Chemical Signals in Vertebrates 16

Lisa M. Schulte
Editor

Chemical Signals in Vertebrates 16

Springer

Editor
Lisa M. Schulte
Wildlife-/Zoo-Animal-Biology and Systematics
Faculty of Biological Sciences
Goethe University Frankfurt
Frankfurt am Main, Hessen, Germany

ISBN 978-3-032-30751-4 ISBN 978-3-032-30752-1 (eBook)
https://doi.org/10.1007/978-3-032-30752-1

This Springer imprint is published by the registered company Springer Nature Switzerland AG
The registered company address is: Gewerbestrasse 11, 6330 Cham, Switzerland

Preface

This volume reports the proceedings of the 16th Chemical Signals in Vertebrates meeting, which was held in Frankfurt am Main, Germany, in August 2024. The meeting took place 48 years after the first meeting of its kind, which was held in Saratoga Springs, New York, USA, in 1976. It was followed, in a triannual pace, by seven more meetings in the USA, three meetings in the UK, two meetings in Germany, one meeting in Poland, and the last meeting, 2021, in France—which unfortunately, due to the COVID pandemic, had to be held online (see footnote for the exact order, places, and years of the meetings[1]). The 16th meeting was hosted by the "Wildlife-/Zoo-animal-biology and Systematics (WZS)" lab at the faculty of Biological Sciences of the Goethe University Frankfurt. It was attended by 56 delegates from 14 countries and 5 continents and covered scientists from all career stages (from students to *emeriti*), all sharing their interest and enthusiasm for vertebrate chemical signalling, and its role in animal behaviour.

Covering all vertebrate classes (except jawless fish), the range of species—and the corresponding chemical substances and behaviours—was extraordinary, spanning sharks to birds, geckos to snow leopards, and poison frogs to humans. As might be expected from such diversity, the program covered an extensive array of topics, including signal detection theory, behavioural responses to con- and heterospecific cues and signals, human chemical communication, chemical communication in non-model systems, glandular sources of chemosignals, disease-related scents, odour-based practices in farm animals, chemical communication-based population management, semiochemicals involved in predation events, odours in urban environments, metabolomics and microbes in chemical communication, chemoreception, genetic identity signals, and many more. The program encompassed various symposia with oral talks, a poster session and three plenary talks on "The evolution of lizard chemical signals across time,

[1]Past CSiV meetings took place in Saratoga Springs, New York, USA (1976); Syracuse, New York, USA (1979); Sarasota, Florida, USA (1982); Laramie, Wyoming, USA (1985); Oxford, England, UK (1988); Philadelphia, Pennsylvania, USA (1991); Tubingen, Germany (1994); Ithaca, New York, USA (1997); Krakow, Poland (2000); Corvallis, Oregon, USA (2003); Chester, England, UK (2006); Urbana-Champaign, Illinois, USA (2014); Cardiff, Wales, UK (2017), Dijon, France (online; 2021).

space, and contexts" by Simon Baeckens, "Avian chemical ecology: Insights from non-model systems" by Leanne A. Grieves, and "Proteins and peptides in vertebrate chemical communication - a retrospective and prospects" by Robert J. Beynon. Together with the three plenary speakers, who contributed insightful reviews, 59 authors were involved in writing chapters for this new volume of the *Chemical Signals in Vertebrates* series. I am very thankful for their devotion and time—without their contributions, this book would not have been possible. I am furthermore grateful for the time and effort of the reviewers—internal and external to our group of authors—for the time and effort, they invested to improve the quality of the contributions.

Besides the people contributing to this book, I am also very thankful to everyone that helped organize the Chemical Signals in Vertebrates meeting. The tireless efforts of all the helpers from the WZS lab—especially Elisa Schulze and Carmen Meyer—made this meeting a success. In addition to all the general logistics, this included the organization of an ice breaker party, a fun BBQ in the botanical garden, and a relaxed conference dinner in a local beer garden. I further want to thank Robert J. Beynon and Jane Hurst for crafting beautiful glass awards and Carsten Müller and Benoist Schaal for donating books for the winners of "Best-Student-Talk" and "Best-Student-Poster." I am furthermore grateful for the financial support from the German Research Foundation (DFG), the "Freunde und Förderer" of the Goethe University, and the European Chemoreception Research Organization (ECRO). Lastly, I would like to thank Benoist Schaal, organizer of the previous meeting, for all the tips I received from him—both for organizing the meeting and for editing this book. I hope I can be the same helpful to the organizers of the upcoming meeting(s).

Frankfurt am Main, Hessen, Germany — Lisa M. Schulte

Contents

Contributors

Peter Apps Botswana Predator Conservation, Wild Entrust, Maun, Botswana
Department of Zoology, Nelson Mandela University, Port Elizabeth, South Africa

Pietro Asproni Tissue Biology and Chemical Communication Department, IRSEA, Institute of Research in Semiochemistry and Applied Ethology, Apt, France

Simon Baeckens Earth and Life Institute, UCLouvain, Louvain-la-Neuve, Belgium

Marie Berčíková Faculty of Science, Charles University, Department of Zoology, Prague, Czech Republic

Robert J. Beynon University of Liverpool, Liverpool, UK

Richard E. Brown Department of Psychology and Neuroscience, Dalhousie University, Halifax, NS, Canada

Sümeyye Carkit-Yilmaz Chair of Aroma and Smell Research, Friedrich-Alexander-Universität Erlangen-Nürnberg, Erlangen, Germany

Manon Chasles Ethology, Physiology and Chemical Communication Department, IRSEA, Institute of Research in Semiochemistry and Applied Ethology, Apt, France

Philippe Chemineau UMR Physiologie de la Reproduction et des Comportements, INRAE, CNRS, IFCE, Université de Tours, Nouzilly, France

Alessandro Cozzi Research and Education Board, IRSEA, Institute of Research in Semiochemistry and Applied Ethology, Apt, France

Javier delBarco-Trillo School of Biological Sciences, Illinois State University, Normal, IL, USA

José-Alberto Delgadillo Centro de Investigación en Reproducción Caprina (CIRCA), Universidad Autónoma Agraria Antonio Narro, Torreón, Coahuila, Mexico

George Priya Doss C School of Bio Sciences and Technology, Vellore Institute of Technology (VIT), Vellore, Tamil Nadu, India

Elisa María Espinosa-López Departamento de Bioquímica y Biología Molecular, Universidad de Córdoba, Córdoba, Spain

Daniel Frynta Faculty of Science, Charles University, Department of Zoology, Prague, Czech Republic

Guadalupe Gómez-Baena Departamento de Bioquímica y Biología Molecular, Universidad de Córdoba, Córdoba, Spain

Mar Gómez-Marín Departamento de Bioquímica y Biología Molecular, Universidad de Córdoba, Córdoba, Spain

Mohanraj Gopikrishnan School of Bio Sciences and Technology, Vellore Institute of Technology (VIT), Vellore, Tamil Nadu, India

Leanne A. Grieves Cornell University, Lab of Ornithology, Macaulay Library, Ithaca, NY, USA

University of Saskatchewan, Department of Biology, Saskatoon, Saskatchewan, Canada

José Guerrero-Casado Departamento de Zoología, Universidad de Córdoba, Córdoba, Spain

Jan Havlíček Department of Zoology, Faculty of Science, Charles University, Prague, Czech Republic

Gnana Thiraviam Annis Inncia School of Bio Sciences and Technology, Vellore Institute of Technology (VIT), Vellore, Tamil Nadu, India

Tolga Karaali Tissue Biology and Chemical Communication Department, IRSEA, Institute of Research in Semiochemistry and Applied Ethology, Apt, France

Matthieu Keller UMR Physiologie de la Reproduction et des Comportements, INRAE, CNRS, IFCE, Université de Tours, Nouzilly, France

Anastasia M. Khrushchova Severtsov Institute of Ecology and Evolution RAS, Moscow, Russia

Laleh Kiavar Chair of Aroma and Smell Research, Friedrich-Alexander-Universität Erlangen-Nürnberg, Erlangen, Germany

Irina V. Kolesnikova Severtsov Institute of Ecology and Evolution RAS, Moscow, Russia

Marlen Kücklich Research Group Behavioural Ecology, Institute of Biology, University of Leipzig, Leipzig, Germany

Research Group Primate Behavioural Ecology, Department of Human Behaviour, Ecology and Culture, Max-Planck-Institute for Evolutionary Anthropology, Leipzig, Germany

German Center for Integrative Biodiversity Research (iDiv), Leipzig, Germany

Erik Kusch Research Group Behavioural Ecology, Institute of Biology, University of Leipzig, Leipzig, Germany

Natural History Museum, University of Oslo, Oslo, Norway

Climate and Nature Risk Research Group, CICERO Centre for International Climate Research, Oslo, Norway

Konrad Lipkowski Museum Wiesbaden, Naturhistorische Sammlung, Wiesbaden, Germany

Wildlife-/Zoo-Animal-Biology and Systematics, Faculty of Biological Sciences, Goethe University Frankfurt, Frankfurt am Main, Hessen, Germany

Helene M. Loos Institute of Nutritional and Food Sciences, Rheinische Friedrich-Wilhelms-Universität Bonn, Bonn, Germany

Sensory Analytics and Technologies, Fraunhofer Institute for Process Engineering and Packaging IVV, Freising, Germany

Chair of Aroma and Smell Research, Friedrich-Alexander-Universität Erlangen-Nürnberg, Erlangen, Germany

Peter Makumbe Oppenheimer Generations Research and Conservation, Shangani Ranch, Shangani, Zimbabwe

Sustainability Research Unit, Nelson Mandela University, Port Elizabeth, South Africa

Miriam Marcet-Rius Ethics, Legislation and Animal Welfare Department, IRSEA, Institute of Research in Semiochemistry and Applied Ethology, Apt, France

J. Weldon McNutt Botswana Predator Conservation, Wild Entrust, Maun, Botswana

Violaine Mechin Tissue Biology and Chemical Communication Department, IRSEA, Institute of Research in Semiochemistry and Applied Ethology, Apt, France

Maxime A. Meunier UMR Physiologie de la Reproduction et des Comportements, INRAE, CNRS, IFCE, Université de Tours, Nouzilly, France

Masao Miyazaki Laboratory of Biomolecular Science, Faculty of Agriculture, Iwate University, Iwate, Japan

Ana María Molina-López Departamento de Anatomía y Anatomía Patológica Comparadas, y Toxicología, Universidad de Córdoba, Córdoba, Spain

Beatriz Ortiz-Guisado Departamento de Bioquímica y Biología Molecular, Universidad de Córdoba, Córdoba, Spain

Patrick Pageat Research and Education Board, IRSEA, Institute of Research in Semiochemistry and Applied Ethology, Apt, France

María Ponte-González Departamento de Zoología, Universidad de Córdoba, Córdoba, Spain

Chantal Porte UMR Physiologie de la Reproduction et des Comportements, INRAE, CNRS, IFCE, Université de Tours, Nouzilly, France

S. Craig Roberts Department of Psychology, University of Stirling, Stirling, UK

Evangelina Rodero-Serrano Departamento de Producción Animal, Universidad de Córdoba, Córdoba, Spain

Veronika Rudolfová Faculty of Science, Charles University, Department of Zoology, Prague, Czech Republic

National Institute of Mental Health, Klecany, Czech Republic

Eliška Rychtecká Faculty of Science, Charles University, Department of Zoology, Prague, Czech Republic

Devaraj Sankarganesh School of Bio Sciences and Technology, Vellore Institute of Technology (VIT), Vellore, Tamil Nadu, India

Sinsha Santhosh School of Bio Sciences and Technology, Vellore Institute of Technology (VIT), Vellore, Tamil Nadu, India

Lisa M. Schulte Wildlife-/Zoo-Animal-Biology and Systematics, Faculty of Biological Sciences, Goethe University Frankfurt, Frankfurt am Main, Hessen, Germany

Dagmar Schwambergová Department of Zoology, Faculty of Science, Charles University, Prague, Czech Republic

Polina D. Semenova Severtsov Institute of Ecology and Evolution RAS, Moscow, Russia

Hana Skalíková Faculty of Science, Charles University, Department of Zoology, Prague, Czech Republic

Iveta Štolhoferová Faculty of Science, Charles University, Department of Zoology, Prague, Czech Republic

Birgit Szabo Centre for Research on Ecology, Cognition and Behaviour of Birds, University of Gent, Gent, Belgium

Hélène Vacher UMR Physiologie de la Reproduction et des Comportements, INRAE, CNRS, IFCE, Université de Tours, Nouzilly, France

Nina Y. Vasilieva Severtsov Institute of Ecology and Evolution RAS, Moscow, Russia

Brigitte M. Weiß Research Group Behavioural Ecology, Institute of Biology, University of Leipzig, Leipzig, Germany

Research Group Primate Behavioural Ecology, Department of Human Behaviour, Ecology and Culture, Max-Planck-Institute for Evolutionary Anthropology, Leipzig, Germany

German Center for Integrative Biodiversity Research (iDiv), Leipzig, Germany

Anja Widdig Research Group Behavioural Ecology, Institute of Biology, University of Leipzig, Leipzig, Germany

Research Group Primate Behavioural Ecology, Department of Human Behaviour, Ecology and Culture, Max-Planck-Institute for Evolutionary Anthropology, Leipzig, Germany

German Center for Integrative Biodiversity Research (iDiv), Leipzig, Germany

Anton P. Zakharov Severtsov Institute of Ecology and Evolution RAS, Moscow, Russia

Part I

Perspectives, Advances and Insights on Vertebrate Chemical Communication

The Evolution of Functional Complexity in the MUP Family: Biochemistry Meets Behaviour

Robert J. Beynon[1](✉)

[1] University of Liverpool, Liverpool, UK

Abstract. Having retired from my academic career (PhD 1974–1977, University of Liverpool 1977–1993, UMIST 1993–1999, and University of Liverpool again 1999–2019), I'm recording a personal and somewhat anecdotal history of my journey through protein chemistry, major urinary proteins, and animal behaviour. In keeping with my plenary at this meeting, this chapter reflects on key milestones in a cross-disciplinary adventure. Because of the extensive timespan and diverse subjects covered, references are necessarily selective, highlighting what I consider to be milestone publications. Over such a long period, including all relevant work would be impossible, and the omission of citations does not reflect a lack of recognition for the many other researchers intrigued by these proteinaceous chemical signals, who are duly acknowledged in our original papers.

Keywords: House mouse · Rat · Major urinary proteins · Darcin · Urine

1 Early History

My career began in biochemistry and enzymology, specifically studying proteolytic enzymes—those that degrade other proteins. While working in Judith Bond's laboratory at the Medical College of Virginia, I was part of the team that discovered meprin, a proteolytic enzyme expressed at remarkably high levels in mouse kidneys (Beynon et al., 1981; Bond & Beynon, 1986). Meprin predominantly binds to the cell membrane of the proximal tubule in the kidney, with its active site facing the external environment of the glomerular filtrate. This raised an intriguing question: why would such an active enzyme be positioned in this space? Our hypotheses included potential roles in the inactivation of peptide hormones that might otherwise negatively affect the urinary system. This focus increased when we identified soluble meprin in mouse urine (Flannery et al., 1990).

To explore the potential substrates for meprin, we examined the protein content of male mouse urine. Simple exploration by gel electrophoresis revealed the presence of very high abundances of proteins at about 15,000–20,000 Da molecular weight (approximately 160 amino acids long). Prior work (Finlayson et al., 1963, 1965; Roy & Neuhaus, 1966) had already recognized this group of proteins since the 1940s, characterized as polymorphic proteins referred to as 'major urinary proteins' (importantly, the

L. M. Schulte (ed.), *Chemical Signals in Vertebrates 16*,
https://doi.org/10.1007/978-3-032-30752-1_1

'M' signifies 'major', not 'mouse'—thus, 'rat MUPs' is correct terminology). The level of MUPs in male mouse urine was remarkably high, at least 1000 times higher than the protein content of normal human urine. Any indication that meprin was responsible for their degradation was rapidly dispelled—MUPs turned out to be remarkably stable proteins, highly resistant to digestion.

By then, we were increasingly intrigued by the MUPs themselves. Emergent cDNA and genomic sequencing technologies had established that multiple MUP genes existed, potentially explaining the heterogeneity observed when MUPs were resolved according to their charge, which was the only method then available for analytical resolution of these proteins.

2 Working in Protein Space: Dissection of Complexity

Around the time of our growing interest in MUPs in the early 1990s, two developments changed the course of our work. First, biological mass spectrometry (MS) was advancing rapidly. We were familiar with 'traditional' gas chromatography/mass spectrometry (GC-MS) for analysis of volatile compounds. The big leap was when we gained the ability to measure protein mass accurately through a combination of soft ionization methods suitable for proteins and new approaches to deconvolution of complex spectra. This allowed us to define the pattern of MUPs as a set of precisely measured masses.

The mass of a protein can be predicted from its gene/cDNA sequence, but there is often a disconnect between predicted and observed mass that needs consideration. A nascent protein typically undergoes changes before reaching its 'mature' form. For example, secreted proteins are proteolyzed to remove a segment of about 20 amino acids (approximately 2000 Da) from the N-terminus—the signal peptide. Many proteins secreted into extracellular space, including MUPs, are stabilized by the formation of covalent disulphide bonds between cysteine residues, reducing mass by 2 Da for each bond. Some proteins are 'decorated' with covalent modifications such as phosphate groups or sugar residues. Finally, mature proteins can be further digested to generate truncated products.

MUPs are present in high concentrations in the urine of many rodent species, making it simple to provide sufficient material for mass spectrometric analysis. When introduced into the ionization source, each protein molecule acquires a variable number of charges, leading to complex, overlapping spectra. At the time we were exploring MUPs by mass spectrometry, a new statistical method (maximum entropy) was developed that could handle protein mixtures and measure protein mass to an accuracy of about 0.01%, or 2 Da in 20 kDa (Evershed et al., 1993). By analysing MUP mixtures with these techniques, we mapped protein masses to predicted cDNA sequences, confirming the loss of the signal peptide and the formation of a single disulphide bond. The concordance of observed and predicted masses confirmed that most MUPs undergo relatively little change after secretion. Armed with this new analytical tool, we were able to reconcile the different MUP mass profiles across three different inbred mouse strains (Robertson et al., 1996, 1997).

We also demonstrated that two of the best-studied low molecular weight pheromones in male mouse urine—2-sec-butyl-4,5-dihydrothiazole ('thiazole') and 3,4-dehydro-exo-brevicomin ('brevicomin')—were bound non-covalently to MUPs (Robertson et al., 1993). This was beautifully confirmed when the three-dimensional structures of these proteins were solved, revealing electron density consistent with bound ligands (Böcskei et al., 1991, 1992; Bacchini et al., 1992). At this time, the speculation was that MUPs acted as carriers of these low molecular weight molecules, either to protect them from chemical degradation or to slow their release into the environment, extending the lifetime of the signal.

3 The Behavioural Connection

Mass spectrometric analysis had given us insights into the molecular composition of MUPs, but we lacked the knowledge or skills to ascribe biological relevance to these observations. The second notable event at this time was my introduction to Jane Hurst, an academic recognized for her studies in mouse scent communication. Jane was investigating deposits left in the environment by free-ranging wild mice and was becoming interested in the role of proteins in creating urinary scent signals. She was working in large poultry farms, where unlimited high-quality food, the warmth provided by the poultry, and plentiful feathers for nests made this an attractive environment for mice. Jane had observed the build-up of large deposits on physical structures, caused by mice repeatedly depositing urine, forming 'urine posts' several centimetres high (Hurst, 1993).

We confirmed that these urine posts contained large quantities of MUPs, creating a sticky binder onto which dust and food particles accumulated (Hurst, 1993). When purified and analysed by GC-MS, these MUPs completely lacked the natural biological ligands (thiazole and brevicomin) expected from laboratory studies. Instead, they contained large amounts of menadione (vitamin K3), a supplement in poultry feed that supports calcium metabolism and shell formation (Robertson et al., 1993). We demonstrated that natural ligands were being displaced by menadione at a rate much higher than their normal slow release, leading to frequent replenishment of urine by the mice. It was also likely that multiple males contributed to these posts to deposit competitive signals. We were able to displace natural ligands from MUPs by addition of exogenous menadione, consistent with this idea (Humphries et al., 1999). Parenthetically, many research groups have subsequently used menadione as a ligand displacer without realizing this was the origin of the displacement approach.

This work, completed in the early to mid-1990s, signalled the start of a hugely rewarding and productive collaboration to establish function and evolution of MUPs in mice and rats as well as identifying scent proteins in other species. Simultaneously, the developments in protein and peptide mass spectrometry were remarkable, allowing increasingly accurate protein mass measurements and enabling the field of 'proteomics'—the study of the entire complement of proteins in a biological sample. It thus became feasible to profile the MUPs in urine samples. The remarkable similarity in most MUP protein sequences makes quantitative analysis particularly challenging, as I will discuss later.

4 From Laboratory to Wild Mice

Previously, we had developed analytical methods using urine samples from laboratory mice, but our collaboration with Jane introduced us to the complexity of wild-derived samples. My previous experience with wild mice had been instructive—when I opened a cage as I would with laboratory mice, the wild mice disappeared within fractions of a second and were never seen again! This was a salutary lesson in working with wild subjects that emphasized the importance of collaborating with qualified behavioural and wildlife experts.

We had previously shown considerable variability in wild mouse urinary patterns using charge-based separation through isoelectric focusing (Hurst et al., 2001; Beynon et al., 2002). Because MUPs differed in their surface charges, we could also purify individual proteins using charge-based ion exchange chromatography. With resolved individual MUPs, we examined the rate of release of natural ligands from MUPs dried onto glass fibre discs. The most anionic (negatively charged) MUP stood out for several reasons beyond its charge. It bound most of the thiazole in urine samples and released this ligand much more slowly than other MUPs, consistent with tight binding between protein and ligand. This protein, known in the lab as the highly memorable 'Peak 4', had a mass of 18,893 Da and exhibited an extremely anomalous migration pattern on gel electrophoresis, appearing to migrate as if it was much smaller than its actual size (Armstrong et al., 2005). There is more on this atypical MUP later.

5 Genomic Insights

Further understanding of MUPs came from genomic sequencing of the inbred mouse strain C57BL/6 J. Although this region of the mouse genome was incomplete (likely due to similarities between genes), the analysis revealed two classes of mouse MUPs—'central' and 'peripheral', referring to their locations in the MUP region of chromosome IV (Mudge et al., 2008). Peripheral MUPs have slightly higher masses, exhibit greater sequence variation (implying different roles), and are expressed in multiple tissues (some in urine and some in nasal tissues). Central MUPs are much more similar in protein sequence (e.g. five MUP genes in the C57BL/6 mouse encode the same mature protein), consistent with rapid gene duplication and expansion of the central region. Examination of MUP profiles across inbred mouse strains revealed limited variation between strains and none within strains, consistent with inbreeding. Notably, some inbred strains lacked the rapidly migrating 'Peak 4' protein (Cheetham et al., 2009).

While earlier studies relied on charge-based separation to establish differences in MUP profiles between laboratory strains or individual wild mice, mass-based analysis had the considerable advantage that measured masses could be reconciled with those predicted from gene sequences. Furthermore, proteomics approaches allowed us to map mass changes to alterations in protein sequence, even without cDNA or genomic data (which is usually unavailable for individual wild mice). This confirmed that wild mice contained novel MUPs, albeit subtly different in sequence from previously known MUPs and those from other *Mus* species (Robertson et al., 1997, 2007). Gradually, we were building a picture of the complexity of the MUP system; some species had multiple MUPs in their urine; other species expressed much simpler patterns (Robertson et al., 2007). Both male and female wild house mice expressed polymorphic protein profiles that differed between individuals. Moreover, the evidence for sexual dimorphism was increasingly clear—while both sexes produced MUPs, there were specific MUPs that were only expressed at an appreciable concentration in the urine of male mice (I use 'only' as a shorthand to emphasize a substantial expression ratio difference between males and females).

6 MUPs as Identity Signals

Why do mice need to express so many different MUP gene products in their urine? Many of the minor amino acid changes were located on the protein surface, unlikely to allow different MUPs to bind and release different ligands. We therefore wondered whether the MUPs themselves were conveying signals.

At this time, it became possible to produce recombinant MUPs in bacteria or fungi. In a male-to-male competitive countermarking study, we demonstrated that adding a single, purified recombinant MUP—without ligands—to a urine scent mark could disrupt recognition of that mark (Hurst et al., 2001). This led us to propose that the pattern of MUPs acted as a molecular barcode encoding individual male identity and suggesting the existence of MUP receptors in the vomeronasal organ. Such a 'barcode' would require multiple MUPs, perhaps explaining the multiplicity of gene products. This suggestion was novel and was explored in the context of the strongly held view that major histocompatibility complex (MHC) haplotype was the main source of genetic signals in mice, used for individual and kin recognition underpinning mate choice. Such studies had largely been conducted with MHC congenic laboratory mice, otherwise genetically identical and with a minimal MUP repertoire. When we explored scent mark recognition in wild house mice using stimuli from MUP- or MHC-controlled individuals on variable genetic backgrounds, it became clear that MHC was not used for individual recognition, but recognition was driven by MUP type (Cheetham et al., 2007). Further, MUP recognition is translated into mate choice. We established a large population of founder house mice, haplotyped for MUP and MHC on a controlled genetic background, which were released into a free-ranging enclosure and allowed to select mates and breed. When the offspring were analysed for MUP and MHC types, it was clear that, while MHC was without influence, there was a strong deficit of mating between MUP-similar mice. MUPs—not MHC—therefore drive kin recognition and inbreeding avoidance in mice (Sherborne et al., 2007). In a subsequent study, we demonstrated that female house mice show a strong preference for MUP heterozygous over MUP homozygous males. Once again, MHC heterozygosity had no influence on this choice (Thom et al., 2008). MUP haplotype, not MHC, also drives the choice of female house mice in selection of nest mates for cooperative breeding (Green et al., 2015).

Indeed, MUP haplotype could even play an important role in speciation. With Guila Ganem and colleagues, we analysed MUPs from *Mus musculus domesticus* and *Mus musculus musculus* from both allopatric and contact zones across Europe. Both subspecies expressed a rich MUP profile with differences, mapped to individual amino acid changes, largely in the central MUPs. The subspecies expressed distinct MUP signatures that varied between allopatric and contact zones (Hurst et al., 2017). At present, MUPs are candidates for the semiochemical barrier to suppress hybridization. Moreover, Michael Sheehan and colleagues conducted an analysis of DNA, RNA, and protein from wild house mice that provides strong confirmation that individual variation is due to different alleles and variability in the expression of each allele, influencing the output of individual central MUPs. Both alleles and expression control are under balancing selection. By contrast, the peripheral MUPs did not seem to undergo this selection (Sheehan et al., 2016).

7 Darcin: A Special MUP

Further evidence for specific roles of some MUPs came from behavioural studies in which MUPs purified from C57BL/6 mouse urine were individually tested for semiochemical properties. One male-specific MUP not only elicited attraction from females but also induced learned attraction to cues associated with the MUP, which persisted even after the stimulus was no longer present (Roberts et al., 2010, 2012). This protein was the 'Peak 4' protein, the peripheral MUP we had previously shown to have unusual properties, including tight binding and slow release of urinary thiazole. To eliminate bound volatiles as the source of the signal, we showed that either aged, ligand-depleted protein or recombinant protein was equally effective in eliciting the same response. Further, male urine that contained darcin at substantially lower levels than that normally expressed by adult male mice was not attractive to females, despite containing normal amounts of typical male volatile components (Roberts et al., 2010).

The potency of this protein and its unusual semiochemical properties distinguished it among MUPs, with specific male-driven effects: initial attraction, followed by learned attraction to the volatile odour signature encountered with the protein (Roberts et al., 2010, 2018), and a remembered attraction to the location where mice had first encountered the protein (Roberts et al., 2012). MUP nomenclature was (and remains) somewhat disorganized. To distinguish this Peak 4 protein and its unusual behavioural role, during a wine-enhanced evening with protein chemist Kathryn Lilley (the responsibility for the name is mine and Jane's however!), we named it 'darcin' (Roberts et al., 2010). Traditionally, many proteins received trivial names ending in 'in' (e.g. pepsin), but the first syllable was a tribute to a model of male attractiveness personified in Jane Austen's novel 'Pride and Prejudice'—the character of Fitzwilliam D'Arcy.

As significant as this scientific discovery (identifying a male protein pheromone in mammals) was, when the work revealing 'darcin' was published, it attracted considerable attention from Jane Austen enthusiasts and science journalists with 'a nose for a good tail'—puns abounded! This response was nicely captured in a BBC Radio interview with Jane [https://www.bbc.co.uk/sounds/play/w3ct2ccf]. Some Jane Austen social media sites were rather offended at equating mouse urine with their beloved Mr. D'Arcy, including detailed discussions about what the actual Mr. D'Arcy would smell like! This proved an object lesson in the connection between hard science and social media, with reactions ranging from 'proof positive that scientists have a sense of humour' to 'what's wrong with you perverts?'!

The darcin story continued to develop. We solved the three-dimensional structure of darcin—it is a 'typical' eight-stranded beta barrel lipocalin but turned out to have remarkable rigidity and structural stability (Phelan et al., 2014). This explained the previously mentioned aberrant migration on gel meprin, for example, as well as the ability to retain pheromonal ligands for a longer time. The latest development in the darcin story was the collaboration with Ebru Demir and Richard Axel, who not only confirmed the behavioural effects of darcin but demonstrated that darcin activates a genetically determined neural circuit extending from the accessory olfactory bulb to the posterior medial amygdala in female mice. This provided a model for the integration of pheromone information with internal state to elicit both innate and learned behaviours (Demir et al., 2020).

Darcin has a rather unique role in mouse semiochemistry. There is a remarkably high degree of sequence conservation across wild mice (more than in central MUPs). The extreme structural rigidity fits well with a role of pheromone binding and slow release. Structural rigidity equated with chemical stability, and darcin is well placed to be a persistent signal. The rigidity might also create a very stable structural motif for recognition by accessory olfactory bulb receptors.

Our current model of the role of darcin in scent communication integrates darcin, other urinary MUPs, and a volatile signature that is conditioned by the urinary MUP profile. Contact with darcin stimulates an attraction response that fixes the location of the scent mark and induces return and a learned preference for the volatile profile of the scent mark (Roberts et al., 2018). Darcin seem to be highly conserved in commensal *Mus* species, consistent with this role in integration of female attraction, recognition, and choice between individual males.

8 The MUPs of the Rat, *Rattus Norvegicus*

It was already known that the rat (*Rattus norvegicus*) also had a relatively complex MUP system. We embarked on detailed molecular phenotypic analysis of rat MUPs using methodologies developed for mouse MUPs. The rat genome sequence was (and remains) less complete but suggested fewer MUP coding genes, which proteomic analysis of rat urine confirmed. One advantage of protein sequence-based mass spectrometric analysis is revealing post-translational modifications—changes occurring after secretion from the liver.

It quickly became clear that the complexity of rat urinary MUPs was partly attributable to post-translational changes (Gómez-Baena et al., 2019). In rat urine, MUPs are often 'trimmed' from each end by exoproteolysis, and some are modified through phosphorylation. We also compared urinary MUP composition with protein content in male preputial glands and female clitoral glands. Unlike house mouse urine, where MUP content in scent marks derives from the liver via urine, MUPs in rat scent marks combine outputs from the liver and from sexual accessory glands (Gómez-Baena et al., 2023). Moreover, the elaborate processing of rat MUPs largely occurs in these secondary sexual glands. Although rats have a complex MUP profile, that complexity arises through completely different mechanisms from that in house mice. Dissecting this post-translational complexity and its functional significance remains a challenge for future work.

9 Cross-Species Collaboration

A particular pleasure of the MUP research has been the interaction with colleagues interested in MUP systems or darcin-like molecules in their own species of interest. Basic mass spectrometric exploration of the protein content of scent secretions is straightforward, and we have always been pleased to share our technology. This work revealed that many species lacked MUP systems, but their urine contained substantial quantities of odorant binding proteins—lipocalins closely related to MUPs and often previously observed by other researchers. Thus, the notion of a generic MUP model was quickly dispelled and the differences between species are particularly interesting.

Beyond providing protein characterization, we proposed names for species-specific proteins: 'roborovskin' for the Roborovski dwarf hamster (*Phodopus roborovskii*) (Turton et al., 2010), 'glareosin' from the bank vole (*Myodes glareolus*) (Loxley et al., 2017; Coombes et al., 2024), and 'vulpeculin' from the common brushtail possum (*Trichosurus vulpecula*) (Loxley et al., 2020). These are all lipocalins with the signature eight-stranded beta barrel and a central cavity potential for ligand binding and are similar to the long known hamster protein, aphrodisin (Singer et al., 1986; Vincent et al., 2001). By contrast, mouse lemurs (*Microcebus murinus* and *Microcebus lehilahytsara*) expressed a male-specific, seasonally regulated urinary protein from the WFDC12 (whey acidic protein 4 disulphide core 12, thought to have antiprotease or antibiotic properties) family rather than a lipocalin (Unsworth et al., 2017). The function of this protein also remains enigmatic.

We should not be surprised by divergent evolution in complex signalling systems; after all, these different species inhabit diverse habitats and exhibit varying life history strategies and requirements for intra-specific communication. Is the observed molecular complexity of some species, specifically mice and rats (Beynon et al., 2015), related to living at high population densities around human resources and the need for degrees of complexity in scent mediated interaction? Is high complexity necessary to encode more complex identity and status signals? Why does one species manage with a single lipocalin while others evolve high molecular complexity? Do single-lipocalin species communicate via complex bouquets of volatile molecules, simply using the lipocalin to bind and control release of the payload? Challenging questions for the future abound.

10 Beyond Molecules: The Behavioural Context

No detailed molecular analysis can fully answer such questions. A highlight of my collaboration with behavioural ecologists has been understanding that molecular knowledge has limited value without behavioural context. Analysing scent marks from wild or wild-derived individuals, isolating key molecules, and conducting controlled behavioural experiments made me realize that while molecular analysis may be technically straightforward, ascribing 'meaning' to molecular data is considerably more challenging. In today's '-omics' era, we too often pursue context-light studies of molecular detail without placing such knowledge in true behavioural or biological contexts. Collaborations like the one with Jane Hurst and her team have been so rewarding precisely because they forced me to think about context while providing tools for biological dissection of meaning—one of my career highlights.

As a further development of our shared interests, my relationship with Jane evolved personally as well—in 2003 we married, fittingly beginning our honeymoon at the CSIV meeting in Corvallis, organized by Bob Mason and colleagues. Our many friends in the field were thrilled for us and made both the conference and our subsequent trip down California's Route 1 memorable! How appropriate that a CSIV conference served as a honeymoon location—and proof that MUPs can influence mate selection even in (some) humans!

11 The Quantitative Dimension

Wearing my analytical chemist hat, I always want to introduce the quantitative dimension to MUP expression. MUP studies present two major challenges. First is the high sequence similarity between different MUPs, making it unlikely that antibodies specific to each MUP could ever be developed, thus precluding immunoquantification. This leaves either intact protein mass analysis (effective with calibration for MUPs of distinct masses; Beynon et al., 2015) or, taking a proteomics approach, digestion of MUPs into precise smaller fragments analysed by mass spectrometry.

Several research groups have used 'label-free quantification' in an attempt to measure individual MUPs, but in practice, this approach faces significant limitations. For example, I have compiled a database of 110 different MUP sequences (from laboratory and wild *Mus musculus domesticus*, *Mus musculus musculus*, and other *Mus* species https://phbuffers.org/MUP/KnownMUPs.fasta). After digestion, these proteins would generate just over 50 unique peptides that could be readily analysed by mass spectrometry. Some peptides appear in over 90 different MUPs, while 32 peptides appear in just 1 or 2 MUPs. This makes simple application of label-free methods to quantification nearly impossible, particularly for the central MUPs—obtaining absolute or even relative quantification of individual MUPs would be extremely difficult. Quantification of the peripheral MUPs is relatively less challenging.

To circumvent this problem, we have used synthetic peptides equivalent to those in MUPs but labelled with stable isotopes, making them 'heavier' in the mass spectrometer.

By comparing variable amounts of peptides in biological samples with known quantities of standard peptides, we can achieve absolute quantification of individual MUPs. Some analytical complexity remains ('from this peptide abundance, reporting on two MUPs, subtract the abundance of a second peptide reporting on one of them to obtain the abundance of the other'), and the complexity of urinary MUP content requires many peptides. It is worth noting that this approach can only work with known MUP sequences of course, but new protein sequences are readily obtained through tandem mass spectrometry.

As a methodological refinement, we developed the QconCAT approach (Pratt et al., 2006; Rivers et al., 2007; Takemori et al., 2017; Johnson et al., 2021). QconCATs are products of artificial genes expressed in bacteria or cell-free systems in the presence of stable-isotope labelled amino acids. When digested, a QconCAT produces peptides that exactly match natural, MUP-derived peptides (Beynon et al., 2015). Because these peptides behave identically in analytical workflows, differing only in mass, the ratio of analyte to standard enables absolute quantification of each MUP (Beynon et al., 2014).

But what does absolute quantification mean for urine-derived signals? Urine concentration varies with water consumption. We've long advocated using creatinine—a metabolic end product that serves as a surrogate measure of body mass—as a reference for urine analysis (Beynon & Hurst, 2004; Cheetham et al., 2009). Without such standardization, attributing changes in urinary MUP concentration to biological responses rather than simple dilution or concentration of urine becomes difficult. As a final complication, urine scent marks typically dry on substrates, effectively rendering MUPs near infinite in concentration—what does concentration actually mean in this scenario?

12 Conclusions

Over the 30 odd years I have been involved with MUPs, there have been many highlights. First, as an analytical protein chemist, I was used to the 'simplicity' and 'cleanliness' of my type of research—clean proteins in, good chemical characterization out. Working with behavioural scientists came as something of an eye-opener. First, the desire for beautifully clean samples was rapidly rejected—scrapings, smears, and other dubious samples became the stuff of everyday analysis, and we quickly developed clean-up techniques to protect our precious instruments from the messy assault of biology! Second, we have been used to essentially unlimited quantities of analytical material. When we first asked Jane and her group for urine samples, we specified about one millilitre as a suitable sample size per animal. The look of dismay from them was followed by intense periods of negotiation and refinement, to the extent that (assisted by developments in instrumentation) we could reduce our demands by a factor of at least 1000. These dialogues—along the lines of '*why can't you do this? – OK we'll try to find a new way to do this*'—were a consistent feature of our interactions and made us better analytical scientists.

Lastly, I have learned so much biology! I attended many more 'biology' seminars and found myself fascinated and stimulated by the ways in which difficult systems might be given a different perspective by analytical science. I remain firmly convinced that the analytical chemistry part of the research is the easy part, and it is the imposition of context and biological meaning that is so challenging.

A final and unexpected consequence of my research path was the connection to the CSIV community. These meetings have always been warm and welcoming and give platforms for younger scientists embarking on their journey. They taught me so much great science, and I now embark myself on 'natural history' vacations with a much greater appreciation of the natural world. I am pleased to recognize my many friends and colleagues in this field.

Acknowledgments. A retrospective like this inevitably makes it impossible to highlight everyone who has contributed to our knowledge over decades. Jane and I have been delighted to see the emergence of 'F1 hybrids'—scientists who move between the 'two towers' of molecular and behavioural science. Rather than risk accidentally omitting important names, I direct readers to our many publications where contributors are properly recognized, including in the previous volumes of this series.

Research requiring access to expensive analytical instrumentation and state-of-the-art behavioural research facilities demands external funding, and we gratefully acknowledge support from the Biotechnology and Biological Sciences Research Council, the Natural Environment Research Council, and the Wellcome Trust. Finally, I dedicate this article to my wife and colleague, Jane, for our collaborations—both in and beyond the laboratory!

References

Armstrong, S. D., Robertson, D. H. L., Cheetham, S. A., Hurst, J. L., & Beynon, R. J. (2005). Structural and functional differences in isoforms of mouse major urinary proteins: A male-specific protein that preferentially binds a male pheromone. *Biochemical Journal, 391*(Pt 2), 343–350.

Bacchini, A., Gaetani, E., & Cavaggioni, A. (1992). Pheromone binding proteins of the mouse, Mus musculus. *Experientia, 48*(4), 419–421.

Beynon, R. J., & Hurst, J. L. (2004). Urinary proteins and the modulation of chemical scents in mice and rats. *Peptides, 25*(9), 1553–1563.

Beynon, R. J., Shannon, J. D., & Bond, J. S. (1981). Purification and characterization of a metallo-endoproteinase from mouse kidney. *Biochemical Journal, 199*(3), 591–598.

Beynon, R. J., Veggerby, C., Payne, C. E., Robertson, D. H. L., Gaskell, S. J., Humphries, R. E., & Hurst, J. L. (2002). Polymorphism in major urinary proteins: Molecular heterogeneity in a wild mouse population. *Journal of Chemical Ecology, 28*(7), 1429–1446.

Beynon, R. J., Armstrong, S. D., Gómez-Baena, G., Lee, V., Simpson, D., Unsworth, J., & Hurst, J. L. (2014). The complexity of protein semiochemistry in mammals. *Biochemical Society Transactions, 42*(4), 837–845.

Beynon, R. J., Armstrong, S. D., Claydon, A. J., Davidosn, A. J., Eyers, C. E., Langridge, J. I., Gómez-Baena, G., Harman, V. M., Hurst, J. L., Lee, V., Mclean, L., Pattison, R., Roberts, S. A., Simpson, D. M., Unsworth, J., Vonderach, M., Williams, J. P., & Woolerton, Y. E. (2015). Mass spectrometry for structural analysis and quantification of the major urinary proteins of the house mouse. *International Journal of Mass Spectrometry, 391*, 146–156.

Böcskei, Z., Findlay, J. B., North, A. C., Phillips, S. E., Somers, W. S., Wright, C. E., Lionetti, C., Tirindelli, R., & Cavaggioni, A. (1991). Crystallization of and preliminary X-ray data for the mouse major urinary protein and rat alpha-2u globulin. *Journal of Molecular Biology, 218*(4), 699–701.

Böcskei, Z., Groom, C. R., Flower, D. R., Wright, C. E., Phillips, S. E., Cavaggioni, A., Findlay, J. B., & North, A. C. (1992). Pheromone binding to two rodent urinary proteins revealed by X-ray crystallography. *Nature, 360*(6400), 186–188.

Bond, J. S., & Beynon, R. J. (1986). Meprin: A membrane-bound metallo-endopeptidase. *Current Topics in Cellular Regulation, 28*, 263–290.

Cheetham, S. A., Thom, M. D., Jury, F., Ollier, W. E. R., Beynon, R. J., & Hurst, J. L. (2007). The genetic basis of individual-recognition signals in the mouse. *Current Biology, 17*(20), 1771–1777.

Cheetham, S. A., Smith, A. L., Armstrong, S. D., Beynon, R. J., & Hurst, J. L. (2009). Limited variation in the major urinary proteins of laboratory mice. *Physiology & Behavior, 96*(2), 253–261.

Coombes, H. A., Prescott, M. C., Stockley, P., Beynon, R. J., & Hurst, J. L. (2024). The role of male scent in female attraction in the bank vole, *Myodes glareolus. Scientific Reports, 14*(1), 4812.

Demir, E., Li, K., Bobrowski-Khoury, N., Sanders, J. I., Beynon, R. J., Hurst, J. L., Kepecs, A., & Axel, R. (2020). The pheromone darcin drives a circuit for innate and reinforced behaviours. *Nature, 578*(7793), 137–141.

Evershed, R. P., Robertson, D. H., Beynon, R. J., & Green, B. N. (1993). Application of electrospray ionization mass spectrometry with maximum-entropy analysis to allelic 'fingerprinting' of major urinary proteins. *Rapid Communications in Mass Spectrometry, 7*(10), 882–886.

Finlayson, J. S., Potter, M., & Runner, C. R. (1963). Electrophoretic variation and sex dimorphism of the major urinary protein complex in inbred mice: A new genetic marker. *Journal of the National Cancer Institute, 31*, 91–107.

Finlayson, J. S., Asofsky, R., Potter, M., & Runner, C. C. (1965). Major urinary protein complex of normal mice: Origin. *Science, 149*(3687), 981–982.

Flannery, A. V., Dalzell, G. N., & Beynon, R. J. (1990). Proteolytic activity in mouse urine: Relationship to the kidney metallo-endopeptidase, meprin. *Biochimica et Biophysica Acta, 1041*(1), 64–70.

Gómez-Baena, G., Armstrong, S. D., Halstead, J. O., Prescott, M., Roberts, S. A., Mclean, L., Mudge, J. M., Hurst, J. L., & Beynon, R. J. (2019). Molecular complexity of the major urinary protein system of the Norway rat, *Rattus norvegicus. Scientific Reports, 9*(1), 10757.

Gómez-Baena, G., Pounder, K. C., Halstead, J. O., Roberts, S. A., Davidson, A. J., Prescott, M., Beynon, R. J., & Hurst, J. L. (2023). Unraveling female communication through scent marks in the Norway rat. *Proceedings of the National Academy of Sciences of the United States of America, 120*(25), e2300794120.

Green, J. P., Holmes, A. M., Davidson, A. J., Paterson, S., Stockley, P., Beynon, R. J., & Hurst, J. L. (2015). The genetic basis of kin recognition in a cooperatively breeding mammal. *Current Biology, 25*(20), 2631–2641.

Humphries, R. E., Robertson, D. H., Beynon, R. J., & Hurst, J. L. (1999). Unravelling the chemical basis of competitive scent marking in house mice. *Animal Behaviour, 58*(6), 1177–1190.

Hurst, J. L. (1993). The priming effects of urine substrate marks on interactions between male house mice, *Mus musculus domesticus* Schwarz & Schwarz. *Animal Behaviour, 45*(1), 55–81.

Hurst, J. L., Payne, C. E., Nevison, C. M., Marie, A. D., Humphries, R. E., Robertson, D. H., Cavaggioni, A., & Beynon, R. J. (2001). Individual recognition in mice mediated by major urinary proteins. *Nature, 414*(6864), 631–634.

Hurst, J. L., Beynon, R. J., Armstrong, S. D., Davidson, A. J., Roberts, S. A., Gómez-Baena, G., Smadja, C. M., & Ganem, G. (2017). Molecular heterogeneity in major urinary proteins of *Mus musculus* subspecies: Potential candidates involved in speciation. *Scientific Reports, 7*, 44992.

Johnson, J., Harman, V. M., Franco, C., Emmott, E., Rockliffe, N., Sun, Y., Liu, L.-N., Takemori, A., Takemori, N., & Beynon, R. J. (2021). Construction of à la carte QconCAT protein standards for multiplexed quantification of user-specified target proteins. *BMC Biology, 19*(1), 195.

Loxley, G. M., Unsworth, J., Turton, M. J., Jebb, A., Lilley, K. S., Simpson, D. M., Rigden, D. J., Hurst, J. L., & Beynon, R. J. (2017). Glareosin: A novel sexually dimorphic urinary lipocalin in the bank vole, *Myodes glareolus*. *Open Biology, 7*(9), 170135.

Loxley, G. M., Hooks, D. O., Antonopoulos, A., Dell, A., Haslam, S. M., Linklater, W. L., Hurst, J. L., & Beynon, R. J. (2020). Vulpeculin: A novel and abundant lipocalin in the urine of the common brushtail possum, *Trichosurus vulpecula*. *Open Biology, 10*(10), 200218.

Mudge, J. M., Armstrong, S. D., Mclaren, K., Beynon, R. J., Hurst, J. L., Nicholson, C., Robertson, D. H., Wilming, L. G., & Harrow, J. L. (2008). Dynamic instability of the major urinary protein gene family revealed by genomic and phenotypic comparisons between C57 and 129 strain mice. *Genome Biology, 9*(5), R91.

Phelan, M. M., Mclean, L., Armstrong, S. D., Hurst, J. L., Beynon, R. J., & Lian, L.-Y. (2014). The structure, stability and pheromone binding of the male mouse protein sex pheromone darcin. *PLoS One, 9*(10), e108415.

Pratt, J. M., Simpson, D. M., Doherty, M. K., Rivers, J., Gaskell, S. J., & Beynon, R. J. (2006). Multiplexed absolute quantification for proteomics using concatenated signature peptides encoded by QconCAT genes. *Nature Protocols, 1*(2), 1029–1043.

Rivers, J., Simpson, D. M., Robertson, D. H. L., Gaskell, S. J., & Beynon, R. J. (2007). Absolute multiplexed quantitative analysis of protein expression during muscle development using QconCAT. *Molecular & Cellular Proteomics, 6*(8), 1416–1427.

Roberts, S. A., Simpson, D. M., Armstrong, S. D., Davidson, A. J., Robertson, D. H., Mclean, L., Beynon, R. J., & Hurst, J. L. (2010). Darcin: A male pheromone that stimulates female memory and sexual attraction to an individual male's odour. *BMC Biology, 8*, 75.

Roberts, S. A., Davidson, A. J., Mclean, L., Beynon, R. J., & Hurst, J. L. (2012). Pheromonal induction of spatial learning in mice. *Science, 338*(6113), 1462–1465.

Roberts, S. A., Prescott, M. C., Davidson, A. J., Mclean, L., Beynon, R. J., & Hurst, J. L. (2018). Individual odour signatures that mice learn are shaped by involatile major urinary proteins (MUPs). *BMC Biology, 16*(1), 48.

Robertson, D. H., Beynon, R. J., & Evershed, R. P. (1993). Extraction, characterization, and binding analysis of two pheromonally active ligands associated with major urinary protein of house mouse (*Mus musculus*). *Journal of Chemical Ecology, 19*(7), 1405–1416.

Robertson, D. H., Cox, K. A., Gaskell, S. J., Evershed, R. P., & Beynon, R. J. (1996). Molecular heterogeneity in the major urinary proteins of the house mouse *Mus musculus*. *Biochemical Journal, 316*, 265–272.

Robertson, D. H., Hurst, J. L., Bolgar, M. S., Gaskell, S. J., & Beynon, R. J. (1997). Molecular heterogeneity of urinary proteins in wild house mouse populations. *Rapid Communications in Mass Spectrometry, 11*(7), 786–790.

Robertson, D. H. L., Hurst, J. L., Searle, J. B., Gündüz, I., & Beynon, R. J. (2007). Characterization and comparison of major urinary proteins from the house mouse, *Mus musculus domesticus*, and the aboriginal mouse, *Mus macedonicus*. *Journal of Chemical Ecology, 33*(3), 613–630.

Roy, A. K., & Neuhaus, O. W. (1966). Identification of rat urinary proteins by zone and immunoelectrophoresis. *Proceedings of the Society for Experimental Biology and Medicine, 121*(3), 894–899.

Sheehan, M. J., Lee, V., Corbett-Detig, R., Bi, K., Beynon, R. J., Hurst, J. L., & Nachman, M. W. (2016). Selection on coding and regulatory variation maintains individuality in major urinary protein scent marks in wild mice. *PLoS Genetics, 12*(3), e1005891.

Sherborne, A. L., Thom, M. D., Paterson, S., Jury, F., Ollier, W. E. R., Stockley, P., Beynon, R. J., & Hurst, J. L. (2007). The genetic basis of inbreeding avoidance in house mice. *Current Biology, 17*(23), 2061–2066.

Singer, A. G., Macrides, F., Clancy, A. N., & Agosta, W. C. (1986). Purification and analysis of a proteinaceous aphrodisiac pheromone from hamster vaginal discharge. *Journal of Biological Chemistry, 261*(28), 13323–13326.

Takemori, N., Takemori, A., Tanaka, Y., Endo, Y., Hurst, J. L., Gómez-Baena, G., Harman, V. M., & Beynon, R. J. (2017). MEERCAT: Multiplexed efficient cell free expression of recombinant QconCATs for large scale absolute proteome quantification. *Molecular & Cellular Proteomics, 16*(12), 2169–2183.

Thom, M. D., Stockley, P., Jury, F., Ollier, W. E. R., Beynon, R. J., & Hurst, J. L. (2008). The direct assessment of genetic heterozygosity through scent in the mouse. *Current Biology, 18*(8), 619–623.

Turton, M. J., Robertson, D. H. L., Smith, J. R., Hurst, J. L., & Beynon, R. J. (2010). Roborovskin, a lipocalin in the urine of the Roborovski hamster, *Phodopus roborovskii. Chemical Senses, 35*(8), 675–684.

Unsworth, J., Loxley, G. M., Davidson, A., Hurst, J. L., Gómez-Baena, G., Mundy, N. I., Beynon, R. J., Zimmermann, E., & Radespiel, U. (2017). Characterisation of urinary WFDC12 in small nocturnal basal primates, mouse lemurs (*Microcebus* spp.). *Scientific Reports, 7*, 42940.

Vincent, F., Löbel, D., Brown, K., Spinelli, S., Grote, P., Breer, H., Cambillau, C., & Tegoni, M. (2001). Crystal structure of aphrodisin, a sex pheromone from female hamster. *Journal of Molecular Biology, 305*(3), 459–469.

Chemical Signals, Olfactory Learning and Individual Differences: 50 Years of Research

Richard E. Brown[1(✉)]

[1] Department of Psychology and Neuroscience, Dalhousie University, Halifax, NS, Canada
Richard.Brown@Dal.Ca

Abstract. This chapter describes 50 years of research in my laboratory into chemical signals in rodents which began with experiments on the role of odour preferences and urine marking in the socio-sexual behaviour of rats. Olfactory communication involves both endocrine and experiential factors as shown by our studies on gonadectomy and isolation rearing. The importance of individual recognition by odours was studied through the rat pups' response to "maternal pheromones" and by examining the role of the major histocompatibility complex in odours of individuality. The initial studies done using habituation-dishabituation tests were continued using an operant olfactometer to study odour discrimination learning using the urine odours of rats and mice. We also investigated the importance of individual recognition in puberty acceleration in juvenile female mice. We then began to examine the use of the olfactometer to study working memory in rats and learning set formation in mice. We are now examining olfactory reversal learning in mice using signal detection methods to identify sex, age and genotype differences in learning strategies. Our studies have attempted to integrate mechanisms of olfactory communication (releaser effects), neuroendocrine responses (primer effects) and learning and memory (cognitive processes) in olfactory signalling in rodents.

Keywords: Rodent olfaction · Social odours · Individual recognition · Scent marking · Major histocompatibility complex · Olfactometer · Olfactory learning and memory · Signal detection analysis

1 From Rat Sexual Behaviour to Olfaction and Urine Marking

My introduction to the Chemical Signals in Vertebrates meetings was to review Volume 2 in 1982 (Brown, 1982a). Since then, my lab has contributed to eight Chemical Signals in Vertebrates meetings, from the third in 1983 to the 16th today. My research in olfaction and behaviour began with my master's thesis on rat sexual behaviour (Brown, 1974), during which I became interested in the role of olfactory communication in social and sexual behaviour. I had read Calhoun's book, *The Ecology and Sociology of the Norway Rat* (Calhoun, 1962), in which he discussed scent marking as an aspect of

L. M. Schulte (ed.), *Chemical Signals in Vertebrates 16*,
https://doi.org/10.1007/978-3-032-30752-1_2

sexual behaviour, saying that "rats were frequently observed to rub their bodies against passages through fences, tile entries into harborage boxes, or over clods of dirt, rocks or entries to burrows. It is assumed that some urination actually occurs during this rubbing process or that scent passed out with the urine and adhering to the fur is rubbed off" (Calhoun, 1962, p. 112).

I decided to examine urine marking by rats using wire-mesh "rocks" wrapped in newsprint and entitled my first presentation at the Animal Behaviour Society meeting: "The fire-hydrant effect: stimuli eliciting urine-marking in the rat" (Brown, 1973). This research involved placing drops of urine collected from unknown male or female rats, the rat's own urine or water in the centre of a paper block and then scoring the time spent investigating each odour and the amount of urine deposited over each block (Brown, 1975). The results showed that male rats preferred to investigate and urine mark over the odours of unfamiliar male and female conspecifics. This research was extended for my PhD thesis which involved using urine odour stimuli from intact and castrated males, intact and ovariectomized females as well as water and the rat's own urine and examining the investigation and urine-marking responses of intact and gonadectomized males and females reared in single sex or mixed sex groups to these odours (Brown, 1977). The results showed that intact males and females investigated conspecific odours and urine marked over them more than gonadectomized rats which showed very little urine marking. Sexual experience increased responses to odours more in males than females.

Following this I tested the effects of gonadal hormone treatment, social isolation and motivational arousal on odour preferences and urine marking in rats in order to examine the importance of hormonal and experiential factors in olfactory signalling. Androgens and oestrogen stimulated urine marking by both male and female rats, while social isolation reduced investigation and scent-marking behaviours (Brown, 1978, 1985a, 1986). I then became interested in trying to identify the types of information contained in the olfactory signals of mammals (Brown, 1979). This review resulted in five classes of identifier odours, which were stable over time (individual, colony, species, age and sex) and four classes of emotional odours, which were correlated with changes in hormone levels (rut, social status, stress and maternal; see Table 1).

2 Individual Recognition by Odour Cues

Michael Leon was the external examiner of my PhD thesis, and we decided to follow up his work on maternal pheromones (Leon & Moltz, 1971; Leon, 1974, 1975) as models of individual recognition by olfactory signals (Table 2). In these experiments we showed that pre-weanling outbred Long-Evans hooded laboratory rats preferred their own nest odour to nests of strange litters and showed low preferences for the nest odours of strange males and virgin females. However, when pups were reared with both their mother and father, nest odours of males became more attractive (Brown, 1982b). The anal excreta odour of the rat pup's own dam was preferred to those of strange dams, and the excreta odour of both own and strange dams was preferred to those of virgin females. Rat pups reared with their dam and sire preferred the anal excreta odour of their own sire to that of other males (Brown & Elrick, 1983). In a third study, my students and I found that male rats which ingested faeces

Table 1. A classification of mammalian social odours (from Brown, 1979).

Identifier odours	Those produced by the body's normal metabolic processes that are stable for long periods of time
Individual	Odours that are unique to each individual animal
Colony	Odours that are characteristic to all members of a colony, nest, den, deme or living group; in some cases family odours
Species typical	Odours that are characteristic to all members of a species or subspecies
Age specific	Odours that distinguish animals of different age classes; such as pre-weanling (infant), postweanling (juvenile) and adult
Sex specific	Odours that distinguish males from females
Emotive odours	Those produced or released only in special circumstances
Rut	Odours distinguishing animals in breeding condition from those that are not
Social status	Odours that distinguish subordinate from dominant animals, usually males
Stress	Odours released in response to fear- or alarm-inducing stimuli; also odours released in frustration situations
Maternal	Odours distinguishing lactating from nonlactating females, and possibly pregnant from nonpregnant females

Reprinted from Brown (1979). Reprinted with permission

Table 2. Social relationships which may involve individual recognition by odour.

1. Parental investment (mother-offspring interaction)
2. Kin selection (kin recognition)
3. Male-male competition (dominance, territory recognition)
4. Mate selection (mate recognition)

of lactating females produced an anal excreta odour which were as attractive to pups as those of lactating females and more attractive than the excreta of males eating a diet of lab chow only or those that consumed faeces of nonlactating females. These results suggested that the "maternal odour" of lactating female rats may contain information on their species and individuality as well as their lactation status and that the consumption of faeces of other rats may result in a "colony odour" that infant rats recognize (Brown, 1983).

3 The MHC and Individual Odours and Habituation-Dishabituation Tests at Cambridge

In 1985, David MacDonald from the Zoology Department at Oxford University and I edited a two-volume book on *Social Odours in Mammals* (Brown & Macdonald, 1985), in which we attempted to identify every source of chemical signals in each genus of mammals, the scent-marking behaviours associated with these chemical signals and the information contained in them. This led to a long series of experiments on the constituents of individual odours in laboratory rats which was done with Bruce Roser and Prim Singh at the Institute of Animal Physiology and Genetics Research, Babraham, Cambridge. These studies began with the influence of genetic differences in the major histocompatibility complex (MHC) on odours of individuality, a study that used the habituation-dishabituation method for examining odour discrimination (Brown et al., 1987) that we published in Nature (Singh et al., 1987). The MHC genes encode proteins which are expressed on the surface of each cell in the body. These proteins display peptide fragments to the T-cells of the immune system, enabling the immune system to distinguish between “self” and “non-self” cells and thus identify pathogens. The MHC proteins also contribute to the chemical signals which produce the odours of individuality. These studies showed that rats could discriminate between the urine odours of rats that differed in their MHC (MHC-congenic strains) and between the odours of two individuals of the same strain, but not between two urine samples from the same individual (Fig. 1).

4 What Determines the Odour of Individuality?

Following these results, we conducted a long series of experiments on the role of class I and class II MHC antigens in determining individual odours (Brown et al., 1989), the effects of rearing MHC-congenic rats in germ-free environments (Singh et al., 1990) and the effects of gonadal hormones on individual odours (Brown, 1988). We also looked at individual recognition by urine odours in gerbils (Brown et al., 1988) and preputial gland odours in mice (Ninomiya & Brown, 1995). Heather Schellinck used the habituation-dishabituation procedure to investigate the effects of gut bacteria (Brown & Schellinck, 1995) and the Y chromosome (Schellinck et al., 1993a) on individual odours in rats and the effects of diet on the individual odours of mice (Schellinck et al., 1992). All of these manipulations influenced the discriminability of individual odours. It has become clear that there are constant genetically mediated factors such as the MHC proteins and the major urinary proteins that form the basics of the individual odour, which can be modulated by diet, gonadal hormones and other factors (Schellinck & Brown; 1999; Roberts, et al., 2018).

5 Using the Operant Olfactometer to Study the Discrimination of Individual Odours in Rats

One of the problems with using the habituation-dishabituation method is that it is difficult to interpret the causes of a non-response to an odour during the dishabituation test, as in Fig. 1c (Schellinck & Brown, 1995), and therefore we started to use the operant olfactometer to study the discrimination of individual odours in rats (Fig. 2; Schellinck et al., 1991). Using this apparatus, we examined the influence of genetic and

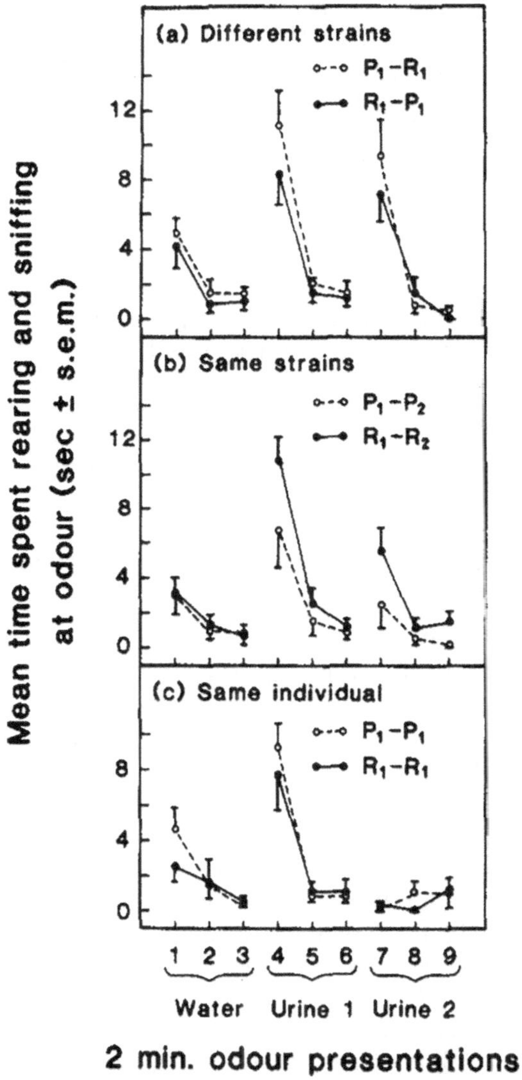

Fig. 1. Mean (± s.e.m.) time rearing and sniffing at urine odours of rats of (**a**) two different genotypes, (**b**) two individuals of the same genotype and (**c**) two samples from the same individual on each 2-min presentation. Abbreviations for stimulus odours: P = PVG and R = PVG.R1 rat urine. "PVG" is an abbreviation for "Piebald Virol Glaxo", a strain of inbred rat. PVG and PVG.R1 rats differ only in the A region (class 1) of the MHC. (From Brown et al., 1987, reprinted with permission).

dietary cues on the individual odours of mice (Brown et al., 1996) and the effect of altering gut bacteria on the individual odours of rats (Schellinck & Brown, 2000). The beauty of the operant method was that we could immediately measure the rat's ability to discriminate between urine odours of two MHC-congenic strains (over 85% correct responses) and their inability to discriminate between two samples from the same individual (50% correct) (Fig. 2).

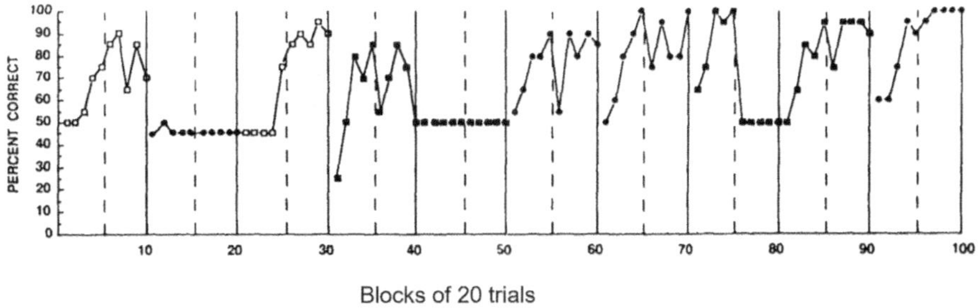

Fig. 2. The learning curves for one rat in the discrimination of the urine odours from individual MHC-congenic (PVG vs. PVG.Rl) rats that were housed in germ-free conditions. After five blocks, both the S+ and the S− urine odours were switched to different aliquots of the same sample (□), to different samples from the same individual (●) or to samples from different individuals (■). Inability to discriminate the two odours shows as 50% correct, and ability to learn to discriminate shows as an increase to 85% correct or better. (From Schellinck et al., 1991, reprinted with permission).

6 Odour Discrimination Is Necessary for Neuroendocrine Responses to Pheromones

Primer pheromone effects involve the activation or inhibition of the hypothalamic-pituitary-gonadal system in mice to accelerate or delay puberty or oestrus in females, for example (Brown, 1985b). We hypothesized that these primer effects involve the discrimination of novel from familiar conspecific odours and tested this in pre-pubertal female mice by exposing them to odours from novel males. The results showed that the odours of unfamiliar males accelerated puberty (Fig. 3) while peppermint odour and odours from the home cages of the females did not (Schellinck et al., 1993b). While there has been research on the differences between "innate versus learned" odour processing in the mouse olfactory system (Kobayakawa et al., 2007), and there is continued interest in the ability of odours from unfamiliar males to accelerate puberty in females (Fu et al., 2022; Wölfl et al., 2023), there has been little research on how the olfactory system that underpins innate responses to odours learns to discriminate between the odours of different individuals in the production of primer pheromone effects.

7 Developing a Theory of the Constituents of Individual Odours

Based on our research, we worked out a theory of the role of the immune system in determining individual odours (Brown, 1995) and then expanded this theory to include the interactions among the genetic, hormonal, immune and metabolic factors contributing to individual odours (Fig. 4; Schellinck & Brown, 1999).

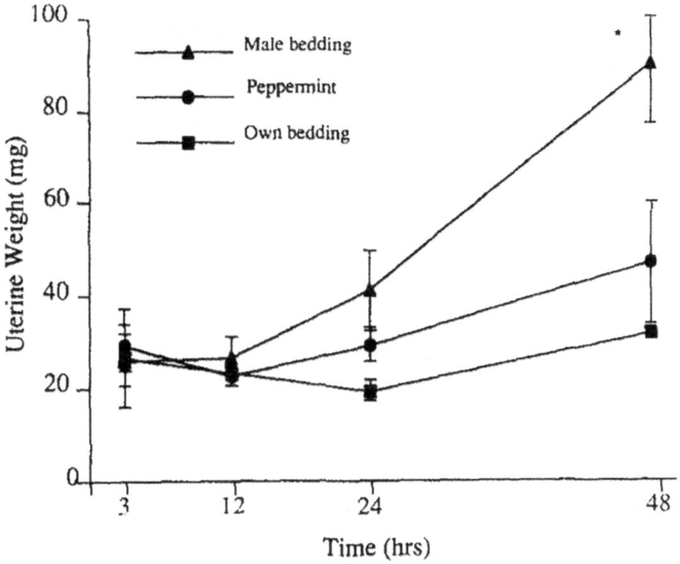

Fig. 3. Mean (± s.e.m.) uterine weights of juvenile female mice exposed to odours from novel male bedding, own bedding or peppermint odour for 3, 12, 24 or 48 h. The asterisk indicates a significant difference in weight for the uteri of those mice exposed to male bedding for 48 h compared with the uteri of mice in all other conditions. (From: Schellinck et al., 1993b, reprinted with permission).

8 Using the Olfactometer to Study Learning, Memory and Odour Detection Sensitivity in Mice: Signal Detection Analysis

Since we had the operant olfactometer, we began to investigate the basic mechanisms of olfactory learning, memory and odour detection sensitivity in rats and mice. Our first study was on olfactory working memory in rats (Winters et al., 2000). Later, when examining cognitive deficits in mouse models of Alzheimer's disease (AD), we used olfactory delayed matching to sample as a test of working memory in the transgenic 5xFAD mouse model (Roddick et al., 2014) and odour discrimination ability in the 3xTg and 5xFAD mouse models of AD (Roddick et al., 2016). These studies showed only small deficits in olfactory learning and memory in these mice. More recently, we began to use signal detection theory to analyse olfactory learning in mice (Roddick et al., 2022). In addition to providing a trial-by-trial learning curve for each mouse, signal detection procedures enable calculation of the sensitivity index (d′) and response bias (c) during learning, which allows us to investigate the decision-making strategies of mice of different genotypes, sexes and ages during olfactory learning.

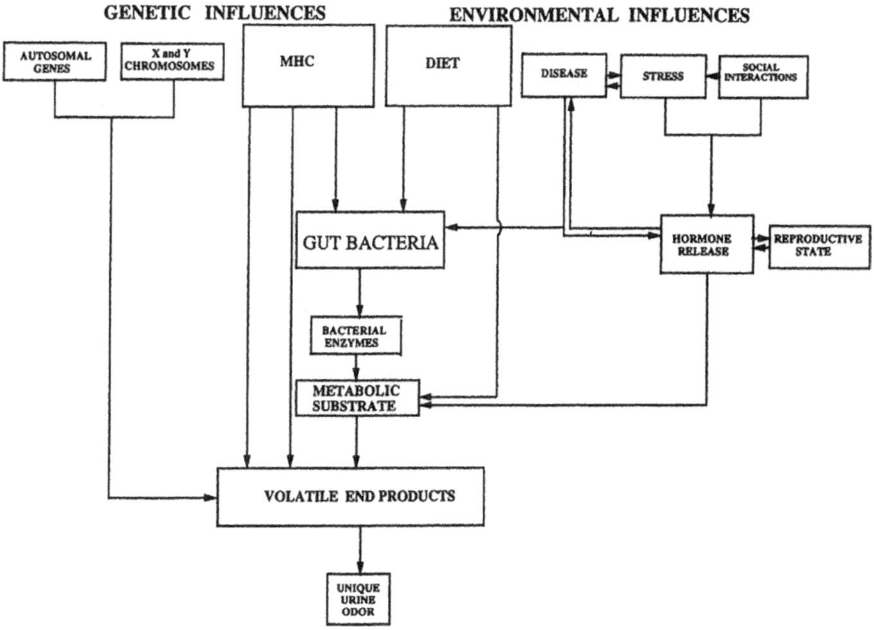

Fig. 4. A hypothetical model for the interactions among the major histocompatibility complex (MHC), gut bacteria, diet and other genetic, environmental and physiological factors which may be involved in the production of urinary odours of individuality in rodents. (From Schellinck & Brown, 1999, reprinted with permission).

9 Using the Olfactometer to Study Reversal Learning in Mice

Slotnick and Katz (1974) showed that rats could form olfactory learning sets which resulted in near errorless discrimination learning, and we found that the same type of learning-to-learn occurred in mice. Using 18 pairs of odours, mice developed a learning-to-learn strategy (Fig. 5) and reached near errorless learning by the sixth odour pair (Roddick et al., 2023).

We are now using signal detection theory to dissociate the phases of reversal learning in which mice are first trained to discriminate a rewarded odour (A+) from a non-rewarded odour (B−), and then, once the discrimination is learned, the rewarded and non-rewarded odours are reversed (B+ vs A−) and the mouse must learn the new discrimination. This is a test of cognitive flexibility. Wyatt Ortibus has trained male and female mice of two genotypes (C57BL 6/J and transgenic Nrxn1+/− mice) on odour discrimination and reversal learning tasks (Ortibus et al., 2024). Figure 6 shows the learning curves of 1 mouse in terms of percent correct responses over blocks of 20 trials (10 S+ and 10 S−). A training odour pair was used to ensure that the mice could learn to discriminate the two odours. The mice were then trained on the discrimination of a new odour pair until a criterion of 85% correct and then presented with the reversal learning phase.

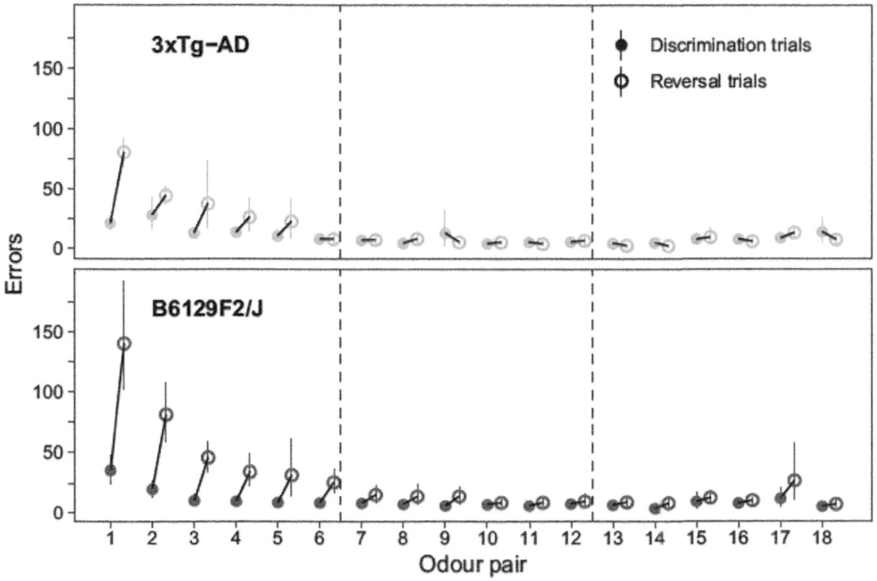

Fig. 5. Mean number of errors (± 95% CI) made by 3xTg-AD and B6129F2/J mice on the odour discrimination and reversal learning trials for each of 18 odour pairs. The difference scores (reversal errors—discrimination errors) are indicated by the lines connecting each odour pair. (From Roddick et al., 2023, reprinted with permission).

There are individual differences in both the odour discrimination learning phase and the reversal learning phase that can be dissociated with signal detection theory (Fig. 6). During the odour discrimination learning phase, graphing hits and false alarms shows how hits (correct responses to the S+) increase over blocks of trials while false alarms (incorrect responses to the S−) decrease over time. During reversal learning, signal detection analysis shows that mice go through a more complex learning process. During the two odour discrimination learning phases, the mouse first responds to both the S+ and S− odours (Respond all), showing both hits and false alarms, and then reduces responding to the S− odour (Respond S+), showing a decrease in false alarms. During reversal learning mice persist in responding to the old S+ as if it were still rewarded as was learned during the discrimination learning phase (Perseverance to old S+), resulting in many false alarms and few hits. They then stop responding altogether (Respond none), showing low numbers of hits and false alarms. In the third phase, they respond to both odours (Respond all) with high numbers of both hits and false alarms. Finally, in the fourth phase (Respond to new S+), they show a high number of hits and fewer false alarms until they reach the criterion of 85% correct. Graphing hits and false alarms shows the changes in response strategy through the two stages of odour discrimination learning and the four stages of reversal learning. This analysis allows us to determine genotype, sex and age differences in learning strategy which will help us to understand cognitive deficiencies in mouse models of neurodegenerative and neurodevelopmental disorders (Brown, 2022).

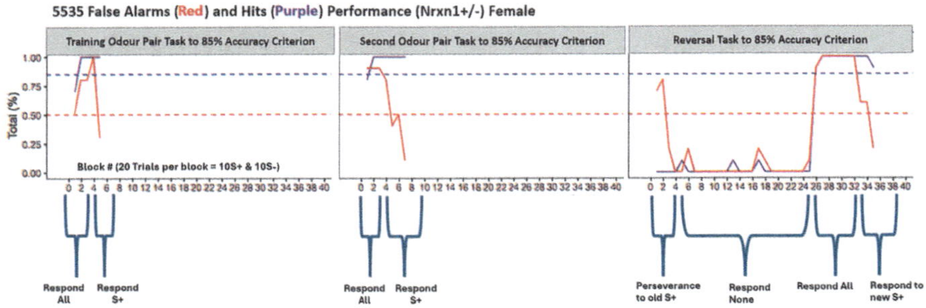

Fig. 6. Learning curves from a single Nrxn1 ± mouse (#5535) on the first (INTRO) and second (OP1) olfactory discrimination tasks followed by reversal learning of the second odour pair (REV1). The red dotted line (50% correct) indicates chance level, and the blue dotted line (85% correct) indicates the criterion for learning the discrimination over blocks of 20 trials (10S+ and 10S−). In a signal detection analysis of the data, the purple line indicates hits (correct responses to the S+), and the red line indicates false alarms (FA, incorrect responding to the S−). During odour discrimination learning, the mouse responds to both the S+ and S− (Respond all) and then reduces responding to the S− (Respond S+). Thus, odour discrimination learning is the result of an increase in hits and false alarms, and then a decrease in false alarms over blocks of 20 trials. Reversal learning has four phases: perseverance in responding to the former S+, responding to neither odour and then responding to both odours before responding only to the new S+. (From: Ortibus et al., 2024, reprinted with permission).

10 Where Does Olfactory Learning and Memory Occur in the Brain?

Much has been written about the neural circuits underlying olfactory learning and memory in the brain (Mori & Sakano, 2021, 2024; Zhang et al., 2024), yet there is still no consensus as to where exactly olfactory memory is located in the brain and how a mouse can learn to discriminate between S+ and S− odours, forming two different types of memories (approach S+, avoid S−) as well as a non-memory to novel odours, which are perceived, but have never been associated with a reward. We hypothesize that olfactory memory consists of complex Hebbian cell assemblies and phase sequences (Langille & Brown, 2018) which are modified by learning and involve all components of the olfactory pathway. These connections can be visualized and analysed using graph theory (Fig. 7) which shows how neurons are synaptically connected into a memory network (Hebbian cell assembly) during learning (Holmaat & Caroni, 2016). One day it might be possible to use graph theory to identify the functional connections for olfactory memory in mice as has been attempted for the human olfactory system, shown in Fig. 8 (Arnold et al., 2020).

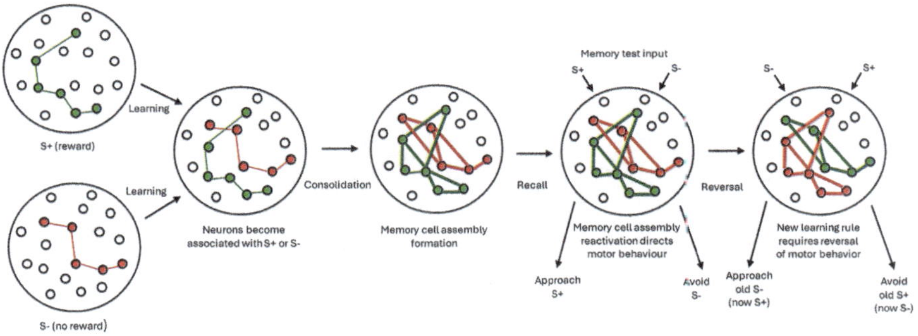

Fig. 7. A graph analysis of odour discrimination learning, memory consolidation, recall and reversal learning for S+ (rewarded) and S− (non-rewarded) odour stimuli. When the S+ odour is associated with a reward, an S+ excitatory cell assembly is formed (green circles), and when the S− odour is associated with non-reward, an S− inhibitory cell assembly is formed (red circles). Both of these cell assemblies (neural networks) exist simultaneously in the olfactory pathways in the brain. During memory consolidation, synaptic connections (green or red lines) within the cell assemblies that represent learning the S+ = reward and S− = no reward rules are modified (stronger connections are symbolized by thicker lines). Re-exposure to the S+ and S− odours in the memory recall test reactivates the S+ and S− cell assemblies, resulting in approach to the S+ and avoidance of the S−. During reversal learning, mice must learn a new set of rules: the old S− is now the new S+ and is associated with reward, while the old S+ is now the new S− and is associated with non-reward. Thus, learning the new association rule results in a reversal of behaviour in which the previously avoided odour is now approached and vice versa. (Drawn by Celia Glenham).

11 Conclusions and Future Directions

How do we integrate our knowledge about the chemical signals used by vertebrates, olfactory sensitivity, odour discrimination, olfactory learning and memory into a coherent theory? That is the job of my students, but one approach is to look back at studies on the role of early experience with odours to determine how the "innate" and learned olfactory systems interact (Kobayakawa et al., 2007; Mori & Sakano, 2024). Remembering Jack Hailman's important paper on how an instinct is learned (Hailman, 1969), we should attend to learning in the innate olfactory system and try to understand how the basic principles of olfactory learning and memory facilitate the use of chemical signals for olfactory communication in vertebrates. We believe that neurodevelopmental and neurodegenerative disorders are caused by the disruption of synaptic connections during development and ageing (Langille & Brown, 2018) and propose that the study of olfactory signalling in rodents will provide a model for understanding synaptic dysfunction in learning, memory and social communication disorders which occur in both Alzheimer's-like and autism spectrum disorders.

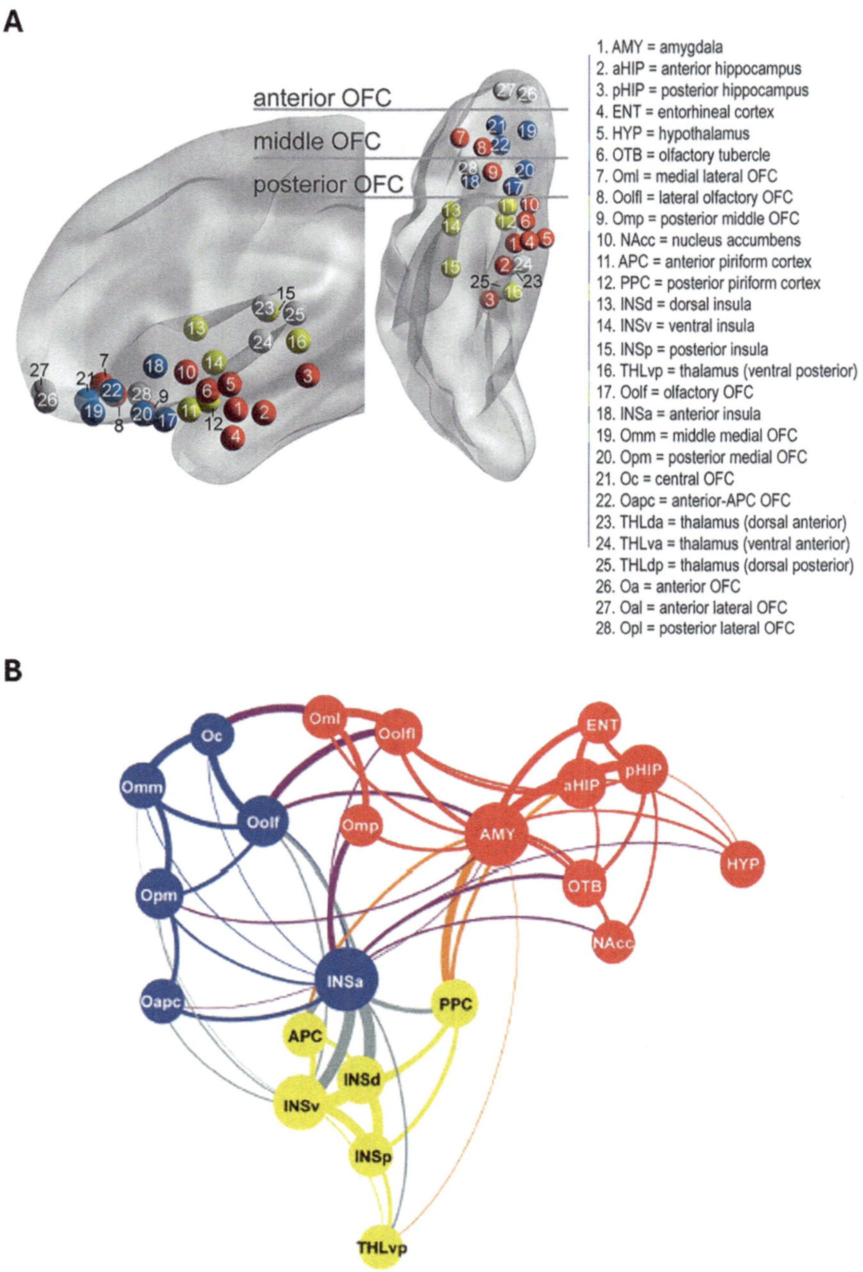

Fig. 8. The human olfactory network exhibits a modular composition of three subnetworks (i.e. the sensory, limbic and frontal) and shows both local segregation (i.e. circuit specialization) and global integration. (**a**) A transparent brain model (in sagittal and axial views) with nodes (regions of interest) for the three subnetwork modules coded in different colours. These are the following: (1) the “olfactory sensory subnetwork” (yellow) consisting of APC, PPC and the ventral, dorsal

Acknowledgements. I would like to thank the many students who worked on the research cited, especially Heather Schellinck, Kyle Roddick and Wyatt Ortibus, Celia Glenham for assistance with the manuscript and the figures and the Natural Sciences and Engineering Research Council of Canada for continued grant support.

References

Arnold, T. C., You, Y., Ding, M., Zuo, X. N., de Araujo, I., & Li, W. (2020). Functional connectome analyses reveal the human olfactory network organization. *eNeuro, 7*(4), 1–14.

Brown, R. E. (1973). The fire-hydrant effect: Stimuli eliciting urine-marking in the rat (*Rattus norvegicus*). *Bulletin of the Ecological Society of America, 54*(44) Abstract.

Brown, R. E. (1974). Sexual arousal, the Coolidge effect and dominance in the rat (*Rattus norvegicus*). *Animal Behaviour, 22*, 634–637.

Brown, R. E. (1975). Object-directed urine-marking by male rats (*Rattus norvegicus*). *Behavioral Biology, 15*, 251–254.

Brown, R. E. (1977). Odor preference and urine-marking scales in male and female rats: Effects of gonadectomy and sexual experience on responses to conspecific odors. *Journal of Comparative and Physiological Psychology, 91*, 1190–1206.

Brown, R. E. (1978). Hormonal control of odor preferences and urine-marking in male and female rats. *Physiology & Behavior, 20*, 21–24.

Brown, R. E. (1979). Mammalian social odors: A critical review. *Advances in the Study of Behavior, 10*, 103–162.

Brown, R. E. (1982a). Chemical signals: Vertebrates and aquatic invertebrates. *Journal of Nervous and Mental Diseases, 170*, 311–312.

Brown, R. E. (1982b). Preferences of pre- and post weanling long-Evans rats for nest odors. *Physiology & Behavior, 29*, 865–874.

Brown, R. E. (1983). Dietary influences in the preferences of pre-weanling Long Evans rats for the anal excreta of adult males. *Physiology & Behavior, 31*, 73–78.

Brown, R. E. (1985a). Effects of social isolation in adulthood on odor preferences and urine-marking in male rats. *Behavioral & Neural Biology, 44*, 139–143.

Brown, R. E. (1985b). The rodents I: Effects of odours on reproductive physiology (primer effects). In R. E. Brown & D. W. Macdonald (Eds.), *Social odours in mammals* (Vol. 2, pp. 245–344). Clarendon Press.

Brown, R. E. (1986). A motivational analysis of odor investigation and urine-marking in male rats (*Rattus norvegicus*). *The Psychological Record, 36*, 69–79.

Fig. 8. (continued) and posterior insula, in addition to the ventral posterior thalamus on the periphery; (2) the "olfactory limbic/paralimbic subnetwork" (red) consisting of the AMY, olfactory tubercle and hippocampus at the centre, in addition to the ENT, HYP, NAcc and three OFC modules (Oml, Oolfl & Omp) on the periphery; and (3) the "olfactory frontal subnetwork" (blue) consisting of the INSa and Oolf at the centre and four additional OFC modules (Oc, Omm, Opm & Oapc) on the periphery. Grey nodes are ROIs not accepted into the olfactory network. (**b**) The topology of the olfactory network. The three subnetworks (sensory, limbic and frontal) are indicated by the yellow, red and blue circles. Line thickness indicates connection strength (based on mean correlation coefficients), and node size reflects connection density (number of connections). (From: Arnold et al., 2020, reprinted with permission).

Brown, R. E. (1988). Individual odors of rats are discriminable independently of changes in gonadal hormone levels. *Physiology & Behavior, 43*, 359–363.
Brown, R. E. (1995). What is the role of the immune system in determining individually distinct body odours? *International Journal of Immunopharmacology, 17*, 655–661.
Brown, R. E. (2022). Genetically modified mice for research on human diseases: A triumph for biotechnology or a work in progress? *The EuroBiotech Journal, 6*, 61–88.
Brown, R. E., & Elrick, D. (1983). Preferences of pre-weanling Long-Evans rats for anal excreta of adult males and females. *Physiology & Behavior, 30*, 567–571.
Brown, R. E., & Macdonald, D. W. (1985). *Social odours in mammals, 2 volumes*. Clarendon Press.
Brown, R. E., & Schellinck, H. M. (1995). Effects of selective depletion of gut bacteria on the odours of individuality in rats. In R. Apfelbach, D. Müller-Schwarze, K. Reutter, & E. Weiler (Eds.), *Chemical signals in vertebrates VII* (pp. 267–271). Pergamon.
Brown, R. E., Singh, P. B., & Roser, B. (1987). The major histocompatibility complex and the chemosensory recognition of individuality in rats. *Physiology & Behavior, 40*, 65–73.
Brown, R. E., Hauschild, M., Holman, S. D., & Hutchison, J. B. (1988). Mate recognition by urine odors in the Mongolian gerbil (*Meriones unguiculatus*). *Behavioral & Neural Biology, 49*, 174–183.
Brown, R. E., Roser, B., & Singh, P. B. (1989). Class I and class II regions of the major histocompatibility complex both contribute to individual odors in congenic inbred strains of rats. *Behavior Genetics, 19*, 659–674.
Brown, R. E., Schellinck, H. M., & West, A. M. (1996). The influence of dietary and genetic cues on the ability of rats to discriminate between the urine odors of MHC congenic mice. *Physiology & Behavior, 60*, 365–372.
Calhoun, J. B. (1962). *The ecology and sociology of the Norway rat*. U.S. Department of Health, Education and Welfare, Public Health Service Publication No. 1008.
Fu, X., Lee, D., Evans, B., & Holy, T. (2022). *The essence of male scent promotes female puberty and estrus*. Unpublished preprint. https://doi.org/10.21203/rs.3.rs-1156787/v1.
Hailman, J. P. (1969). How an instinct is learned. *Scientific American, 221*, 98–108.
Holtmaat, A., & Caroni, P. (2016). Functional and structural underpinnings of neuronal assembly formation in learning. *Nature Neuroscience, 19*(12), 1553–1562.
Kobayakawa, K., Kobayakawa, R., Matsumoto, H., Oka, Y., Imai, T., Ikawa, M., Okabe, M., Ikeda, T., Itohara, S., Kikusui, T., Mori, K., & Sakano, H. (2007). Innate versus learned odour processing in the mouse olfactory bulb. *Nature, 450*(7169), 503–508.
Langille, J. J., & Brown, R. E. (2018). The synaptic theory of memory: A historical survey and reconciliation of recent opposition. *Frontiers in Systems Neuroscience, 12*, 52.
Leon, M. (1974). Maternal pheromone. *Physiology & Behavior, 13*, 441–453.
Leon, M. (1975). Dietary control of maternal pheromone in the lactating rat. *Physiology & Behavior, 14*, 311–319.
Leon, M., & Moltz, H. (1971). Maternal pheromone discrimination by pre-weaning albino rats. *Physiology & Behavior, 7*, 265–267.
Mori, K., & Sakano, H. (2021). Olfactory circuitry and behavioral decisions. *Annual Review of Physiology, 83*, 231–256.
Mori, K., & Sakano, H. (2024). Circuit formation and sensory perception in the mouse olfactory system. *Frontiers in Neural Circuits, 18*, 1342576.
Ninomiya, K., & Brown, R. E. (1995). Removal of the preputial glands alters the individual odors of male MHC-congenic mice and the preferences of females for these odours. *Physiology and Behavior, 58*, 191–194.
Ortibus, W., Roddick, K. M., & Brown, R. E. (2024). Olfactometer methodologies in rodent olfaction research: What can we test about learning and memory? In A. Spink, G. Riedel,

K. Truong, & L. Robinson (Eds.), *Proceedings of the 13th international conference on methods and techniques in behavioral research* (pp. 89–94).
Roberts, S. A., Prescott, M. C., Davidson, A. J., McLean, L., Beynon, R. J., & Hurst, J. L. (2018). Individual odour signatures that mice learn are shaped by involatile major urinary proteins (MUPs). *BMC Biology, 16*(1), 48.
Roddick, K. M., Schellinck, H. M., & Brown, R. E. (2014). Olfactory delayed matching to sample performance in mice: Sex differences in the 5XFAD mouse model of Alzheimer's disease. *Behavioural Brain Research, 270*, 165–170.
Roddick, K. M., Roberts, A. D., Schellinck, H. S., & Brown, R. E. (2016). Sex and genotype differences in odour detection in the 3xTg-AD and 5XFAD mouse models of Alzheimer's disease at 6 months of age. *Chemical Senses, 41*, 433–440.
Roddick, K. M., Fertan, E., Schellinck, H. M., & Brown, R. E. (2022). A signal detection analysis of olfactory learning in 12-month-old 5xFAD mice. *Journal of Alzheimer's Disease, 88*, 37–44.
Roddick, K. M., Schellinck, H. M., & Brown, R. E. (2023). Serial reversal learning in an olfactory discrimination task in 3xTg-AD mice. *Learning & Memory, 30*, 310–319.
Schellinck, H. M., & Brown, R. E. (1995). Methodological questions in the study of the rat's ability to discriminate between the odours of individual conspecifics. In R. Apfelbach, D. Müller-Schwarze, K. Reutter, & E. Weiler (Eds.), *Chemical signals in vertebrates 7* (pp. 427–436). Pergamon.
Schellinck, H. M., & Brown, R. E. (1999). Searching for the source of urinary odors of individuality in rodents. In R. E. Johnston, D. Müller-Schwartze, & P. W. Sorenson (Eds.), *Advances in chemical signals in vertebrates* (pp. 267–280). Kluwer Academic/Plenum Publishers.
Schellinck, H. M., & Brown, R. E. (2000). Selective depletion of gut bacteria alters but does not eliminate odors of individuality in *Rattus norvegicus*. *Physiology & Behavior, 70*, 261–270.
Schellinck, H. M., Brown, R. E., & Slotnick, B. M. (1991). Training rats to discriminate between the odors of individual conspecifics. *Animal Learning & Behavior, 19*, 223–233.
Schellinck, H. M., West, A. M., & Brown, R. E. (1992). Rats can discriminate between the urine odors of genetically identical mice maintained on different diets. *Physiology & Behavior, 51*, 1079–1082.
Schellinck, H. M., Monahan, E., Brown, R. E., & Maxson, S. C. (1993a). A comparison of the contribution of the major histocompatibility complex (MHC) and Y chromosomes to the discriminability of individual urine odors of mice by long-Evans rats. *Behavior Genetics, 23*, 257–263.
Schellinck, H. M., Smyth, C., Brown, R. E., & Wilkinson, M. (1993b). Odor-induced sexual maturation and expression of c-fos in the olfactory system of juvenile female mice. *Developmental Brain Research, 74*, 138–141.
Singh, P. B., Brown, R. E., & Roser, B. (1987). MHC antigens in urine as olfactory recognition cues. *Nature, 327*, 161–164.
Singh, P. B., Herbert, J., Roser, B., Arnott, L., Tucker, D. K., & Brown, R. E. (1990). Rearing rats in a germ-free environment eliminates their odours of individuality. *Journal of Chemical Ecology, 16*, 1667–1682.
Slotnick, B. M., & Katz, H. M. (1974). Olfactory learning-set formation in rats. *Science, 185*, 796–798.
Winters, B., Mathieson, W. R., McGregor, I. S., & Brown, R. E. (2000). An automated two-choice test of olfactory working memory in the rat: Effect of scopolamine. *Psychobiology, 28*, 21–31.
Wölfl, S., Zala, S. M., & Penn, D. J. (2023). Male scent but not courtship vocalizations induce estrus in wild female house mice. *Physiology & Behavior, 259*, 114053.
Zhang, Y. J., Lee, J. Y., & Igarashi, K. M. (2024). Circuit dynamics of the olfactory pathway during olfactory learning. *Frontiers in Neural Circuits, 18*, 1437575.

Avian Chemical Ecology: History and Future Directions

Leanne A. Grieves[1,2] (✉)

[1] Cornell University, Lab of Ornithology, Macaulay Library, Ithaca, NY, USA
[2] University of Saskatchewan, Department of Biology, Saskatoon, Saskatchewan, Canada
leanne.grieves@usask.ca

Abstract. It is now well established that birds have a fully functional olfactory system and use chemical information in a wide variety of contexts, including navigation, foraging, predator detection, individual recognition, and mate choice. However, many unanswered questions in avian chemical ecology remain. For example, the origin and function of avian chemosignals involved in intraspecific chemical communication and their role in sexual signalling and olfactory crypsis are unresolved. Many reviews have been published on avian chemical ecology compared to the number of empirical studies. The goal of the present review, then, is not to recapitulate what has already been summarized in the literature, but to place this information in the broader historical and technical context of chemical ecology and bring together recommendations that will advance our understanding of chemically mediated interactions in birds. I first provide a brief overview and history of chemical ecology as a whole and avian chemical ecology in particular, including an overview of relevant contributions to Chemical Signals in Vertebrates. Next, leveraging guiding themes in chemical ecology, I provide recommendations for future research, emphasizing methodological development, study design, comparative approaches, experimental manipulation, and the importance of collaboration between biologists and chemists. While this review considers avian chemical ecology specifically, the recommendations outlined are broadly applicable to any study species or system.

Keywords: Bird odour · Chemical analysis · Chemical communication · Comparative methods · Infochemical · Olfaction · Preen oil · Uropygial gland

1 Introduction

The field of chemical ecology, the study of chemically mediated interactions between organisms and their environment, is generally considered to have originated in the 1950s. This timeline corresponds with the recognition of plant secondary metabolites as key components of plant-insect interactions (Hartmann, 2008) and early studies on

L. M. Schulte (ed.), *Chemical Signals in Vertebrates 16*,
https://doi.org/10.1007/978-3-032-30752-1_3

olfaction in vertebrates (Tucker & Smith, 1976). However, the importance of chemical interactions among plants and between plants and insects was noted nearly 100 years earlier, in the 1870s through 1900s, by pioneering plant and insect chemical ecologists (Hartmann, 2008). The well-developed olfactory anatomy of birds had been noted as early as 1789 (Strong, 1911; Wenzel, 2007), and a review of sight and smell in vertebrates was published in 1881 (Allen, 1881). Of course, Indigenous peoples have understood the importance of plant and animal chemical interactions since time immemorial; for example, Western explorers travelling to South America in the 1820s documented the long-standing use of toxins collected from dendrobatid frogs by Indigenous hunters (Myers & Daly, 1983).

The famous artist, naturalist, and ornithologist John James Audubon is often given the dubious credit of originating the misconception that birds are either anosmic or have an underdeveloped sense of smell. In 1826, Audubon published research findings from which he concluded that black vultures (*Coragyps atratus*) rely on sight rather than smell to find food (Audubon, 1826). His experimental design has since been criticized (Stager, 1967), and follow-up experiments on turkey vultures (*Cathartes aura*) provided evidence of olfactorily mediated food location (Stager, 1964). While Audubon may have popularized the myth that olfaction is an unimportant sensory modality for birds, it seems more likely that this misconception was perpetuated by an overall bias among researchers emphasizing the stunning visual and acoustic displays of birds, trade-offs among sensory modalities, and the presumed importance of the overall size of sensory anatomical structures.

In his review of sight and smell in vertebrates, Canadian science writer and novelist Grant Allen began by stating "Sight and smell stand to one another to some extent in rough inverse ratio. Where sight is the main intellectual sense, smell is generally deficient or unimportant; where smell is the main intellectual sense, sight is generally deficient or unimportant". Allen's arguments against the importance of smell in birds were based on the large size of the optic nerve, the brilliant colours of many birds, and their apparent visual acuity, contrasted with the small size of the olfactory lobes, apparent lack of evidence for scent glands as sexual ornaments, and lack of behavioural evidence for sniffing (Allen, 1881). However, Allen also acknowledged the likelihood of variation in olfactory capabilities among species, noting that birds of prey may have a better sense of smell than granivores and pointing to the fact that, while similar in body mass, vultures have larger olfactory nerves than turkeys (Allen, 1881).

2 Chemical Signals in Vertebrates

The first Chemical Signals in Vertebrates (CSiV) meeting was held in Saratoga Springs, New York, USA, in 1976. Since then, meetings have typically been held every 3 years, alternating between North American and European host institutions. On completion of each meeting, a conference book of submissions from meeting attendees is published. Mammals have consistently been a major focus of the research presented at CSiV, with

Table 1. Number of papers published in Chemical Signals in Vertebrates (CSiV) volumes that focused on birds.

CSiV	Num. papers	Topic	Authors	Year
1	0			1977
2	1	Foraging	Hutchison and Wenzel	1980
3	0			1983
4	2	Chemoreception	Mason and Clark; Wenzel	1986
5	0			1990
6	8	Anatomy, chemoreception, foraging, navigation	Benvenuti et al.; Berkhoudt; Clark and Shah; Dolbeer et al.; Mason et al.; Shah et al.; Wenzel; Wiltschko and Wiltschko	1992
7	0			1995
8	1	Chemoreception	Clark	1999
9	1	Foraging	Nevitt	2001
10	1	Chemoreception	Cunningham and Nevitt	2005
11	0			2008
12	4	Chemoreception, communication	Bonadonna and Mardon; Caspers and Krause; Gwinner; Hagelin et al.	2013
13	6	Communication, defence, diet, methods, microbes	Azzani et al.; Gabirot et al.; Kohlwey et al.; Krause et al.; Röder et al.; Whittaker and Theis	2016
14	0			2019
15	2	Chemical composition, communication	Alves Soares et al.; Pineaux et al.	2023

avian chemical ecology papers making up just 3.9% (26/675) of contributions across the 15 volumes published between 1977 and 2023 (Table 1).

Among the avian chemical ecology papers published in CSiV collections are eight reviews spanning olfaction in Procellariiform seabirds (Wenzel, 1986; Cunningham & Nevitt, 2005), olfactory sensitivity (Wenzel 1992), taste buds (Berkhoudt, 1992), the olfactory map of homing pigeons (Benvenuti et al., 1992), olfactory foraging in seabirds (Nevitt, 2001), chemical communication (Bonadonna & Mardon, 2013), and chemical signatures (Alves Soares et al., 2021). The research article themes include chemical repellents and chemical defence (Mason & Clark, 1986; Dolbeer et al., 1992; Mason et al., 1992; Shah et al., 1992; Clark, 1999; Gwinner, 2013; Röder et al., 2016), navigation (Wiltschko & Wiltschko, 1992), foraging (Clark & Shah, 1992; Krause et al., 2016), communication (Caspers & Krause, 2013; Hagelin et al., 2013; Pineaux et al., 2023), diet (Azzani et al., 2016), chemical signatures (Kohlwey et al., 2016), microbe-chemical interactions (Whittaker & Theis, 2016), and methods (Gabirot et al., 2016).

In general, the themes represented in the CSiV collections mirror those in the avian chemical ecology literature more broadly. I conducted a non-exhaustive literature search using Web of Science (search terms, preen* OR uropygial* OR feather OR skin)

AND (gland* OR secret* OR oil* OR wax* OR chemical*) AND (communication* OR ecology*) and Google Scholar (search terms, avian OR bird AND chemical communication OR chemical ecology OR chemical OR avian chemical communication review) and found 51 reviews published from 1967 to 2025. The review topics include chemoreception and olfaction (23), navigation and foraging (11), communication and reproduction (9), chemical composition (3), chemical defence (2), microbe-chemical

Table 2. Review papers on topics in avian chemical ecology from 1967 to 2025.

Author	Year	Publication	Topic
1. Stager	1967	Am Zool	Chemoreception
2. Wenzel	1967	Olfaction and Taste	Chemoreception
3. Wenzel	1971	Handbook Sens Physiol	Chemoreception
4. Wenzel	1980	Marine Birds	Chemoreception
5. Wenzel	1982	Int Symp Avian Navigation	Chemoreception
6. Papi	1982	Int Symp Avian Navigation	Navigation, foraging
7. Wenzel	1986	CSiV	Chemoreception
8. Papi	1986	Ital J Zool	Navigation, foraging
9. Wenzel	1987	Ann NY Acad Sci	Chemoreception
10. Papi	1989	Ethol Ecol Evol	Navigation, foraging
11. Waldvogel	1989	Curr Ornithol	Navigation, foraging
12. Benvenuti	1992	CSiV	Navigation, foraging
13. Wenzel	1992	CSiV	Chemoreception
14. Able	1996	J Exp biol	Navigation, foraging
15. Dumbacher and Pruett-Jones	1996	Curr Ornithol	Chemical defence
16. Jones and Roper	1997	Physiol Behav	Chemoreception
17. Malakoff	1999	Science	Chemoreception
18. Roper	1999	Adv Study Behav	Chemoreception
19. Nevitt	2001	CSiV	Navigation, foraging
20. Wallraff	2004	Anim Behav	Navigation, foraging
21. Nevitt and Bonadonna	2005	Mar Ecol Prog Ser	Chemoreception
22. Cunningham and Nevitt	2005	CSiV	Chemoreception
23. Hagelin and Jones	2007	Auk	Communication
24. Hagelin	2007	J Ornithol	Communication
25. Wenzel	2007	J Ornithol	Chemoreception
26. Rajchard	2008	Vet Med	Chemoreception
27. Nevitt	2008	J Exp Biol	Chemoreception
28. DeBose and Nevitt	2008	J Chem Ecol	Chemoreception
29. Balthazart and Taziaux	2009	Behav Brain Res	Communication
30. Caro and Balthazart	2010	J Comp Physiol A	Communication
31. Castro et al.	2010	J Avian Biol	Chemoreception
32. Campagna et al.	2012	Chem Senses	Chemical composition
33. Caspers and Krause	2013	CSiV	Communication
34. Bonadonna and Mardon	2013	CSiV	Communication
35. Gagliardo	2013	J Exp Biol	Navigation, foraging
36. Caro et al.	2015	Horm Behav	Communication

(continued)

Table 2. (continued).

Author	Year	Publication	Topic
37. Wallraff	2015	J Exp Biol	Navigation, foraging
38. Wiltschko and Wiltschko	2017	J Exp Biol	Navigation, foraging
39. Maraci et al.	2018	Genes	Microbe-chemical
40. Krause et al.	2018	Adv Study Behav	Chemoreception
41. Prada and Furton	2018	Front Vet Sci	Chemoreception
42. Shutler	2019	J Avian Biol	Nest ecology
43. Abankwah et al.	2020	Comp Cogn Behav Rev	Chemoreception
44. Potier	2020	Zool J Linn Soc	Chemoreception
45. Whittaker and Hagelin	2020	J Chem Ecol	Communication
46. Alves Soares et al.	2021	CSiV	Chemical composition
47. Grieves, Gilles et al.	2022	Biol Rev	Chemoreception
48. Praveenkumar et al.	2023	Avian Biol Res	Microbe-chemical
49. Weldon	2023	Ibis	Chemical defence
50. Alves Soares et al.	2024	Biol Rev	Chemical composition
51. Creece et al.	2025	Ibis	Chemoreception

interactions (2), and nest ecology (1) (Table 2). A recent review of empirical studies in avian chemical ecology reported 151 studies of avian olfaction published between 1964 and 2024 (Creece et al., 2025), suggesting that 1 review paper has been published for every 3 empirical studies!

Given the large number of reviews on avian chemical ecology, there seems at first glance no need to produce—or read—yet another. However, in this review I have endeavoured to provide a unique perspective by summarizing a brief history of research in avian chemical ecology and providing recommendations to advance the field, situating both history and recommendations within the context of chemical ecology as a whole.

3 History of Avian Chemical Ecology

As with early work on chemical ecology in insects and plants, research on olfaction and chemical ecology in vertebrates generally seems to have stalled from the early 1900s until its resurgence in the 1950s and 1960s. Foremost among the avian chemical ecologists of this mid-twentieth-century renaissance were Betsy Bang and Bernice Wenzel. Betsy Bang (1912–2003) was an American scientific and medical illustrator and biologist who made major contributions to our understanding of olfaction in birds through her detailed illustrations, descriptions, and comparative studies of avian olfactory anatomy, which included descriptions from at least 23 orders and 108 species of birds (Bang, 1960, 1964, 1971; Bang & Cobb, 1968). Bang's studies highlighted the extreme variation in olfactory bulb size among bird species (including variation among species within the same order): the percent ratio of the size of the olfactory bulb to the size of the cerebral hemisphere ranges from 3% in the black-capped chickadee (*Poecile atricapillus*, order Passeriformes) to 37% in the snow petrel (*Pagodroma nivea*, order Procellariiformes) (Bang & Cobb, 1968).

Concurrently, American psychologist and physiologist Bernice Wenzel (1921–2018) published sophisticated studies providing electrophysiological evidence of olfactory perception in birds by measuring changes in heart rate and respiratory rate in response to odour stimuli (Wenzel, 1967; Sieck & Wenzel, 1969; Wenzel & Sieck, 1972). Wenzel also contributed some of the first neuroanatomical evidence for olfaction in birds (Wenzel et al., 1969; Rieke & Wenzel, 1978; Wenzel & Meisami, 1987), and behavioural evidence of olfactorily guided foraging in Procellariiformes (Hutchinson et al., 1984; Wenzel, 1986) and kiwis (Wenzel, 1968). Wenzel's neuroanatomical study revealed that northern fulmars (*Fulmarus glacialis*, order Procellariiformes) have roughly twice as many mitral cells (neurons located in the olfactory bulb that receive information from olfactory receptor neurons and transfer that information to other brain regions) as rats and rabbits and that rock pigeons (*Columba livia*) have roughly the same number of mitral cells as mice (Wenzel & Meisami, 1987; Wenzel, 2007), further solidifying the evidence for keen olfactory capabilities in at least some bird species.

In the 1970s, the Italian biologist and World War II veteran Floriano Papi (1926–2016) began a series of influential studies demonstrating evidence of olfactory navigation in homing pigeons (Papi et al., 1971, 1972; Papi, 1976, 1989). Papi showed that anosmic pigeons did not return from unfamiliar release sites while those with an intact sense of smell did (Papi et al., 1971). This work led Papi to develop the olfactory navigation hypothesis, which posited that pigeons learn windborn environmental odours associated with wind direction at their home roosts and use information about both release site and roost odours to determine their direction of displacement and return home (Papi et al., 1972; Benvenuti & Wallraff, 1985; Gagliardo, 2013). Considering the wealth of evidence, the use of olfactory navigation by homing pigeons in particular and by birds in general remains surprisingly controversial (Papi, 1986; Able, 1996; Gagliardo, 2013; Kishkinev, 2015; Wallraff, 2015; Wiltschko & Wiltschko, 2017).

Influential work on avian olfaction and chemical communication continued throughout the 1980s and 1990s with additional studies in anatomy and physiology (Jacob & Ziswiler, 1982; Abalain et al., 1984a; Downing, 1986; Healy & Guilford, 1990; Bohnet et al., 1991; Asnani & Ramachandran, 1993; Clark et al., 1993), chemical repellents and chemical defence (Mason & Clark, 1986; Clark & Mason, 1987; Dolbeer et al., 1992; Mason et al., 1992; Shah et al., 1992; Clark, 1999; Gwinner, 2013; Röder et al., 2016), olfactory navigation (Gagliardo et al., 1988; Benvenuti & Gagliardo, 1996; Guilford et al., 1998), and olfactory foraging (Nevitt, 1994, 1999a, b; Nevitt et al., 1995).

In the 2000s, the advent of affordable, high-throughput genomics techniques revealed that birds have a large number of olfactory receptor genes (with high variability among species), the majority of which are predicted to be functional. The estimated total number of olfactory receptor genes (not merely those estimated to be functional) correlates positively with relative olfactory bulb size, which has been used as an anatomical correlate of olfactory capability in birds since the early days of work on avian olfaction pioneered by Bang and Wenzel (Steiger et al., 2008, 2009; Khan et al., 2015). Strikingly, birds also possess their own unique clade of olfactory receptor genes, the gamma-c clade, which constitutes the majority of olfactory receptor genes in the bird species studied to date (Steiger et al., 2008; Driver & Balakrishnan, 2021).

During this same period, research on interspecific and intraspecific chemical communication took off, with researchers exploring olfactory crypsis (Reneerkens et al., 2005), antipredator behaviour (Amo et al., 2008, 2017; Roth et al., 2008; Mahr & Hoi, 2018, but see Amo et al., 2018; Blackwell et al., 2018), species recognition (Zhang et al., 2013; Krause et al., 2014; Grieves et al., 2019a; Van Huynh & Rice, 2019), sexual signalling (Bonadonna et al., 2009; Hirao et al., 2009; Whittaker et al., 2010, 2011a; Grieves et al., 2019a; Krause et al., 2023; Rivers & DuVal, 2024), individual and kin recognition (Bonadonna & Nevitt, 2004; Coffin et al., 2011; Bonadonna & Sanz-Aguilar, 2012; Caspers et al., 2017; Schlatmann et al., 2022), genetic discrimination (Leclaire et al., 2017b; Grieves et al., 2019c; Pineaux et al., 2023, but see Amo et al., 2022), and parental care (Whittaker et al., 2009; Gwinner, 2013; Leclaire et al., 2017a). The 1980s–2000s also marked a surge of research interest in the avian uropygial gland, chemical analysis of its secretions, and the factors that affect the chemical composition of uropygial gland secretions, also called preen wax or preen oil.

3.1 The Uropygial Gland

The uropygial gland is a large holocrine integumentary gland located near the base of the tail in most bird species (Jacob & Ziswiler, 1982; Salibian & Montalti, 2009). The gland is present in the embryonic stages of all bird species that have been studied, but it is absent in the adults of some species in the orders Struthioniformes, Piciformes, and Psittaciformes, and in some varieties of rock pigeon (*Columba livia*; order Columbiformes) (Moreno-Rueda, 2017). In birds, the entire skin is lipogenic and acts as a sebaceous secretory organ, with the uropygial gland acting as a specialized part (Menon & Menon, 2000; Salibian & Montalti, 2009). Avian odours may be derived not only from the uropygial gland but also from faeces, blood, stomach oils, powder down, plumage, and secretions of the anal gland, salt gland, salivary gland, ear glands, and sebokeratocytes in the skin (reviewed in Hagelin & Jones, 2007). With the exception of plumage, these other odour sources have rarely been studied.

Uropygial gland secretions are typically monoester waxes comprised of a fatty acid esterified to an alcohol moiety. These secretions usually consist of a mixture of fatty acids and alcohols with varying chain lengths and branching patterns, resulting in complex mixtures of potentially hundreds of individual wax esters of variable molecular weight (Haahti et al., 1964; Livezey et al., 1986; Dekker et al., 2000; reviewed in Campagna et al., 2012). Higher molecular weight diester waxes have been identified in the secretions of some Charadriiform shorebirds (Piersma et al., 1999). More recently, researchers have focused on the volatile fraction of preen oil, which in passerines contains predominantly a mix of alkanes, alkenes, alcohols, aldehydes, ketones, carboxylic acids, aromatic and cyclic compounds, amines, amides, and sulphides (Alves Soares et al., 2021, 2024a).

Research from the 2000s onward has tended to focus on understanding the diverse functions of preen oil (and other avian odour sources, such as plumage and whole-body odour) and their role in interspecific and intraspecific communication. Preen oil secretions serve multiple non-mutually exclusive functions in birds including waterproofing, feather maintenance, protection against ectoparasites, pollutant depuration (reviewed in Moreno-Rueda, 2017), olfactory crypsis (Reneerkens et al., 2002, 2005; reviewed in Grieves, Gilles et al., 2022), and social communication via cosmetic colouration (Amat et al., 2011, 2018) and chemical cues (reviewed in Caro et al., 2015; Moreno-Rueda, 2017; Whittaker & Hagelin, 2020; Grieves, Gilles et al., 2022).

The chemical composition of preen oil is dynamic, differs among species (Krause et al., 2014; Grieves et al., 2019a; Van Huynh & Rice, 2019; Fischer et al., 2020), and can be affected by diverse factors such as diet (Thomas et al., 2010; Leclaire et al., 2019), food stress (Reneerkens et al., 2007a, b; Grieves et al., 2020), time of year (Bhattacharyya & Chowdhury, 1995; Soini et al., 2007; Martín-Vivaldi et al., 2009; Fischer et al., 2017; reviewed in Grieves, Gilles et al., 2022), age (Shaw et al., 2011; Grieves et al., 2019b), sex (Jacob et al., 1979; Mardon et al., 2010; Whittaker et al., 2010; Tuttle et al., 2014; reviewed in Grieves, Gilles et al., 2022), levels of circulating sex steroids (Abalain et al., 1984a, b; Bohnet et al., 1991; Asnani & Ramachandran, 1993; Whittaker et al., 2011b, 2018), genotype (Leclaire et al., 2014; Slade et al., 2016; Grieves et al., 2019c; Whittaker et al., 2019a), and host microbiota (Martín-Vivaldi et al., 2010; Jacob et al., 2014; Whittaker et al., 2019b).

Avian preen oil thus has the potential to act as a chemical cue that can convey a wealth of information to receivers, and there is growing evidence that birds use preen oil cues in a variety of social contexts including discrimination of species (Grieves et al., 2019a), sexes (reviewed in Grieves, Gilles et al., 2022), and individuals (Coffin et al., 2011). Other odour sources have also been investigated; for example, birds can discriminate among individuals using whole-body odour (Bonadonna & Nevitt, 2004; Caspers et al., 2017; Fracasso et al., 2018).

Creece et al. (2025) recently reviewed studies testing the behavioural responses of birds to odour cues. While not restricted to studies involving preen oil, of 151 studies the authors considered, 34% (52/151) tested how birds use olfaction during foraging, followed by nesting (28%, 42/151), social interactions (25%, 38/151), predator detection (14%, 21/151), and navigation (13%, 20/151). Foraging studies most often used seabirds, navigation studies used pigeons followed by seabirds, social interaction and nesting studies mostly used seabirds and songbirds, and songbirds were most commonly used in predator detection studies.

While evidence of olfaction was found in all areas evaluated, much remains to be learned. For example, little is known about the use of olfaction in nocturnal species (predicted to be more likely to use olfactory information than diurnal birds), olfactory foraging in landbirds is rarely studied, little is known about olfactory navigation in species other than the domestic pigeon, the role of olfaction in speciation and hybridization is poorly understood, and evidence that birds detect predator odour cues is mixed (Creece et al., 2025). Moreover, current data provide support for both the olfactory crypsis and sex semiochemical hypotheses, but do not allow these hypotheses to be disentangled (Grieves, Gilles et al. 2022). Most avian behavioural studies have focused on just two avian orders, Procellariiformes (seabirds) and Passeriformes (songbirds), and predominantly on species that breed in the Global North (Grieves, Gilles et al., 2022; Creece et al., 2025). Given the relatively small number of behavioural studies overall (approx. 150), and the lack of taxonomic diversity of study species, there remains much to be learned about the ways in which birds use their sense of smell.

4 Recommendations

The field of chemical ecology has largely been guided by four major themes that can be applied to any study system or species. These are to understand the following: (i) diversity of chemicals, (ii) factors driving variation in chemicals, (iii) costs of producing chemicals, and (iv) mechanisms by which chemicals influence interactions among individuals, species, and communities (Dyer et al., 2018). The diversity of and factors influencing avian chemical cues have been relatively well studied for at least a handful of species (Salibian & Montalti, 2009; Whittaker & Hagelin, 2020; Alves Soares et al., 2024a), but our understanding of the costs of chemical production and mechanisms of chemical interactions is limited (though some mechanistic work has been done on the effect of sex steroid hormones on chemical cues, e.g. Abalain et al., 1984b; Whittaker et al., 2011a, b). The field of chemical ecology already has a well-developed theoretical framework; thus, the best way to advance avian chemical ecology is to focus on using modern methodological and statistical approaches in chemical ecology and collecting empirical data to test hypotheses (Dyer et al., 2018).

Most organisms, including birds, produce complex mixtures of diverse, multifunctional compounds. According to the screening hypothesis originally developed for plant chemical ecology, unless they are too costly, higher levels of chemical diversity should increase the probability of some functional chemical activity against a novel actor (e.g. a competitor, predator, or pathogen), similar to how an immune system works. Thus, metabolic traits that favour chemical diversity are expected to be favoured by natural selection (Jones & Firn, 1991). Due to the potential for additive or synergistic effects among chemicals, high compound diversity and plasticity have been suggested to be functional in their own right (Raguso et al., 2015a). Our current understanding of the selection pressures affecting avian chemical diversity is in its infancy, but comparative methods combined with explicit hypothesis testing and experimental manipulation, along with integrated field and laboratory studies, can improve our ability to make generalizations about the biological relevance of avian semiochemicals and allow us to identify exceptions to developing rules. In the following sections, I provide some key considerations and recommendations for research in vertebrate chemical ecology in general and avian chemical ecology in particular, which are summarized in Box 1.

Box 1 Key Considerations and Recommendations for Research in Avian Chemical Ecology

Sample type	Commonly used sample types include preen oil, feathers from different body regions, and whole-body volatile emissions Choice of sample type should be based on the study goals. A priori chemical characterization and comparison of sample types may be required to determine the appropriate sample type for a given study Characterizing and comparing the chemical composition of different sample types can itself help identify biologically relevant compounds used in avian chemical communication
Methodological development	Methodological studies to evaluate the effect of sampling and extraction methods and analytical techniques on study results, including evaluation of sterilization methods and sources of contamination using blanks, replicates, and standards are essential Standardization is required before meaningful biological comparisons of results across studies can be made Key methodological considerations include sample type; sample collection and storage; extraction type, time, and temperature; solvent type (where applicable); analytical technique, including instrument type and settings
Hypothesis testing	Clearly framing a priori hypotheses and predictions, testing alternative hypotheses or otherwise allowing for reliable evaluation of support for non-mutually exclusive hypotheses where necessary, and avoiding post hoc hypothesis fitting
Experimentation	Characterizing chemical profiles and diversity and identifying variation among individuals or groups of interest provides a valuable starting point for the development of experimental methods to test whether chemical variation has any biological function Carefully designed experimental manipulations (typically behavioural assays) are required to demonstrate functional consequences of variation in chemical profiles among individuals or groups of interest

4.1 Choosing the Right Sample

While sampling and study design have been reviewed elsewhere (e.g. Whittaker & Hagelin, 2020; Grieves, Gilles et al., 2022; Alves Soares et al., 2024a, b), the choice of sample type is worth further consideration. The most common sample types collected in avian chemical ecology research are preen oil, feathers, and whole-body volatile emissions. Only a handful of studies have analysed the chemical composition of preen oil and feathers collected simultaneously from the same individuals, with some studies finding substantial overlap in preen oil and feather chemicals (Zhang et al., 2009, 2010; Mardon et al., 2011; Alves Soares et al., 2024b) and others finding substantial differences (Bolliger & Varga, 1961; Jacob & Grimmer, 1975; Jacob & Pomeroy, 1979;

Jacob & Ziswiler, 1982; Grieves et al., 2024). Characterizing and comparing the chemical composition of these and other biological sample types from birds is important in its own right and can help researchers identify biologically relevant compounds involved in avian chemical communication (Alves Soares et al., 2024a). While routine collection and analysis of multiple sample types may be ideal, this is not always possible for both technical and financial reasons, so the choice of sample type(s) must often be made strategically based on the study goals.

For example, in species such as the crested auklet (*Aethia cristatella*), which are known to produce a fragrant tangerine-like plumage odour associated with courtship displays (Hagelin et al., 2003) or other species with notable plumage odour (e.g. Leach's storm petrels, *Oceanodroma leucorhoa*; Jennings & Ebeler, 2020), collecting feather or whole-body volatile samples seems an appropriate choice. In contrast, many songbird species lack a distinctive odour (Grieves, unpublished data); in these species, collecting preen oil may be more appropriate. It must also be noted that not all extant bird species possess a uropygial gland (e.g. some species in the orders Struthioniformes, Piciformes, and Psittaciformes), so collecting other sample types is necessary for these species.

4.1.1 Preen Oil. A major benefit of working with preen oil is its relative ease of collection and storage. For smaller species, a micropipette or nonheparinized capillary tube can be used to gently probe the papilla of the uropygial gland, causing preen oil to be expressed directly into the tube. Samples can be stored on ice in the field and then transferred to a − 20 or − 80 °C freezer for storage pending analysis (Whittaker et al., 2010; Grieves et al., 2019b). Another advantage of this method is that frozen preen oil can be easily pushed from the capillary tube into a glass vial using a plunger (e.g. from a Hamilton syringe) and then weighed, enabling quantitative analysis.

For species with larger uropygial glands, preen oil is often collected by rubbing the gland gently with sterile cotton until oil is expressed onto the cotton (Reneerkens et al., 2005; Gilles et al., 2024), which can then be stored in a glass vial (Fischer et al., 2017, 2020). A drawback of this method is the difficulty in obtaining the mass of preen oil to allow for quantitative analysis, so it is worth considering the development of methods in which the oil can be expressed from the gland and then collected into a capillary tube or other container so the sample can be weighed. Technically, individual cotton swabs could be weighed and labelled prior to sterilization and use for collection, but this is not recommended because it introduces an impractical requirement to individually package, label, and track each piece of cotton, so it can be linked to a particular sample, increasing logistical difficulties and the risk of labelling errors.

Baking capillary tubes and aluminium foil to sterilize them prior to sample collection and then storing foil-wrapped tubes in sterile WhirlPaks should minimize contamination (Grieves et al., 2024). Similarly, for cotton wool, solvent washing followed by baking can be used to minimize contamination (Nonnamaker et al., 2023). Methodological tests for contamination using these methods are warranted, and field blanks should always be collected to test for contamination during data collection. For whole preen oil (i.e. non-volatile) analysis, the chemical composition of preen oil remains stable over long-term storage (Grieves, unpublished data), but volatiles may be more sensitive to long-term storage so repeated time course studies, along with comparative studies focusing on a range of non-volatile, "semi-volatile", and volatile compounds, should be performed to test the effects of storage in capillary tubes and on cotton wool or other substrates.

One drawback to using preen oil is that this sample type imperfectly reflects the whole bird odour that individuals likely perceive. The chemical composition of preen oil may be altered by environmental conditions such as temperature, precipitation, humidity, and wind after application onto the bird's body and symbiotic microbes living within or surrounding the uropygial gland or on the skin or other body regions (Whittaker et al. 2019b). However, observational studies have shown that preen oil contains info-chemicals (Whittaker et al., 2010; Grieves et al., 2018, 2019b, 2020) and behavioural experiments demonstrate that birds can detect and respond to odour cues in preen oil (Whittaker et al., 2011a; Grieves et al., 2019a, c), so despite this limitation, preen oil appears to be a reasonable proxy for odour, at least in some species.

4.1.2 Feathers. As with preen oil, feathers are easy to collect and store, and the methods described above can be replicated using feather samples, enabling the standardization essential for comparative work evaluating chemical composition between sample types. Alternatively, feather samples can be immediately placed into vials and frozen pending analysis (Jennings & Ebeler, 2020). A complication of feather sampling then becomes the choice of sampling site. Researchers have sampled feathers from directly around the uropygial gland, which likely more closely reflects the chemical composition of preen oil (Leclaire et al., 2011; Mardon et al., 2011; Jennings & Ebeler, 2020); body feathers, which can be collected from a variety of body regions that may differ in their chemical composition (Hagelin et al., 2003; Leclaire et al., 2019; Grieves et al., 2024); and wing feathers (Zhang et al., 2009, 2010).

Another source of variation is in the method of feather collection, including plucking feathers (Jennings & Ebeler, 2020; Grieves et al., 2024), cutting feathers (Mardon et al., 2011), and rolling a stir bar over the feather surface (Soini et al., 2007). There is a paucity of comparative data on the chemical composition of feathers collected from different body regions, which should be addressed (Alves Soares et al., 2024a). While the method of feather plucking versus cutting seems unlikely to affect analytical results, rolling a stir bar over feathers may be less likely to capture the entirety of chemicals on the feather surface, and the amount captured will depend on the duration and method of rolling (e.g. pressure, body region). In addition, some methods may be more prone to contamination, and this should be evaluated experimentally. Standardizing the feather collection method would facilitate the comparison of results across studies, but the choice of feather sample type may vary depending on study goals and should be carefully considered during study design.

Another important consideration associated with feather collection is quantification. Researchers should either weigh feathers prior to analysis (Grieves et al., 2024) or note the number of feathers collected and perform extractions on a similar number or volume of feathers (Gabirot et al., 2016, 2018). For solvent extraction, unlike preen oil which dissolves completely, feathers do not dissolve, making quantification more complicated. Ideally, the goal is to extract chemicals only from the feather surface, so it is worth evaluating whether any degradation of the feather occurs during solvent extraction that might affect downstream analyses (Bortolotti, 2010). Lastly, as with preen oil, feathers likely imperfectly reflect the whole bird odour that individuals perceive. To the best of my knowledge, no behavioural experiments have tested bird responses to feather odour.

4.1.3 Whole-Body Volatiles. The major advantage of collecting whole-body emissions is that this sample type likely best represents what receivers perceive. A major challenge is to effectively capture whole-body emissions under in situ field conditions. Relatively few studies have collected whole-body odours from birds for chemical analysis; however, whole-body volatile emissions can be collected by holding birds in a clean fibre container (e.g. a nylon sock, Krause et al., 2014; Griebel & Dawson, 2020, or cotton bag, Bonadonna & Nevitt, 2004; Bonadonna et al., 2009) for some period of time, then storing and extracting chemicals from the bag (Krause et al., 2014). Critically, the exact same type and colour of material must be used for all birds within a study to avoid confounding effects of different contaminants that might be extracted from different fibres and dyes (Krause et al., 2014). Preliminary tests to evaluate the range of chemicals that can be captured with a given material (e.g. with respect to volatility and polarity) and the type of contaminants associated with different materials (e.g. nylon versus cotton or nylon socks of different colours) should be conducted prior to choosing a container type, and blank (clean) containers should also be collected in the field and analysed alongside samples.

Another option is to use an air sampling apparatus whereby birds are placed in an airtight glass chamber with controlled airflow that is connected to an activated charcoal filter and an adsorbent tube for volatile compound collection (Douglas, 2006; Gabirot et al., 2016; Alves Soares et al., 2024b). This method may be less likely to result in contamination, but it is critical to ensure the apparatus remains airtight and that a constant airflow is maintained. Preliminary analysis should be performed to determine how frequently the activated charcoal filter should be replaced (Raguso & Pellmyr, 1998), and blank samples should be collected regularly to assess contamination. In addition, the apparatus must be thoroughly cleaned and dried between uses, which may pose difficulties under field conditions, depending on the frequency of bird sampling and capture.

A third option is to sample air in the immediate vicinity of the study subject, such as by placing tubing filled with adsorbent material and connected to an air pump within a burrow or nest cavity to collect volatiles emitted from both the environment and the bird (Zidat et al., 2023). This method may be particularly useful for burrow- or cavity-nesting species that can be sampled without disturbing the bird, e.g. during incubation at the nest, but may not be effective for open-cup or ground-nesting species that are unlikely to remain in situ during sampling.

For all methods, care should be taken to minimize stress to birds during sampling, as stress can affect body odour (Inagaki et al., 2014; Grieves et al., 2020; Parr-Cortes et al., 2024). Placing birds in a dark, quiet, and confined space like a cotton bag is a commonly used approach in ornithology to reduce stress and keep birds calm prior to handling, which may present one advantage of this odour collection method over a glass container sampling method. It may be possible to reduce stress using the container method by making the container small enough to limit the birds' movement and by tinting the outside of the glass or placing a box over the entire apparatus so birds can be kept in the dark during sampling. The container method may also present a greater risk of birds escaping, so care should be taken during the design phase to minimize this risk.

For any new study, preliminary data should be collected to determine the length of time birds must be held in either fibre containers (bags), glass containers, or by sampling within a burrow or cavity, and a balance must be struck between ensuring sufficient sample is collected for analysis while minimizing handling and disturbance time for the birds. Suitable collection time will likely vary across species, as some species are more odorous than others (Weldon & Rappole, 1997; Röder et al., 2016).

4.2 Methodological Development

Chemical ecologists have long recognized the importance of using comparative approaches to evaluate the effect of different sampling methods and analytical techniques on study results. This is crucial because without standardization, it is impossible to make meaningful biological comparisons of results across studies (Raguso & Pellmyr, 1998). For example, Agelopoulos and Pickett evaluated the effects of different dynamic headspace sampling methods (solid-phase microextraction [SPME], thermal desorption with the trapping sorbents Porapak Q and Tenax TA, and solvent desorption with the same sorbents, followed by gas-chromatography with flame ionization detection [GC-FID] and mass spectroscopy [GC-MS]) on volatiles emitted from both a standardized chemical mixture and ripe banana. They found that the ratios of some compounds in both the standardized mixture and ripe banana differed between SPME and the polymers using both thermal and solvent desorption while the ratios of the compounds did not substantially differ among polymers or with desorption type (Agelopoulos & Pickett, 1998).

Raguso and Pellmyr also evaluated dynamic headspace sampling methods, considering the effects of different trapping sorbents (Porapak Q, Tenax TA, Tenax + activated charcoal), elution solvents (diethyl ether, hexane, and pentane), and vacuum pump flow rates (500 mL/min, 1 L/min) on the collection of volatiles using both a standardized chemical mixture and living flowers (Raguso & Pellmyr, 1998). Porapak Q again trapped more volatiles than Tenax TA, solvent type affected volatile recovery in complex ways, with more volatiles detected in flower samples eluted with hexane, and a higher flow rate was associated with better collection of lower volatility compounds but also more contamination (Raguso & Pellmyr, 1998). Comparing these two studies reveals different results for different standardized mixtures and natural products, highlighting the importance of investing time in comparative method development for projects involving new chemical mixtures (e.g. species).

Similar methodological tests have been performed by mammalian chemical ecologists, for example, by comparing different sample collection types (cotton wool, viscose), swab sterilization methods ("baked", "solvent-soaked", "soaked-and-baked), and extraction methods (solventless, SPME, thermal desorption [TD], mobile thermal desorption and GC-MS; solvent, hexane, diethyl ether; Birkemeyer et al., 2016; Kücklich et al., 2017; Nonnamaker et al., 2023). General findings from these and related studies suggest that TD outperforms SPME (Kücklich et al., 2017; Weiß et al., 2018; Nonnamaker et al., 2023), and, while some studies report issues with cotton wool for volatile sampling (Birkemeyer et al., 2016), others find that cotton rinsed in methanol followed by pentane ("soaked") and then heated (e.g. at 190 °C for 210 min with a pressure of 30 psi of high purity nitrogen; "baked") substantially reduced contamination and allowed for the recovery of a wide range of volatile compounds (Nonnamaker et al., 2023). Sterilization methods should be performed close to the time of sample collection to avoid contamination post-sterilization (Birkemeyer et al., 2016). These comparative studies in plant and mammal chemical ecology indicate the importance of performing methodological comparisons in the early stages of research on any new species or system.

Comparative methodological studies in avian chemical ecology are sorely needed. Two reviews provide excellent starting points for ornithologists new to chemical ecology (for a detailed review of methodological considerations broadly applicable to vertebrate chemical ecology, see Drea et al., 2013; for a review of general sample preparation methods for GC analysis, see de Koning et al., 2009). Nevitt and Prada provide an introduction to best practices for avian chemical sample collection, over-

Table 3. Examples of methodological variation among published studies of avian preen oil.

Extraction solvent	Analytical technique	Study
Acetone	HPLC	Amat et al. (2018)
Acetone	GC-MS	Livezey et al. (1986)
Acetone	GLC	Livezey et al. (1986)
Acetone	TLC	Livezey et al. (1986)
Chloroform	GC-FID	Thomas et al. (2010), Leclaire et al. (2011), Slade et al. (2016), Grieves et al. (2018, 2019a, b, 2020), To (2024)
Chloroform	GC-MS	Jacob et al. (1979), Pan et al. (1979), Kolattukudy et al. (1985, 1987), Thomas et al. (2010), Leclaire et al. (2011), Slade et al. (2016), Grieves et al. (2019b)
Chloroform	TLC	Jacob et al. (1979), Kolattukudy et al. (1985, 1987), Bhattacharyya and Chowdhury (1995)
Chloroform	TLC-FID	Thomas et al. (2010)
Dichloromethane	GC-FID	Burger et al. (2004), Bonadonna et al. (2007), Grieves et al. (2018), Caspers et al. (2022), Gilles et al. (2024)
Dichloromethane	GC-MS	Burger et al. (2004), Haribal et al. (2005, 2011), Bonadonna et al. (2007), Zhang et al. (2009, 2010, 2013), Mardon et al. (2010), Martin-Vivaldi et al. (2010), Shaw et al. (2011), Amo et al. (2012), Jacob et al. (2014), Gabirot et al. (2016), Potier (2020), Van Huynh and Rice (2019), Gilles et al. (2024), Grieves et al. (2024), Jeanjean et al. (2024)
Dichloromethane	EI-MS	Burger et al. (2004)
Ethyl acetate	GC-FID	Azzani et al. (2016), Fluen (2008)
Ethyl acetate	GC-MS	Piersma (1999), Damste et al. (2000), Reneerkens et al. (2002, 2006, 2007a, b), Rasmussen (2013), Fischer et al. (2017, 2020), Thomas et al. (2010)
Ethyl acetate	TLC	Thomas et al. (2010)
Hexane	GC-MS	Mardon et al. (2010), Jacob et al. (2014), Gabirot (2016), Lopez-Perea et al. (2019), Díez-Fernández (2020)

(continued)

Table 3. (continued).

Extraction solvent	Analytical technique	Study
Hexane	EI-MS	Lopez-Perea et al. (2019)
Methanol	GC-MS	Pan et al. (1979), Haribal et al., 2005, 2011; Montalti et al., 2005
Methanol	TLC	Bhattacharyya and Chowdhury (1995)
Solventless		
SPME	GC-MS	Burger et al. (2004), Whelan (2010), Shaw et al. (2011), Mihailova (2014), Leclaire et al. (2019)
SBSE	GC-MS	Soini et al. (2007), Whittaker et al. (2010, 2011a, b, 2013, 2016, 2018, 2019a, b), Tuttle et al. (2014), Forrette (2018), Jennings and Ebeler (2020), Talbott et al. (2022)
SBSE	GC-AED	Soini et al. (2007), Whittaker et al. (2010)
TD	GC-MS	Gabirot et al. (2018)

SBSE stir bar sorptive extraction, *TD* thermal desorption, *HPLC* high-performance liquid chromatography, *GLC* gas-liquid chromatography, *TLC* thin-layer chromatography, *EI* electron ionization, *AED* atomic emission detector

viewing sample types, collection methods, extraction, and analysis (Nevitt & Prada, 2015). More recently, Alves Soares et al. provide in-depth best practices for volatile sample collection, extraction, and analysis, noting that there is substantial variation in the avian chemical ecology literature with respect to sample type, storage, transport, extraction, and analysis (Alves Soares et al., 2024a). Even within a given sample type, for example, preen oil, the methodological variation is extensive (Table 3), precluding our ability to compare results across studies to better understand the composition and diversity of avian preen oil.

To the best of my knowledge, there is only one comparative study of different sample collection and extraction methods on avian chemical samples. Gabirot et al. collected feathers and rubbed cotton wool on body feathers of blue petrels (*Halobaena caerulea*) in situ under field conditions, comparing results from solvent extraction with dichloromethane (DCM) and hexane and TD on feathers and TD and SPME on cotton rubbed on body feathers (Gabirot et al., 2016). TD outperformed solvent extraction for analysis of feather volatiles, and TD and SPME performed similarly for analysis of volatiles from cotton swabs rubbed on body feathers (Gabirot et al., 2016). The authors also note the importance of standardizing all sample collection, handling and storage, preparation, and analytical procedures to allow for robust methodological comparison (Gabirot et al., 2016), consistent with standard best practices in chemical ecology.

4.3 Omics Approaches

In chemical ecology, the major or most abundant compounds are not always biologically active and may not be the targets of selection. By contrast, minor or low abundance compounds often have important biological functions. This creates a challenge for chemical ecologists attempting to parse signal from noise when analysing complex chemical blends—such as those derived from avian uropygial secretions, feathers, or whole-body volatile emissions—especially because modern analytical methods can detect large numbers of compounds in minute quantities. Screening protocols such as bioassay-guided fractionation can help researchers distinguish signal from noise when analysing complex chemical datasets (Raguso et al., 2015b).

Thus, two key aspects of chemical ecology are the development of bioassays to test the activity of extracts or pure compounds and the identification of chemical stimuli responsible for observed interactions. Bioassay-guided fractionation begins with collection of an extract from the organism producing compounds that may elicit the interaction, followed by fractionation to separate compounds and purify the active metabolite(s), then conducting additional bioassays to confirm biological activity and relevance of the purified compound(s). This approach has been the gold standard in chemical ecology since the 1950s, but a major drawback is how time-consuming and costly it is.

Metabolomics, the study of all primary and secondary metabolites produced by an organism, presents a promising alternative. Metabolic profiling uses standard analytical techniques in chemical ecology such as GC-MS, LC-MS, and NMR spectroscopy coupled with multivariate statistical techniques that allow for the visualization of patterns and identification of relevant chemicals. In metabolomics-guided fractionation, once extracts are collected, they are profiled using spectroscopy techniques; then, multivariate statistical methods are used to identify metabolic features that correlate with bioactivity in the extracts. Next, structural identification is made via NMR spectroscopy or database matching. Following identification, commercially available or synthetic metabolites are tested using bioassays to confirm biological activity, and the results are compared to predictions derived from the multivariate modelling performed earlier in the process. While commercial and synthetic metabolites are expensive, compared to bioassay-guided fractionation, metabolomics-guided fractionation allows for more rapid and cost-effective identification of complex mixtures of metabolically active candidate compounds that may elicit biological interactions of interest (Poulin & Pohnert, 2019). Thus, metabolomics and other "-omics" approaches will undoubtedly prove useful in the identification of biologically relevant compounds that influence chemically mediated interactions in birds.

4.4 Hypothesis Testing and Experimentation

Clear hypothesis testing with empirical data is another key to advancing the field of avian chemical ecology, and several recent reviews have provided important recommendations for future research. Whittaker and Hagelin provide a comprehensive list of hypothesized functions of female chemosignals and associated predictions, focusing on intersexual attraction (priming, receptivity, mate choice, quality), intraspecific competition (quality, competitive ability, scent marking, reproductive suppression), and parental behaviour (protection/crypsis, recognition) (Table 3 in Whittaker & Hagelin, 2020). Grieves, Gilles et al. make recommendations for testing predictions of the olfactory crypsis and sex semiochemical hypotheses (Grieves, Gilles et al., 2022). Alves Soares et al. emphasize the importance of determining the origin and function of avian volatiles and characterizing variation in the chemical profiles of different sample types (e.g. uropygial and other glandular secretions, feathers, and feathers from different body regions, skin, and whole-body emissions) (Alves Soares et al., 2024a). These reviews also highlight key considerations for sampling and study design and the need to increase the phylogenetic and geographic diversity of bird species sampled to gain a more comprehensive understanding of avian chemical diversity and function (Whittaker & Hagelin, 2020; Grieves, Gilles et al., 2022; Alves Soares et al., 2024a). Care should also be taken to design studies that allow testing and discrimination of alternative hypotheses. For example, a comparative analysis by Grieves, Gilles et al. (2022) revealed support for both the sex semiochemical and olfactory crypsis hypotheses, non-mutually exclusive hypotheses that cannot be differentiated using data currently in the literature. The authors outline a series of testable predictions that would allow researchers to better differentiate these hypotheses (Grieves, Gilles et al., 2022).

Crucially, characterizing chemical profiles and chemical diversity and identifying differences among individuals, groups, or populations does not demonstrate a function or role for behaviour (Nevitt & Prada, 2015). Carefully designed behavioural assays (i.e. experiments) taking into account the natural history of a given study species or system are required to demonstrate a link between variation and function. Moving beyond the observation of chemical variation to experimentally demonstrate a functional consequence of that variation is essential to advancing our understanding of the role odours play in the behaviour and survival of birds.

4.5 Interdisciplinary Collaboration

Where ecological approaches alone might reveal patterns or context dependency, chemical ecology allows us to identify the mechanisms driving interactions, integrating the functional analysis of chemical traits with fundamental evolutionary questions (Raguso et al., 2015a). Funding agencies are increasingly recognizing the power and potential of interdisciplinary research. Thus, a major opportunity in chemical ecology is that successful, productive research in this field relies on interdisciplinary collaboration between chemists and biologists. Ecologists interested in using the power of chemical ecology to understand behaviour and communication must be able to understand chemical principles and analytical techniques sufficiently to develop robust study design and accurate reporting. Chemists interested in applying evolutionary theory to chemical interactions must be able to understand the need for statistical replication and complex modelling in biology (Raguso et al., 2015b). Bridging these gaps in our knowledge and training through constructive consultation prior to data collection is key to producing robust results, leading to higher-impact research.

5 Conclusions

Chemical ecology has flourished since the 1950s, and employing modern methodological and technological approaches to chemical ecology research will advance our understanding of chemically mediated interactions among individuals, species, communities, and ecosystems. By holding in mind the major themes in chemical ecology, developing robust interdisciplinary collaborations, employing careful study design informed by natural history, emphasizing a combination of chemical characterization, identification, bioassays, and behavioural validation (i.e. experimental manipulation), we are well-poised to gain a deeper understanding of ecological and evolutionary principles in biology. For avian chemical ecologists, as with any other taxon-specific focus, methodological development coupled with comparative approaches, empirical data collection, explicit hypothesis testing, and experimentation will undoubtedly lead to many exciting discoveries about the breadth and depth of chemically mediated interactions among birds and their environment.

In avian chemical ecology, comparative methodological studies that evaluate the effects of different sample types, collection, storage, extraction, analytical, and statistical methods are lacking. Such studies are essential to promote standardization and facilitate our ability to compare results across species and studies. Our understanding of the origin and function of avian chemical cues and signals is also lacking, as are studies characterizing chemical variation and testing behavioural responses to chemical information across species from diverse phylogenetic and geographic backgrounds. Numerous hypotheses in avian chemical ecology have been tested, but results are generally equivocal; for example, hypotheses about the use of chemical cues or signals in speciation, individual recognition, intraspecific competition, intersexual attraction, mate choice, parental behaviour, crypsis, predator detection, chemical defence, foraging, homing, and navigation all warrant further testing. Explicit a priori hypothesis testing coupled with behavioural assays will allow researchers to draw more robust conclusions about the depth and breadth of ways in which birds use chemical information.

Acknowledgements. This review was written with support from the Rose Postdoctoral Fellowship program at the Cornell University Lab of Ornithology. The author thanks editor L. M. Schulte, an anonymous reviewer, K. Keleher, R. Raguso, M. Webster, and A. Wilder for providing constructive feedback on this manuscript, and the organizers of the 16th Chemical Signals in Vertebrates meeting in Frankfurt, Germany, for the invitation to be part of this international research community. This review is dedicated to the memory of Jeremy McNeil (1944–2024), British-Canadian biologist, zoologist, Helen Battle professor of Chemical Ecology at Western University, teacher, mentor, role model, and friend.

References

Abalain, J. H., Amet, Y., Daniel, J. Y., & Floch, H. H. (1984a). Androgen regulation of secretions in the sebaceous-like uropygial gland of the male Japanese quail. *Journal of Endocrinology, 103*, 147–153.

Abalain, J. H., Amet, Y., Daniel, J. Y., & Floch, H. H. (1984b). Androgen control of the secretion in the sebaceous-like preen gland. *Journal of Steroid Biochemistry, 20*, 529–531.

Abankwah, V., Deeming, D. C., & Pike, T. W. (2020). Avian olfaction: A review of the recent literature. *Comparative Cognition & Behavior Reviews, 15*, 149–161.

Able, K. P. (1996). The debate over olfactory navigation by homing pigeons. *Journal of Experimental Biology, 199*, 121–124.

Agelopoulos, N. G., & Pickett, J. A. (1998). Headspace analysis in chemical ecology: Effects of different sampling methods on ratios of volatile compounds present in headspace samples. *Journal of Chemical Ecology, 24*, 1161–1172.

Allen, G. (1881). Sight and smell in vertebrates. *Mind, 6*, 453–471.

Alves Soares, T., Caspers, B. A., & Loos, H. M. (2021). Avian chemical signatures: An overview. In B. Schaal, D. Rekow, M. Keller, & F. Damon (Eds.), *Chemical signals in vertebrates* (Vol. 15, pp. 113–137). Springer.

Alves Soares, T., Caspers, B. A., & Loos, H. M. (2024a). Volatile organic compounds in preen oil and feathers – A review. *Biological Reviews, 99*, 1085–1099. https://doi.org/10.1111/brv.13059

Alves Soares, T., Caspers, B. A., Veit, D., & Loos, H. M. (2024b). Analytical characterization of volatiles present in the whole body odour of zebra finches. *Analytical and Bioanalytical Chemistry, 416*, 5335–5349. https://doi.org/10.1007/s00216-024-05466-8

Amat, J. A., Rendón, M. A., Garrido-Fernández, J., Garrido, A., Rendón-Martos, M., & Pérez-Gálvez, A. (2011). Greater flamingos *Phoenicopterus roseus* use uropygial secretions as make-up. *Behavioral Ecology and Sociobiology, 65*, 665–673.

Amat, J. A., Garrido, A., Portavia, F., Rendón-Martos, M., Pérez-Gálvez, A., Garrido-Fernández, J., Gómez, J., Béchet, A., & Rendón, M. A. (2018). Dynamic signalling using cosmetics may explain the reversed sexual dichromatism in the monogamous greater flamingo. *Behavioral Ecology and Sociobiology, 72*, 1–10.

Amo, L., Galván, I., Tomás, G., & Sanz, J. J. (2008). Predator odour recognition and avoidance in a songbird. *Functional Ecology, 22*, 289–293. https://doi.org/10.1111/j.1365-2435.2007.01361.x

Amo, L., Avilés, J. M., Parejo, D., Peña, A., Rodríguez, J., & Tomás, G. (2012). Sex recognition by odour and variation in the uropygial gland secretion in starlings. *Journal of Animal Ecology, 81*, 605–613. https://doi.org/10.1111/j.1365-2656.2011.01940.x

Amo, L., Tomás, G., & López-García, A. (2017). Role of chemical and visual cues of mammalian predators in nest defense in birds. *Behavioral Ecology and Sociobiology, 71*, 49.

Amo, L., Tomás, G., Saavedra, I., & Visser, M. E. (2018). Wild great and blue tits do not avoid chemical cues of predators when selecting cavities for roosting. *PLoS One, 13*, e0203269.

Amo, L., Amo de Paz, G., Kabbert, J., & Machordom, A. (2022). House sparrows do not exhibit a preference for the scent of potential partners with different MHC-I diversity and genetic distances. *PLoS One, 17*, e0278892.

Asnani, M. V., & Ramachandran, A. V. (1993). Roles of adrenal and gonadal steroids and season in uropygial gland function in male pigeons, *Columba livia. General and Comparative Endocrinology, 92*, 213–224.

Audubon, J. J. (1826). Account of the habits of the Turkey buzzard (*Vultur aura*) particularly with the view of exploding the opinion generally entertained of its extraordinary power of smelling. *Edinburgh New Philosophical Journal, 2*, 172–184.

Azzani, L., Rasmussen, J. L., Gieseg, S. P., & Briskie, J. V. (2016). An experimental test of the effect of diet on preen wax composition in New Zealand silvereyes (*Zosterops lateralis*). In B. A. Schulte, T. E. Goodwin, & M. H. Ferkin (Eds.), *Chemical signals in vertebrates* (Vol. 13, pp. 511–525). Springer.

Balthazart, J., & Taziaux, M. (2009). The underestimated role of olfaction in avian reproduction? *Behavioral Brain Research, 200*, 248–259. https://doi.org/10.1016/j.bbr.2008.08.036

Bang, B. G. (1960). Anatomical evidence for olfactory function in some species of birds. *Nature, 188*, 547–549.

Bang, B. G. (1964). The nasal organs of the black and turkey vultures; a comparative study of the cathartid species *Coragyps atratus* atratus and *Cathartes aura septentrionalis* (with notes on *Cathartes aura falklandica*, *Pseudogyps bengalensis*, and *Neophron percnopterus*). *Journal of Morphology, 115*, 153–183.

Bang, B. G. (1971). Functional anatomy of the olfactory system in 23 orders of birds. *Acta Anatomica, 58*, 1–76.

Bang, B. G., & Cobb, S. (1968). The size of the olfactory bulb in 108 species of birds. *The Auk, 85*, 55–61. https://doi.org/10.2307/4083624

Benvenuti, S., & Gagliardo, A. (1996). Homing behaviour of pigeons subjected to unilateral zinc sulphate treatment of their olfactory mucosa. *Journal of Experimental Biology, 199*, 2531–2535.

Benvenuti, S., & Wallraff, H. G. (1985). Pigeon navigation: Site simulation by means of atmospheric odours. *Journal of Comparative Physiology A, 156*, 737–746.

Benvenuti, S., Ioalè, P., & Papi, F. (1992). The olfactory map of homing pigeons. In R. L. Doty & D. Müller-Schwarze (Eds.), *Chemical signals in vertebrates* (Vol. 6, pp. 429–434). Springer.

Berkhoudt, H. (1992). Avian taste buds: Topography, structure and function. In R. L. Doty & D. Müller-Schwarze (Eds.), *Chemical signals in vertebrates* (Vol. 6, pp. 15–20). Springer.

Bhattacharyya, S. P., & Chowdhury, S. R. (1995). Seasonal variation in the secretory lipids of the uropygial gland of a sub-tropical wild passerine bird, *Pycnonotus cafer* (L) in relation to the testicular cycle. *Biological Rhythm Research, 26*, 79–87. https://doi.org/10.1080/09291019509360326

Birkemeyer, C. S., Thomsen, R., Jänig, S., Kücklich, M., Slama, A., Weiß, B. M., & Widdig, A. (2016). Sampling the body odor of primates: Cotton swabs sample semivolatiles rather than volatiles. *Chemical Senses, 41*, 525–535.

Blackwell, B. F., Seamans, T. W., Pfeiffer, M. B., & Buckingham, B. N. (2018). European Starling (*Sturnus vulgaris*) reproduction undeterred by predator scent inside nest boxes. *Canadian Journal of Zoology, 96*, 980–986.

Bohnet, S., Rogers, L., Sasaki, G., & Kolattukudy, P. E. (1991). Estradiol induces proliferation of peroxisome-like microbodies and the production of 3-hydroxy fatty acid diesters, the female pheromones, in the uropygial glands of male and female mallards. *Journal of Biological Chemistry, 266*, 9795–9804.

Bolliger, A., & Varga, D. (1961). Feather lipids. *Nature, 190*, 1125–1125.

Bonadonna, F., & Mardon, J. (2013). Besides colours and songs, odour is the new black of avian communication. In M. L. East & M. Dehnhard (Eds.), *Chemical signals in vertebrates* (Vol. 12, pp. 325–339). Springer.

Bonadonna, F., & Nevitt, G. A. (2004). Partner-specific odor recognition in an Antarctic seabird. *Science, 306*, 835–835. https://doi.org/10.1126/science.1103001

Bonadonna, F., & Sanz-Aguilar, A. (2012). Kin recognition and inbreeding avoidance in wild birds: The first evidence for individual kin-related odour recognition. *Animal Behaviour, 84*, 509–513. https://doi.org/10.1016/j.anbehav.2012.06.014

Bonadonna, F., Miguel, E., Grosbois, V., Jouventin, P., & Bessiere, J. M. (2007). Individual odor recognition in birds: An endogenous olfactory signature on petrels' feathers? *Journal of Chemical Ecology, 33*, 1819–1829. https://doi.org/10.1007/s10886-007-9345-7

Bonadonna, F., Caro, S. P., & Brooke, M. L. (2009). Olfactory sex recognition investigated in Antarctic prions. *PLoS One, 4*, e4148.

Bortolotti, G. R. (2010). Flaws and pitfalls in the chemical analysis of feathers: Bad news–good news for avian chemoecology and toxicology. *Ecological Applications, 20*, 1766–1774. https://doi.org/10.1890/09-1473.1

Burger, B. V., Reiter, B., Borzyk, O., & Du Plessis, M. A. (2004). Avian exocrine secretions. I. Chemical characterization of the volatile fraction of the uropygial secretion of the green woodhoopoe, *Phoeniculus purpureus*. *Journal of Chemical Ecology, 30*, 1603–1611.

Campagna, S., Mardon, J., Celerier, A., & Bonadonna, F. (2012). Potential semiochemical molecules from birds: A practical and comprehensive compilation of the last 20 years studies. *Chemical Senses, 37*, 3–25. https://doi.org/10.1093/chemse/bjr067

Caro, S. P., & Balthazart, J. (2010). Pheromones in birds: Myth or reality? *Journal of Comparative Physiology A, 196*, 751–766. https://doi.org/10.1007/s00359-010-0534-4

Caro, S. P., Balthazart, J., & Bonadonna, F. (2015). The perfume of reproduction in birds: Chemosignaling in avian social life. *Hormones and Behavior, 68*, 25–42. https://doi.org/10.1016/j.yhbeh.2014.06.001

Caspers, B. A., & Krause, E. T. (2013). Intraspecific olfactory communication in zebra finches (*Taeniopygia guttata*): Potential information apart from visual and acoustic cues. In M. L. East & M. Dehnhard (Eds.), *Chemical signals in vertebrates* (Vol. 12, pp. 341–351). Springer.

Caspers, B. A., Hagelin, J. C., Paul, M., Bock, S., Willeke, S., & Krause, E. T. (2017). Zebra finch chicks recognise parental scent, and retain chemosensory knowledge of their genetic mother, even after egg cross-fostering. *Scientific Reports, 7*, 12859. https://doi.org/10.1038/s41598-017-13110-y

Caspers, B. A., Marfull, R., Dannenhaus, T., Komdeur, J., & Korsten, P. (2022). Chemical analysis reveals sex differences in the preen gland secretion of breeding blue tits. *Journal of Ornithology, 163*, 191–198. https://doi.org/10.1007/s10336-021-01921-w

Castro, I., Cunningham, S. J., Gsell, A. C., Jaffe, K., Cabrera, A., & Liendo, C. (2010). Olfaction in birds: A closer look at the kiwi (Apterygidae). *Journal of Avian Biology, 41*, 213–218.

Clark, L. (1999). Bird repellents: Interactions of agents in mixture. In R. E. Johnston, D. Müller-Schwarze, & P. W. Sorensen (Eds.), *Advances in chemical signals in vertebrates* (Vol. 8, pp. 623–632). Springer.

Clark, L., & Mason, J. R. (1987). Olfactory discrimination of plant volatiles by the European starling. *Animal Behaviour, 35*, 227–235.

Clark, L., & Shah, P. S. (1992). Information content of prey odor plumes: What do foraging Leach's storm petrels know? In R. L. Doty & D. Müller-Schwarze (Eds.), *Chemical signals in vertebrates* (Vol. 6, pp. 421–427). Springer.

Clark, L., Avilova, K. V., & Bean, N. J. (1993). Odor thresholds in passerines. *Comparative Biochemistry and Physiology Part A: Physiology, 104*, 305–312.

Coffin, H. R., Watters, J. V., & Mateo, J. M. (2011). Odor-based recognition of familiar and related conspecifics: A first test conducted on captive Humboldt penguins (*Spheniscus humboldti*). *PLoS One, 6*, e25002. https://doi.org/10.1371/journal.pone.0025002

Creece, D., Freire, R., & Massaro, M. (2025). Past research and future directions in understanding how birds use their sense of smell. *Ibis, 133*, 98. https://doi.org/10.1111/ibi.13398

Cunningham, G. B., & Nevitt, G. A. (2005). The sense of smell in procellariiforms: An overview and new directions. In R. T. Mason, M. P. LeMaster, & D. Müller-Schwarze (Eds.), *Chemical signals in vertebrates* (Vol. 10, pp. 403–408). Springer.

Damsté, J. S., Dekker, M. H. A., & Piersma, T. (2000). Molecular analysis of intact preen waxes of *Calidris canutus* (Aves: Scolopacidae) by GC/MS and GC/MS/MS. *Lipids, 35*, 533–542.

de Koning, S., Janssen, H.-G., & Brinkman, U. A. T. (2009). Modern methods of sample preparation for GC analysis. *Chroma, 69*, 33–78. https://doi.org/10.1365/s10337-008-0937-3

DeBose, J. L., & Nevitt, G. A. (2008). The use of odors at different spatial scales: Comparing birds with fish. *Journal of Chemical Ecology, 34*, 867–881.

Dekker, M. H., Piersma, T., & Damsté, J. S. S. (2000). Molecular analysis of intact preen waxes of *Calidris canutus* (Aves: Scolopacidae) by gas chromatography/mass spectrometry. *Lipids, 35*, 533–541. https://doi.org/10.1007/s11745-000-553-7

Díez Fernández, A. (2020). *The role of uropygial secretion and birds' body odour on their interaction with mosquitoes and parasites*. PhD Thesis, Universidad de Seville.

Dolbeer, R. A., Woronecki, P. P., & Bullard, R. W. (1992). Visual cue fails to enhance bird repellency of methiocarb in ripening sorghum. In R. L. Doty & D. Müller-Schwarze (Eds.), *Chemical signals in vertebrates* (Vol. 6, pp. 323–330). Springer.

Douglas, H. D. (2006). Measurement of chemical emissions in crested auklets (*Aethia cristatella*). *Journal of Chemical Ecology, 32*, 2559–2567.

Downing, D. T. (1986). Preen gland and scent gland lipids. In J. Bereiter-Hahn, A. G. Matoltsy, & K. S. Richards (Eds.), *Biology of the integument* (pp. 833–840). Springer.

Drea, C. M., Boulet, M., delBarco-Trillo, J., Greene, L. K., Sacha, C. R., Goodwin, T. E., & Dubay, G. R. (2013). The "secret" in secretions: Methodological considerations in deciphering primate olfactory communication. *American Journal of Primatology, 75*, 621–642. https://doi.org/10.1002/ajp.22143

Driver, R. J., & Balakrishnan, C. N. (2021). Highly contiguous genomes improve the understanding of avian olfactory receptor repertoires. *Integrative and Comparative Biology, 61*, 1281–1290.

Dumbacher, J. P., & Pruett-Jones, S. (1996). Avian chemical defense. In V. Nolan & E. D. Ketterson (Eds.), *Current ornithology* (pp. 137–174). Springer.

Dyer, L. A., Philbin, C. S., Ochsenrider, K. M., Richards, L. A., Massad, T. J., Smilanich, A. M., Forister, M. L., Parchman, T. L., Galland, L. M., Hurtado, P. J., Espeset, A. E., Glassmire, A. E., Harrison, J. G., Mo, C., Yoon, S., Pardikes, N. A., Muchoney, N. D., Jahner, J. P., Slinn, H. L., Shelef, O., Dodson, C. D., Kato, M. J., Yamaguchi, L. F., & Jeffrey, C. S. (2018). Modern approaches to study plant–insect interactions in chemical ecology. *Nature Reviews Chemistry, 2*, 50–64.

Fischer, I., Haliński, L. P., Meissner, W., Stepnowski, P., & Knitter, M. (2017). Seasonal changes in the preen wax composition of the herring gull *Larus argentatus*. *Chemoecology, 27*, 127–139. https://doi.org/10.1007/s00049-017-0239-z

Fischer, I., Meissner, W., Haliński, Ł. P., & Stepnowski, P. (2020). Preen oil chemical composition in herring gull *Larus argentatus,* common gull *Larus canus* and black-headed gull *Chroicocephalus ridibundus* confirms their status as two separate genera. *Biochemical Systematics and Ecology, 89*, 103987. https://doi.org/10.1016/j.bse.2019.103987

Fluen, T. (2008). *A comparative analysis of evolutionary changes in island birds*. MSc Thesis, University of Canterbury.

Forrette, L. M. (2018). *A description and analysis of preen oil volatiles in wild white-throated sparrows (Zonotrichia albicollis)*. PhD Thesis, Indiana State University.

Fracasso, G., Tuliozi, B., Hoi, H., & Griggio, M. (2018). Can house sparrows recognize familiar or kin-related individuals by scent? *Current Zoology, 65*, 53–59.

Gabirot, M., Mardon, J., Campagna, S., West, N., Bonadonna, F., & Saunders, S. M. (2016). Guidelines for collecting and extracting avian odors in a remote field: Case study of a subantarctic seabird. In B. A. Schulte, T. E. Goodwin, & M. H. Ferkin (Eds.), *Chemical signals in vertebrates* (Vol. 13, pp. 435–460). Springer.

Gabirot, M., Buatois, B., Müller, C. T., & Bonadonna, F. (2018). Odour of King Penguin feathers analysed using direct thermal desorption discriminates between individuals but not sexes. *Ibis, 160*, 379–389.

Gagliardo, A. (2013). Forty years of olfactory navigation in birds. *Journal of Experimental Biology, 216*, 2165–2171. https://doi.org/10.1242/jeb.070250

Gagliardo, A., Fiaschi, V., & Benvenuti, S. (1988). Pigeon homing: Olfactory experiments with young inexperienced birds. *Naturwissenschaften, 75*, 211–213.

Gilles, M., Fokkema, R. W., Korsten, P., Caspers, B. A., & Schmoll, T. (2024). Preen oil composition of pied flycatchers is similar between partners but differs between sexes and breeding stages. *Ibis, 166*, 171–186. https://doi.org/10.1111/ibi.13246

Griebel, I. A., & Dawson, R. D. (2020). Nestling tree swallows (*Tachycineta bicolor*) alter begging behaviour in response to odour of familiar adults, but not their nests. *Ethology, 126*, 630–636. https://doi.org/10.1111/eth.13015

Grieves, L. A., Kelly, T. R., Bernards, M. A., & MacDougall-Shackleton, E. A. (2018). Malarial infection alters wax ester composition of preen oil in songbirds: Results of an experimental study. *The Auk, 135*, 767–776. https://doi.org/10.1642/AUK-17-242.1

Grieves, L. A., Bernards, M. A., & MacDougall-Shackleton, E. A. (2019a). Behavioural responses of songbirds to preen oil odour cues of sex and species. *Animal Behaviour, 156*, 57–65. https://doi.org/10.1016/j.anbehav.2019.06.035

Grieves, L. A., Bernards, M. A., & MacDougall-Shackleton, E. A. (2019b). Wax ester composition of songbird preen oil varies seasonally and differs between sexes, ages, and populations. *Journal of Chemical Ecology, 45*, 37–45. https://doi.org/10.1007/s10886-018-1033-2

Grieves, L. A., Gloor, G. B., Bernards, M. A., & MacDougall-Shackleton, E. A. (2019c). Songbirds show odour-based discrimination of similarity and diversity at the major histocompatibility complex. *Animal Behaviour, 158*, 131–138. https://doi.org/10.1016/j.anbehav.2019.10.005

Grieves, L. A., Bottini, C. L. J., Branfireun, B. A., Bernards, M. A., MacDougall-Shackleton, S. A., & MacDougall-Shackleton, E. A. (2020). Food stress, but not experimental exposure to mercury, affects songbird preen oil composition. *Ecotoxicology, 29*, 275–285. https://doi.org/10.1007/s10646-020-02171-x

Grieves, L. A., Gilles, M., Cuthill, I. C., Székely, T., MacDougall-Shackleton, E. A., & Caspers, B. A. (2022). Olfactory camouflage and communication in birds. *Biological Reviews, 97*, 1193–1209. https://doi.org/10.1111/brv.12837

Grieves, L. A., Brady, A. L., Slater, G. F., & Quinn, J. S. (2024). Chemical profiles differ between communal breeding groups in a highly social bird. *The American Naturalist, 203*, 490–502. https://doi.org/10.1086/729221

Guilford, T., Gagliardo, A., Chappell, J., Bonadonna, F., Perera, T. B. D., & Holland, R. (1998). Homing pigeons use olfactory cues for navigation in England. *Journal of Experimental Biology, 201*, 895–900.

Gwinner, H. (2013). Male European starlings use odorous herbs as nest material to attract females and benefit nestlings. In M. L. East & M. Dehnhard (Eds.), *Chemical signals in vertebrates* (Vol. 12, pp. 353–362). Springer.

Haahti, E., Lagerspetz, K., Nikkari, T., & Fales, H. M. (1964). Lipids of the uropygial gland of birds. *Comparative Biochemistry and Physiology, 12*(4), 35–437.

Hagelin, J. C. (2007). The citrus-like scent of crested auklets: Reviewing the evidence for an avian olfactory ornament. *Journal of Ornithology, 148*, 195–201.

Hagelin, J. C., & Jones, I. L. (2007). Bird odors and other chemical substances: A defense mechanism or overlooked mode of intraspecific communication? *The Auk, 124*, 741–761. https://doi.org/10.1642/0004-8038(2007)124[741:BOAOCS]2.0.CO;2

Hagelin, J. C., Jones, I. L., & Rasmussen, L. E. L. (2003). A tangerine-scented social odour in a monogamous seabird. *Proceedings of the Royal Society of London B: Biological Sciences, 270*, 1323–1329.

Hagelin, J. C., Simonet, J. C., & Lyson, T. R. (2013). Embryonic domestic chickens can detect compounds in an avian chemosignal before breathing air. In M. L. East & M. Dehnhard (Eds.), *Chemical signals in vertebrates* (Vol. 12, pp. 363–377). Springer.

Haribal, M., Dhondt, A. A., Rosane, D., & Rodriguez, E. (2005). Chemistry of preen gland secretions of passerines: Different pathways to same goal? Why? *Chemoecology, 15*, 251–260.

Haribal, M., Proctor, H., Dhondt, A. A., & Rodriguez, E. (2011). Biology of house finch feather mites, *Proctophyllodes pinnatus* (Acari: Proctophyllodidae), parallels variation in preen gland secretions. *International Journal of Acarology, 37*, 75–90.

Hartmann, T. (2008). The lost origin of chemical ecology in the late 19th century. *Proceedings of the National Academy of Sciences USA, 105*, 4541–4546. https://doi.org/10.1073/pnas.0709231105

Healy, S., & Guilford, T. (1990). Olfactory-bulb size and nocturnality in birds. *Evolution, 44*, 339–346. https://doi.org/10.1111/j.1558-5646.1990.tb05203.x

Hirao, A., Aoyama, M., & Sugita, S. (2009). The role of uropygial gland on sexual behavior in domestic chicken, *Gallus gallus domesticus*. *Behavioural Processes, 80*, 115–120.

Hutchinson, L. V., Wenzel, B. M., Stager, K. E., & Tedford, B. L. (1984). Further evidence for olfactory foraging by sooty shearwaters and northern fulmars. In *Marine birds: Their feeding ecology and commercial fisheries relationships* (pp. 72–77).

Inagaki, H., Kiyokawa, Y., Tamogami, S., Watanabe, H., Takeuchi, Y., & Mori, Y. (2014). Identification of a pheromone that increases anxiety in rats. *Proceedings of the National Academy of Sciences USA, 111*, 18751–18756. https://doi.org/10.1073/pnas.1414710112

Jacob, J., & Grimmer, G. (1975). Gefiederlipide der Ringeltaube *Columba palumbus* plumage lipids from the ring dove (*Columba palumbus*). *Zeitschrift für Naturforschung C, 30*, 363–368.

Jacob, J., & Pomeroy, D. E. (1979). The feather lipids of the marabou stork (*Leptoptilos crumeniferus*). *Comparative Biochemistry and Physiology Part B: Comparative Biochemistry, 64*, 301–303.

Jacob, J., & Ziswiler, V. (1982). The uropygial gland. In D. S. Farner, J. R. King, & K. C. Parkes (Eds.), *Avian biology* (pp. 199–324). Academic Press.

Jacob, J., Balthazart, J., & Schoffeniels, E. (1979). Sex differences in the chemical composition of uropygial gland waxes in domestic ducks. *Biochemical Systematics and Ecology, 7*, 149–153.

Jacob, S., Immer, A., Leclaire, S., Parthuisot, N., Ducamp, C., Espinasse, G., & Heeb, P. (2014). Uropygial gland size and composition varies according to experimentally modified microbiome in great tits. *BMC Evolutionary Biology, 14*, 134. https://doi.org/10.1186/1471-2148-14-134

Jeanjean, L., Caspers, B. A., Schmoll, T., & Gilles, M. (2024). Chemical signatures in the preen oil of pied flycatchers: Testing reproducibility and exploring ontogeny. *Journal of Avian Biology, 2025*(3), e03365. https://doi.org/10.1111/jav.03365

Jennings, S. L., & Ebeler, S. E. (2020). Individual chemical profiles in the Leach's storm-petrel. *Journal of Chemical Ecology, 46*, 845–864.

Jones, C. G., & Firn, R. D. (1991). On the evolution of plant secondary chemical diversity. *Philosophical Transactions of the Royal Society of London Series B: Biological Sciences, 333*, 273–280. https://doi.org/10.1098/rstb.1991.0077

Jones, R. B., & Roper, T. J. (1997). Olfaction in the domestic fowl: A critical review. *Physiology & Behavior, 62*, 1009–1018.

Khan, I., Yang, Z., Maldonado, E., Li, C., Zhang, G., Gilbert, M. T. P., Jarvis, E. D., O'Brian, S. J., Johnson, W. E., & Antunes, A. (2015). Olfactory receptor subgenomes linked with broad ecological adaptations in Sauropsida. *Molecular Biology and Evolution, 32*, 2832–2843.

Kishkinev, D. (2015). Sensory mechanisms of long-distance navigation in birds: A recent advance in the context of previous studies. *Journal of Ornithology, 156*, 145–161.

Kohlwey, S., Krause, E. T., Baier, M. C., et al. (2016). Chemical analyses reveal family-specific nest odor profiles in zebra finches (*Taeniopygia guttata*): A pilot study. In B. A. Schulte, T. E. Goodwin, & M. H. Ferkin (Eds.), *Chemical signals in vertebrates* (Vol. 13, pp. 167–175). Springer.

Kolattukudy, P. E., Bohnet, S., & Rogers, L. (1985). Disappearance of short chain acids from the preen gland wax of male mallard ducks during eclipse. *Journal of Lipid Research, 26*, 989–994.

Kolattukudy, P. E., Bohnet, S., & Rogers, L. (1987). Diesters of 3-hydroxy fatty acids produced by the uropygial glands of female mallards uniquely during the mating season. *Journal of Lipid Research, 28*, 582–588.

Krause, E. T., Brummel, C., Kohlwey, S., Baier, M. C., Müller, C., Bonadonna, F., & Caspers, B. A. (2014). Differences in olfactory species recognition in the females of two Australian songbird species. *Behavioral Ecology and Sociobiology, 68*, 1819–1827. https://doi.org/10.1007/s00265-014-1791-y

Krause, E. T., Kabbert, J., & Caspers, B. A. (2016). Exploring the use of olfactory cues in a nonsocial context in zebra finches (*Taeniopygia guttata*). In B. A. Schulte, T. E. Goodwin, & M. H. Ferkin (Eds.), *Chemical signals in vertebrates* (Vol. 13, pp. 177–187). Springer.

Krause, E. T., Bischof, H.-J., Engel, K., Golueke, S., Maraci, O., Mayer, U., Sauer, J., & Caspers, B. A. (2018). Olfaction in the zebra finch (*Taeniopygia guttata*): What is known and further perspectives. In *Advances in the study of behavior* (pp. 37–85). Elsevier.

Krause, E. T., Paul, M., Krüger, O., & Caspers, B. A. (2023). Olfactory sex preferences in six Estrildid Finch species. *Frontiers in Ecology and Evolution, 11*, 1000531.

Kücklich, M., Möller, M., Marcillo, A., Einspanier, A., Weiß, B. M., Birkemeyer, C., & Widdig, A. (2017). Different methods for volatile sampling in mammals. *PLoS One, 12*(8), e0183440. https://doi.org/10.1371/journal.pone.0183440

Leclaire, S., Merkling, T., Raynaud, C., Giacinti, G., Bessière, J. M., Hatch, S. A., & Danchin, É. (2011). An individual and a sex odor signature in kittiwakes? Study of the semiochemical composition of preen secretion and preen down feathers. *Naturwissenschaften, 98*, 615–624. https://doi.org/10.1007/s00114-011-0809-9

Leclaire, S., van Dongen, W. F. D., Voccia, S., Merkling, T., Ducamp, C., Hatch, S. A., Blanchard, P., Danchin, É., & Wagner, R. H. (2014). Preen secretions encode information on MHC similarity in certain sex-dyads in a monogamous seabird. *Scientific Reports, 4*, 1–6. https://doi.org/10.1038/srep06920

Leclaire, S., Bourret, V., & Bonadonna, F. (2017a). Blue petrels recognize the odor of their egg. *Journal of Experimental Biology, 220*(17), 3022–3025.

Leclaire, S., Strandh, M., Mardon, J., Westerdahl, H., & Bonadonna, F. (2017b). Odour-based discrimination of similarity at the major histocompatibility complex in birds. *Proceedings of the Royal Society of London B: Biological Sciences, 284*, 20162466. https://doi.org/10.1098/rspb.2016.2466

Leclaire, S., Chatelain, M., Pessato, A., Buatois, B., Frantz, A., & Gasparini, J. (2019). Pigeon odor varies with experimental exposure to trace metal pollution. *Ecotoxicology, 28*, 76–85.

Livezey, B. C., Jacob, J., & Humphrey, P. S. (1986). Biochemical composition of secretions from uropygial glands of steamer-ducks. *Biochemical Systematics and Ecology, 14*, 445–450.

López-Perea, J. J., & Mateo, R. (2019). Wax esters of uropygial gland secretion as biomarkers of endocrine disruption in birds exposed to treated sewage water. *Environmental Pollution, 250*, 323–330.

Mahr, K., & Hoi, H. (2018). Red-legged partridges perceive the scent of predators and alarm scents of an avian heterospecific. *Animal Behaviour, 144*, 109–114.

Malakoff, D. (1999). Following the scent of avian olfaction. *Science, 286*, 704–705. https://doi.org/10.1126/science.286.5440.704

Maraci, Ö., Engel, K., & Caspers, B. A. (2018). Olfactory communication via microbiota: What is known in birds? *Genes, 9*, 387. https://doi.org/10.3390/genes9080387

Mardon, J., Saunders, S. M., Anderson, M. J., Couchoux, C., & Bonadonna, F. (2010). Species, gender, and identity: Cracking petrels' sociochemical code. *Chemical Senses, 35*, 309–321.

Mardon, J., Saunders, S. M., & Bonadonna, F. (2011). From preen secretions to plumage: The chemical trajectory of blue petrels' *Halobaena caerulea* social scent. *Journal of Avian Biology, 42*, 29–38.

Martín-Vivaldi, M., Ruiz-Rodríguez, M., Soler, J. J., Peralta-Sánchez, M. J., Méndez, M., Valdivia, E., Martín-Platero, A. M., & Martínez-Bueno, M. (2009). Seasonal, sexual and developmental differences in hoopoe *Upupa epops* preen gland morphology and secretions: Evidence for a role of bacteria. *Journal of Avian Biology, 40*, 191–205.

Martín-Vivaldi, M., Peña, A., Peralta-Sánchez, J. M., Sánchez, L., Ananou, S., Ruiz-Rodríguez, M., & Soler, J. J. (2010). Antimicrobial chemicals in hoopoe preen secretions are produced by symbiotic bacteria. *Proceedings of the Royal Society of London B: Biological Sciences, 277*, 123–130. https://doi.org/10.1098/rspb.2009.1377

Mason, J. R., & Clark, L. (1986). Chemoreception and the selection of green plants as nest fumigants by starlings. In D. Duvall, D. Müller-Schwarze, & R. M. Silverstein (Eds.), *Chemical signals in vertebrates* (Vol. 4, pp. 369–384). Springer.

Mason, J. R., Clark, L., & Shah, P. S. (1992). Taxonomic differences between birds and mammals in their responses to chemical irritants. In R. L. Doty & D. Müller-Schwarze (Eds.), *Chemical signals in vertebrates* (Vol. 6, pp. 311–317). Springer.

Menon, G. K., & Menon, J. (2000). Avian epidermal lipids: Functional considerations and relationship to feathering. *American Zoologist, 40*, 540–552.

Mihailova, M. (2014). *Olfactory signaling in an avian species complex: The crimson rosella.* PhD Thesis, Deakin University.

Montalti, D., Gutiérrez, A. M., Reboredo, G., & Salibián, A. (2005). The chemical composition of the uropygial gland secretion of rock dove *Columba livia. Comparative Biochemistry and Physiology Part A: Molecular & Integrative Physiology, 140*, 275–279.

Moreno-Rueda, G. (2017). Preen oil and bird fitness: A critical review of the evidence. *Biological Reviews, 92*, 2131–2143. https://doi.org/10.1111/brv.12324

Myers, C. W., & Daly, J. W. (1983). Dart-poison frogs. *Scientific American, 248*, 120–133.

Nevitt, G. A. (1994). Antarctic procellariiform seabirds can smell krill. *Antarctic Journal of the United States, 29*, 168–169.

Nevitt, G. (1999a). Olfactory foraging in Antarctic seabirds: A species-specific attraction to krill odors. *Marine Ecology Progress Series, 177*, 235–241.

Nevitt, G. (1999b). Foraging by seabirds on an olfactory landscape. *American Scientist, 87*, 46.

Nevitt, G. A. (2001). Mechanisms of olfactory foraging by Antarctic procellariiform seabirds. In A. Marchlewska-Koj, J. J. Lepri, & D. Müller-Schwarze (Eds.), *Chemical signals in vertebrates* (Vol. 9, pp. 27–33). Springer.

Nevitt, G. A. (2008). Sensory ecology on the high seas: The odor world of the procellariiform seabirds. *Journal of Experimental Biology, 211*, 1706–1713.

Nevitt, G. A., & Bonadonna, F. (2005). Seeing the world through the nose of a bird: New developments in the sensory ecology of procellariiform seabirds. *Marine Ecology-Progress Series, 287*, 292–295.

Nevitt, G. A., & Prada, P. A. (2015). The chemistry of avian odors: An introduction to best practices. In R. L. Doty (Ed.), *Handbook of olfaction and gustation* (pp. 565–578). Wiley.

Nevitt, G. A., Veit, R. R., & Kareiva, P. (1995). Dimethyl sulphide as a foraging cue for Antarctic procellariiform seabirds. *Nature, 376*, 680–682.

Nonnamaker, E., Muszynska, M., Lightcap, I., Drea, C. M., & Archie, E. A. (2023). From swab to spectrum: A comprehensive guide to optimizing simple gas chromatography-mass spectrometry analysis of vertebrate odorants. In B. Schaal, D. Rekow, M. Keller, & F. Damon (Eds.), *Chemical signals in vertebrates* (Vol. 15, pp. 11–38). Springer.

Pan, P. R., Dilworth, B. C., Day, E. J., & Chen, T. C. (1979). Effect of season of the year, sex, and dietary fats on broiler performance, abdominal fat, and preen gland secretion. *Poultry Science, 58*, 1564–1574. https://doi.org/10.3382/ps.0581564

Papi, F. (1976). The olfactory navigation system of the homing pigeon. *Verhandlungen der Deutschen Zoologischen Gesellschaft, 69*, 184–205.

Papi, F. (1982). Olfaction and homing in pigeons: Ten years of experiments. In F. Papi & H. G. Wallraff (Eds.), *Avian navigation* (pp. 149–159). Springer.

Papi, F. (1986). Pigeon navigation: Solved problems and open questions. *Monitore Zoologico Italiano-Italian Journal of Zoology, 20*, 471–517.

Papi, F. (1989). Pigeons use olfactory cues to navigate. *Ethology Ecology & Evolution, 1*, 219–231. https://doi.org/10.1080/08927014.1989.9525511

Papi, F., Fiore, L., Fiaschi, V., & Benvenuti, S. (1971). The influence of olfactory nerve section on the homing capacity of carrier pigeons. *Monitore Zoologico Italiano-Italian Journal of Zoology, 5*, 265–267.

Papi, F., Fiore, L., Fiaschi, V., & Benvenuti, S. (1972). Olfaction and homing in pigeons. *Monitore Zoologico Italiano-Italian Journal of Zoology, 6*, 85–95.

Parr-Cortés, Z., Müller, C. T., Talas, L., Mendl, M., Guest, C., & Rooney, N. J. (2024). The odour of an unfamiliar stressed or relaxed person affects dogs' responses to a cognitive bias test. *Scientific Reports, 14*, 15843.

Piersma, T., Dekker, M., & Sinninghe Damsté, J. S. (1999). An avian equivalent of make-up? *Ecology Letters, 2*, 201–203.

Pineaux, M., Blanchard, P., Ribeiro, L., Hatch, S. A., & Leclaire, S. (2023). A gull species recognizes MHC-II diversity and dissimilarity using odor cues. In B. Schaal, D. Rekow, M. Keller, & F. Damon (Eds.), *Chemical signals in vertebrates* (Vol. 15, pp. 139–151). Springer.

Potier, S. (2020). Olfaction in raptors. *Zoological Journal of the Linnean Society, 189*, 713–721.

Poulin, R. X., & Pohnert, G. (2019). Simplifying the complex: Metabolomics approaches in chemical ecology. *Analytical and Bioanalytical Chemistry, 411*, 13–19.

Prada, P. A., & Furton, K. G. (2018). Birds and dogs: Toward a comparative perspective on odor use and detection. *Frontiers in Veterinary Science, 5*, 188.

Praveenkumar, D., Vinothkumar, A., Saravanan, G., Selvakumar, M., Vijayakumar, A. S., Kolanchinathan, P., Kamalakkannan, S., & Achiraman, S. (2023). Symbiotic microbes play a role more important than preen gland in avian pheromone production – A review. *Avian Biology Research, 16*, 32–41. https://doi.org/10.1177/17581559221137503

Raguso, R. A., & Pellmyr, O. (1998). Dynamic headspace analysis of floral volatiles: A comparison of methods. *Oikos, 81*, 238–254.

Raguso, R. A., Agrawal, A. A., Douglas, A. E., Jander, G., Kessler, A., Poveda, K., & Thaler, J. S. (2015a). The raison d'être of chemical ecology. *Ecology, 96*, 617–630. https://doi.org/10.1890/14-1474.1

Raguso, R. A., Thompson, J. N., & Campbell, D. R. (2015b). Improving our chemistry: Challenges and opportunities in the interdisciplinary study of floral volatiles. *Natural Product Reports, 32*, 893–903.

Rajchard, J. (2008). Exogenous chemical substances in bird perception: A review. *Veterinární Medicína, 53*, 412–419.

Rasmussen, J. L. (2013). *Investigations of evolutionary arms races and host diversity in avian brood parasite systems*. PhD Thesis, University of Canterbury.

Reneerkens, J., Piersma, T., & Damsté, J. S. S. (2002). Sandpipers (*Scolopacidae*) switch from monoester to diester preen waxes during courtship and incubation, but why? *Proceedings of the Royal Society of London B: Biological Sciences, 269*, 2135–2139. https://doi.org/10.1098/rspb.2002.2132

Reneerkens, J., Piersma, T., & Damsté, J. S. S. (2005). Switch to diester preen waxes may reduce avian nest predation by mammalian predators using olfactory cues. *Journal of Experimental Biology, 208*, 4199–4202. https://doi.org/10.1242/jeb.01872

Reneerkens, J., Piersma, T., & Damsté, J. S. S. (2006). Discerning adaptive value of seasonal variation in preen waxes: Comparative and experimental approaches. *Acta Zoologica Sinca, 52*, 272–275.

Reneerkens, J., Almeida, J. B., Lank, D. B., Jukema, J., Lanctot, R. B., Morrison, R. I. G., Rijpstra, W. I. C., Schamel, D., Schekkerman, H., Damsté, J. S. S., Tomkovich, P. S., Tracy, D. M., Tulp, I., & Piersma, T. (2007a). Parental role division predicts avian preen wax cycles. *Ibis, 149*, 721–729.

Reneerkens, J., Piersma, T., & Damsté, J. S. S. (2007b). Expression of annual cycles in preen wax composition in red knots: Constraints on the changing phenotype. *Journal of Experimental Zoology Part A: Ecological Genetics and Physiology, 307A*, 127–139. https://doi.org/10.1002/jez.a.347

Rieke, G. K., & Wenzel, B. M. (1978). Forebrain projections of the pigeon olfactory bulb. *Journal of Morphology, 158*, 41–55. https://doi.org/10.1002/jmor.1051580105

Rivers, P. R., & DuVal, E. H. (2024). Chemical cues in the mating behavior of a highly polygynous bird. *Behavioral Ecology and Sociobiology, 78*, 69.

Röder, G., Baglione, V., Bolopo, D., Marcos, J. M., Trnka, A., & Canestrari, D. (2016). Small emissions with major consequences: Specialized malodorous defenses in birds. In B. A. Schulte, T. E. Goodwin, & M. H. Ferkin (Eds.), *Chemical signals in vertebrates* (Vol. 13, pp. 281–302). Springer.

Roper, T. J. (1999). Olfaction in birds. *Advances in the Study of Behavior, 28*, 247–247.

Roth, T. C., Cox, J. G., & Lima, S. L. (2008). Can foraging birds assess predation risk by scent? *Animal Behaviour, 76*, 2021–2027.

Salibian, A., & Montalti, D. (2009). Physiological and biochemical aspects of the avian uropygial gland. *Brazilian Journal of Biology, 69*, 437–446. https://doi.org/10.1590/S1519-69842009000200029

Schlatmann, A. A., Salazar, S. M., Yu, G., Baas, K. B., Versteegh, M. A., Komdeur, J., Caspers, B. A., & Korsten, P. (2022). Olfactory kin discrimination in begging blue tit nestlings? In *European conference on behavioural biology 2022: All of life is social!*

Shah, P. S., Mason, J. R., & Clark, L. (1992). Avian chemical repellency: A structure-activity approach and implications. In R. L. Doty & D. Müller-Schwarze (Eds.), *Chemical signals in vertebrates* (Vol. 6, pp. 291–296). Springer.

Shaw, C. L., Rutter, J. E., Austin, A. L., Garvin, M. C., & Whelan, R. J. (2011). Volatile and semi-volatile compounds in gray catbird uropygial secretions vary with age and between breeding and wintering grounds. *Journal of Chemical Ecology, 37*, 329–339. https://doi.org/10.1007/s10886-011-9931-6

Shutler, D. (2019). Some important overlooked aspects of odors in avian nesting ecology. *Journal of Avian Biology, 50*, e02003.

Sieck, M. H., & Wenzel, B. M. (1969). Electrical activity of the olfactory bulb of the pigeon. *Electroencephalography and Clinical Neurophysiology, 26*, 62–69.

Slade, J. W. G., Watson, M. J., Kelly, T. R., Gloor, G. B., Bernards, M. A., & MacDougall-Shackleton, E. A. (2016). Chemical composition of preen wax reflects major histocompatibility complex similarity in songbirds. *Proceedings of the Royal Society of London B: Biological Sciences, 283*, 20161966. https://doi.org/10.1098/rspb.2016.1966

Soini, H. A., Schrock, S. E., Bruce, K. E., Wiesler, D., Ketterson, E. D., & Novotny, M. V. (2007). Seasonal variation in volatile compound profiles of preen gland secretions of the dark-eyed junco (*Junco hyemalis*). *Journal of Chemical Ecology, 33*, 183–198. https://doi.org/10.1007/s10886-006-9210-0

Stager, K. E. (1964). The role of olfaction in food location by the turkey vulture (*Cathartes aura*). *Wilson Bulletin, 76*, 390–392.

Stager, K. E. (1967). Avian olfaction. *American Zoologist, 7*, 415–420. https://doi.org/10.1093/icb/7.3.415

Steiger, S. S., Fidler, A. E., Valcu, M., & Kempenaers, B. (2008). Avian olfactory receptor gene repertoires: Evidence for a well-developed sense of smell in birds? *Proceedings of the Royal Society of London B: Biological Sciences, 275*, 2309–2317. https://doi.org/10.1098/rspb.2008.0607

Steiger, S. S., Kuryshev, V. Y., Stensmyr, M. C., Kempenaers, B., & Mueller, J. C. (2009). A comparison of reptilian and avian olfactory receptor gene repertoires: Species-specific expansion of group γ genes in birds. *BMC Genomics, 10*, 446. https://doi.org/10.1186/1471-2164-10-446

Strong, R. M. (1911). On the olfactory organs and the sense of smell in birds. *Journal of Morphology, 22*, 619–661.

Talbott, K. M., Becker, D. J., Soini, H. A., Higgins, B. J., Novotny, M. V., & Ketterson, E. D. (2022). Songbird preen oil odour reflects haemosporidian parasite load. *Animal Behaviour, 188*, 147–155.

Thomas, R. H., Price, E. R., Seewagen, C. L., Mackenzie, S. A., Bernards, M. A., & Guglielmo, C. G. (2010). Use of TLC-FID and GC-MS/FID to examine the effects of migratory state, diet and captivity on preen wax composition in white-throated sparrows (*Zonotrichia albicollis*). *Ibis, 152*, 782–792. https://doi.org/10.1111/j.1474-919X.2010.01050.x

To, J. (2024). *The correspondence between genotype at bacterial and viral-facing toll-like receptors and preen oil chemical composition in song sparrows*. PhD Thesis, Western University.

Tucker, D., & Smith, J. C. (1976). Vertebrate olfaction. In R. B. Masterton, M. E. Bitterman, C. B. G. Campbell, & N. Hotton (Eds.), *Evolution of brain and behavior in vertebrates*. Routledge.

Tuttle, E. M., Sebastian, P. J., Posto, A. L., Soini, H. A., Novotny, M. V., & Gonser, R. G. (2014). Variation in preen oil composition pertaining to season, sex, and genotype in the polymorphic white-throated sparrow. *Journal of Chemical Ecology, 40*, 1025–1038. https://doi.org/10.1007/s10886-014-0493-2

Van Huynh, A., & Rice, A. M. (2019). Conspecific olfactory preferences and interspecific divergence in odor cues in a chickadee hybrid zone. *Ecology and Evolution, 9*, 9671–9683. https://doi.org/10.1002/ece3.5497

Waldvogel, J. A. (1989). Olfactory orientation by birds. In D. M. Power (Ed.), *Current ornithology* (pp. 269–321). Springer.

Wallraff, H. G. (2004). Avian olfactory navigation: Its empirical foundation and conceptual state. *Animal Behaviour, 67*, 189–204. https://doi.org/10.1016/j.anbehav.2003.06.007

Wallraff, H. G. (2015). An amazing discovery: Bird navigation based on olfaction. *Journal of Experimental Biology, 218*, 1464–1466.

Weiß, B. M., Marcillo, A., Manser, M., Weiß, B. M., Marcillo, A., Manser, M., Holland, R., Birkemeyer, C., & Widdig, A. (2018). A non-invasive method for sampling the body odour of mammals. *Methods in Ecology and Evolution, 9*, 420–429. https://doi.org/10.1111/2041-210X.12888

Weldon, P. J. (2023). Chemical aposematism: The potential for non-host odours in avian defence. *Ibis, 165*, 1054–1067. https://doi.org/10.1111/ibi.13165

Weldon, P. J., & Rappole, J. H. (1997). A survey of birds odorous or unpalatable to humans: Possible indications of chemical defense. *Journal of Chemical Ecology, 23*, 2609–2633. https://doi.org/10.1023/B:JOEC.0000006670.79075.92

Wenzel, B. M. (1967). Olfactory perception in birds. *Olfaction and Taste, 11*, 203–217.

Wenzel, B. M. (1968). Olfactory prowess of the kiwi. *Nature, 220*, 1133–1134. https://doi.org/10.1038/2201133a0

Wenzel, B. M. (1971). Olfaction in birds. In L. M. Beidler (Ed.) Olfaction. Handbook of sensory physiology, volume 4 (pp. 432–448). Springer.

Wenzel, B. M. (1980). Chemoreception in seabirds. In J. Burger, B. L. Olla, & H. E. Winn (Eds.), *Behavior of marine animals* (pp. 41–67). Springer.

Wenzel, B. M. (1982). Functional status and credibility of avian olfaction. In F. Papi & H. G. Wallraff (Eds.), *Avian navigation* (pp. 352–361). Springer.

Wenzel, B. M. (1986). The ecological and evolutionary challenges of Procellariiform olfaction. In D. Duvall, D. Müller-Schwarze, & R. M. Silverstein (Eds.), *Chemical signals in vertebrates* (Vol. 4, pp. 357–368). Springer.

Wenzel, B. M. (1987). The olfactory and related systems in birds. *Annals of the New York Academy of Sciences, 519*, 137–149. https://doi.org/10.1111/j.1749-6632.1987.tb36293.x

Wenzel, B. M. (1992). The puzzle of olfactory sensitivity in birds. In R. L. Doty & D. Müller-Schwarze (Eds.), *Chemical signals in vertebrates* (Vol. 6, pp. 443–448). Springer.

Wenzel, B. M. (2007). Avian olfaction: Then and now. *Journal of Ornithology, 148*, 191–194.

Wenzel, B. M., & Meisami, E. (1987). Number, size, and density of mitral cells in the olfactory bulbs of the northern fulmar and rock dove. *Annals of the New York Academy of Sciences, 510*, 700–702.

Wenzel, B. M., & Sieck, M. H. (1972). Olfactory perception and bulbar electrical activity in several avian species. *Physiology & Behavior, 9*, 287–293.

Wenzel, B. M., Albritton, P. F., Salzman, A., & Oberjat, T. E. (1969). Behavioral changes in pigeons after olfactory nerve section or bulb ablation. *Olfaction and Taste, 3*, 278–287.

Whelan, R. J., Levin, T. C., Owen, J. C., & Garvin, M. C. (2010). Short-chain carboxylic acids from gray catbird (*Dumetella carolinensis*) uropygial secretions vary with testosterone levels and photoperiod. *Comparative Biochemistry and Physiology Part B: Biochemistry and Molecular Biology, 156*, 183–188.

Whittaker, D. J., & Hagelin, J. C. (2020). Female-based patterns and social function in avian chemical communication. *Journal of Chemical Ecology, 47*, 53–62. https://doi.org/10.1007/s10886-020-01230-1

Whittaker, D. J., & Theis, K. R. (2016). Bacterial communities associated with junco preen glands: Preliminary ramifications for chemical signaling. In B. A. Schulte, T. E. Goodwin, & M. H. Ferkin (Eds.), *Chemical signals in vertebrates* (Vol. 13, pp. 105–117). Springer.

Whittaker, D. J., Reichard, D. G., Dapper, A. L., & Ketterson, E. D. (2009). Behavioral responses of nesting female dark-eyed juncos (*Junco hyemalis*) to hetero- and conspecific passerine preen oils. *Journal of Avian Biology, 40*, 579–583.

Whittaker, D. J., Soini, H. A., Atwell, J. W., Hollars, C., Novotny, M. V., & Ketterson, E. D. (2010). Songbird chemosignals: Volatile compounds in preen gland secretions vary among individuals, sexes, and populations. *Behavioral Ecology, 21*, 608–614. https://doi.org/10.1093/beheco/arq033

Whittaker, D. J., Richmond, K. M., Miller, A. K., Kiley, R., Burns, C. B., Atwell, J. W., & Ketterson, E. D. (2011a). Intraspecific preen oil odor preferences in dark-eyed juncos (*Junco hyemalis*). *Behavioral Ecology, 22*, 1256–1263. https://doi.org/10.1093/beheco/arr122

Whittaker, D. J., Soini, H. A., Gerlach, N. M., Posto, A. L., Novotny, M. V., & Ketterson, E. D. (2011b). Role of testosterone in stimulating seasonal changes in a potential avian chemosignal. *Journal of Chemical Ecology, 37*, 1349–1357. https://doi.org/10.1007/s10886-011-0050-1

Whittaker, D. J., Gerlach, N. M., Soini, H. A., Novotny, M. V., & Ketterson, E. D. (2013). Bird odour predicts reproductive success. *Animal Behaviour, 86*, 697–703.

Whittaker, D. J., Gerlach, N. M., Slowinski, S. P., Corcoran, K. P., Winters, A. D., Soini, H. A., Novotny, M. V., Ketterson, E. D., & Theis, K. R. (2016). Social environment has a primary influence on the microbial and odor profiles of a chemically signaling songbird. *Frontiers in Ecology and Evolution, 4*, 1–15. https://doi.org/10.3389/fevo.2016.00090

Whittaker, D. J., Rosvall, K. A., Slowinski, S. P., Soini, H. A., Novotny, M. V., & Ketterson, E. D. (2018). Songbird chemical signals reflect uropygial gland androgen sensitivity and predict aggression: Implications for the role of the periphery in chemosignaling. *Journal of Comparative Physiology A, 204*, 5–15.

Whittaker, D. J., Kuzel, M., Burrell, M. J. E., Soini, H. A., Novotny, M. V., & DuVal, E. H. (2019a). Chemical profiles reflect heterozygosity and seasonality in a tropical lekking passerine bird. *Animal Behaviour, 151*, 67–75. https://doi.org/10.1016/j.anbehav.2019.03.005

Whittaker, D. J., Slowinski, S. P., Greenberg, J. M., Alian, O., Winters, A. D., Ahmad, M. M., Burrell, M. J. E., Soini, H. A., Novotny, M. V., Ketterson, E. D., & Theis, K. R. (2019b). Experimental evidence that symbiotic bacteria produce chemical cues in a songbird. *Journal of Experimental Biology, 222*, jeb.202978. https://doi.org/10.1242/jeb.202978

Wiltschko, W., & Wiltschko, R. (1992). Pigeon homing: The effect of temporary anosmia on orientation behavior. In R. L. Doty & D. Müller-Schwarze (Eds.), *Chemical signals in vertebrates* (Vol. 6, pp. 435–442). Springer.

Wiltschko, R., & Wiltschko, W. (2017). Considerations on the role of olfactory input in avian navigation. *Journal of Experimental Biology, 220*, 4347–4350.

Zhang, J.-X., Sun, L., & Zuo, M. X. (2009). Uropygial gland volatiles may code for olfactory information about sex, individual, and species in Bengalese finches (*Lonchura striata*). *Current Zoology, 55*, 357–365.

Zhang, J.-X., Wei, W., Zhang, J.-H., & Yang, W.-H. (2010). Uropygial gland-secreted alkanols contribute to olfactory sex signals in budgerigars. *Chemical Senses, 35*, 375–382. https://doi.org/10.1093/chemse/bjq025

Zhang, Y.-H., Du, Y.-F., & Zhang, J.-X. (2013). Uropygial gland volatiles facilitate species recognition between two sympatric sibling bird species. *Behavioral Ecology, 24*, 1271–1278. https://doi.org/10.1093/beheco/art068

Zidat, T., Gabirot, M., Bonadonna, F., & Müller, C. T. (2023). Homing and nest recognition in nocturnal blue petrels: What scent may attract birds to their burrows? *Journal of Chemical Ecology, 49*, 384–396. https://doi.org/10.1007/s10886-023-01424-3

The Evolution of Lizard Chemical Signals Across Time, Space, and Contexts

Simon Baeckens[1](✉)

[1] Earth and Life Institute, UCLouvain, Louvain-la-Neuve, Belgium
simon.baeckens@uclouvain.be

Abstract. Research on the evolution and diversity of animal signals has long been dominated by studies of visual and acoustic modalities, leaving chemical communication comparatively overlooked. Moreover, within the existing chemical signaling literature, research has been strongly skewed toward invertebrates—leaving chemical signaling in vertebrates, especially nonmammalian groups, relatively underexplored. In this chapter, I argue that lizards represent a particularly promising group for advancing our understanding of chemical signal evolution in vertebrates. I highlight major macro-evolutionary advances over the past decade, during which large-scale phylogenetic comparative studies have begun to uncover how ecological, social, and phylogenetic factors shape chemical signal diversity across lizard species. In contrast, micro-evolutionary research—investigating how chemical signals evolve within species across space and time—is lagging behind, not just in lizards, but across vertebrates more broadly. Encouragingly, recent longitudinal and experimental studies are beginning to demonstrate the power of this approach. I also draw attention to the largely untapped potential of studying chemical communication under major anthropogenic perturbations, which are increasingly exploited in evolutionary research but remain rarely applied to chemical signaling systems. These coming years are especially promising for scholars seeking to connect macro-evolutionary patterns with fine-scale, contemporary evolutionary processes—a critical next step in understanding the evolution of vertebrate chemical signaling.

Keywords: Chemical communication · Signal evolution · Lacertid lizards · Macroevolution · Microevolution · Anthropocene · Phylogenetic comparative methods

1 The Skewed Landscape of Animal Signal Research

The study of animal communication has captivated naturalists and scientists for centuries—from the early writings of Aristotle and Pliny the Elder (Fögen, 2014) to Darwin's seminal work on emotional expression (Ekman, 2009), and the landmark discoveries of Tinbergen (1953), von Frisch (1967), and Wilson (1975). Building on this rich legacy,

L. M. Schulte (ed.), *Chemical Signals in Vertebrates 16*,
https://doi.org/10.1007/978-3-032-30752-1_4

the evolution and function of animal signals remain actively studied, driven in part by the sheer richness and complexity of signaling diversity across taxa. However, many aspects of animal communication remain unevenly explored. Much of our current understanding has been shaped by a strong emphasis on visual and acoustic modalities: signals that are conspicuous, easily detected and tracked, and intuitive to human observers. As a result, dominant theories of animal communication have largely emerged from systems where displays are seen or heard (e.g., Andersson, 1982; Caro, 1995; Sinervo & Lively, 1996; Zuk & Kolluru, 1998).

In contrast, chemical communication—though evolutionary ancient and taxonomically widespread—has received comparatively less attention (Müller-Schwarze & Mozzel, 1977; Wyatt, 2014). This neglect may stem from human sensory biases and the technical complexity associated with identifying and manipulating chemical signals. And even when chemical communication has been studied in detail, the focus remained overwhelmingly narrow. Research has centered largely on invertebrates, particularly insects (Fig. 1), driven by applied interests in pest control and biological management (Hartmann, 2008; Symonds & Elgar, 2008; Witzgall et al., 2010). As a fortunate byproduct, these applied studies have produced some of the most detailed fundamental knowledge of chemical signal structure and function (Yew & Chung, 2015). However, the insights they offer are often limited in their relevance to vertebrates, where signal production and reception operate under vastly different physiological and ecological constraints.

Although recent years have seen growing recognition of the importance of chemical signals in vertebrate biology (Apps et al., 2015; Müller-Schwarze, 2006), research efforts remain heavily skewed. Mammals have received disproportionate attention (Fig. 1)—likely due to their prominence in biomedical research (e.g., rodents) and broader trends in science where charismatic fauna (e.g., elephants) tend to attract more attention than less "appealing" taxa (Rosenthal et al., 2017; Troudet et al., 2017). Meanwhile, nonmammalian vertebrates remain markedly underrepresented in the chemical communication literature, despite many of these groups relying heavily on chemosensory cues (Müller-Schwarze, 2006). Among them, the neglect of squamate reptiles, notably lizards, is particularly striking. Their deep reliance on chemical information, actively sampled from the environment via specialized vomeronasal systems and, in many species, forked tongues, has been widely acknowledged for decades (Bissinger & Simon, 1979; Cooper, 1994; Schwenk, 1994, 1995). This trait is so central to their biology that it has even shaped how they are portrayed in mythology, folklore, and popular media (Le Busque & Litchfield, 2023; Morgan, 2008). Yet, their chemical signaling systems remain surprisingly understudied—especially when compared to the wealth of research on mammals and, overwhelmingly, on insects.

2 Lizards as Models for Studying Chemical Signal Evolution

Despite their historical neglect, lizards have increasingly emerged as a promising vertebrate group for studying the evolution of chemical communication, for several reasons.

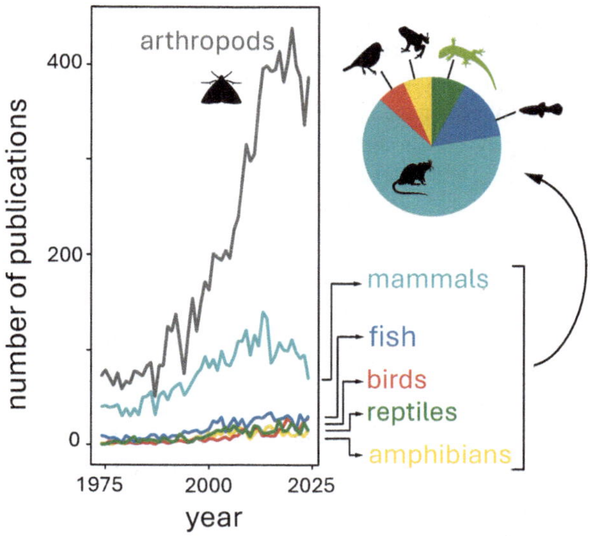

Fig. 1. Taxonomic bias in research on animal chemical signals. The graph shows the number of articles published on chemical signals over a 50-year period for all major vertebrate clades, alongside arthropods as a representative invertebrate group. A pie chart (top right) further illustrates the relative proportion of publications focused on each vertebrate group over this half-century of research. Two patterns emerge: first, studies on vertebrate chemical signaling are vastly outnumbered by those on arthropods—a group that itself represents only a portion of invertebrate diversity—highlighting a profound imbalance in research effort. Second, within vertebrates, chemical communication studies are overwhelmingly concentrated on mammals (64.1%), with far fewer publications addressing fishes (14.3%), reptiles (8.3%), birds (6.7%), and amphibians (6.6%). Data were obtained from a Scopus search (conducted on 03/04/2025), which queried peer-reviewed articles (excluding reviews) published between 1974 and 2024. Searches were restricted to title, abstract, and keywords using the terms (pheromone* OR "chemical signal*" OR "scent mark*" OR semiochemical* OR chemosignal*), in combination with taxon-specific scientific and common names[†] similar to Symonds and Elgar (2008). While this literature search was designed as a coarse, exploratory assessment to capture broad patterns in research focus rather than serve as an exhaustive bibliometric analysis, it nonetheless underscores a long-standing and persistent taxonomic bias in chemical communication research. [†]Search terms used for each group were as follows: reptiles = (reptilia OR reptile OR lepidosauria OR lepidosaur* OR testudines OR turtle* OR crocodilia* OR crocodil* OR squamata OR squamate* OR lizard* OR snake*); fishes = (fish OR fishes OR agnatha OR chondrichthyes OR osteichthyes); amphibians = (amphibia OR amphibian* OR anura* OR frog* OR caudata* OR salamander* OR gymnophiona OR caecilian*); mammals = (mammalia OR mammal* OR mice OR mouse OR rat OR rats); arthropods = (arthropoda OR arthropod* OR insect*).

First, many lizards possess specialized epidermal glands (e.g., femoral, precloacal, and generation glands), which are typically active in adult males and secrete waxy compounds onto environmental surfaces as scent marks during the reproductive season (Mayerl et al., 2015; Mouton et al., 2010). These secretions—chemically complex mixtures of lipophilic and proteinaceous compounds (Mangiacotti et al., 2023a, b)—can be detected by conspecifics and heterospecifics via tongue-flicking. Recent integrative research combining behavioral experiments, chemical analyses, and ecological data has revealed substantial diversity and specificity in these glandular secretions. Variations in composition and relative abundance of specific compounds can convey details about an individual's age, health, parasite load, hormonal state, immune condition, and even access to thermal resources. These chemical signals function not only in species, sex, and population recognition but also in mediating social interactions, including competitive and sexual encounters. This rapidly growing body of work has been extensively reviewed in recent years (e.g., Weldon et al., 2008; Mason & Parker, 2010; Font et al., 2012; Apps et al., 2015; Martín & López, 2015; Martín & López, 2024; Fleishman & Font, 2019). While epidermal gland secretions are generally considered the primary source of chemical signals in lizards, there is also evidence that other sources—such as feces, cloacal secretions, and skin lipids—may convey socially relevant chemical cues (Cooper & Vitt, 1984; Labra, 2008; Mason & Parker, 2010; Moreira et al., 2008).

Second, to efficiently collect and process complex chemical cues from the environment, many lizards have evolved a highly specialized chemosensory system: the vomeronasal-lingual system. This system consists of a tongue, sometimes highly forked to facilitate the detection of chemical gradients, which samples airborne and substrate-bound chemicals and delivers them to paired vomeronasal organs located in the roof of the mouth (Filoramo & Schwenk, 2009). Unlike in other vertebrates, the vomeronasal organs in squamates are anatomically and functionally independent from the main olfactory system (Graves, 1993; Keverne, 1999), enabling the precise detection and processing of a wide array of social signals and environmental cues.

Third, although all lizards share a broadly similar chemical communication system, there is considerable variation in the extent and specialization of both their signaling and receiving structures. While many lizards possess epidermal glands, others lack them altogether (Mayerl et al., 2015). In species with glands, these structures are typically active only in adult males; however, in some lineages, both sexes maintain functional glands (García-Roa et al., 2016; Martín et al., 2016). Moreover, gland size and number vary substantially across species, potentially reflecting differences in the degree of investment or reliance on chemical communication (Baeckens, 2017). Similarly, the morphology of the tongue and vomeronasal organs exhibits broad interspecific diversity (Baeckens et al., 2017a, 2017c; Cooper, 1996; Schwenk, 1993, 1994; Smith et al., 2008). In parallel, the composition of glandular secretions varies widely among species and populations—both in the specific compounds present and in their relative abundances (Martín & López, 2024; Mayerl et al., 2015). These layers of diversity offer a rich foundation for investigating the ecological and evolutionary forces shaping chemical communication.

Fourth, lizards have radiated across nearly all terrestrial ecosystems, except the coldest extremes, occupying a wide range of ecological niches and displaying remarkable phenotypic diversity (Huey et al., 2007; Roll et al., 2017). With over 7800 recognized species, lizards are more species-rich than all mammals (currently ~6760 species; Burgin et al., 2018). Their well-resolved phylogenies (Tonini et al., 2016) make them especially well-suited for macro-evolutionary studies using phylogenetic comparative methods (Meiri, 2018). Lizards are also abundant and diverse on islands around the globe, where geographic isolation has promoted evolutionary divergence along unique ecological trajectories (Novosolov et al., 2013). These circumstances make them particularly valuable for investigating the ecological drivers of phenotypic evolution through island comparisons (Losos & Ricklefs, 2009).

Finally, from a practical perspective, lizards offer numerous experimental advantages. Many populations occur at high densities (Novosolov et al., 2016), have short generation times, and reach sexual maturity relatively quickly (Mancini et al., 2025), facilitating experimental designs both in the field and in controlled environments. This enables longitudinal and even multigenerational studies under seminatural or natural conditions.

In the sections that follow, I highlight recent research efforts that have capitalized on these advantageous traits to deepen our understanding of chemical signal evolution in lizards—across spatial and temporal scales, and a range of socio-ecological contexts.

3 Macro-evolutionary Patterns in Lizard Chemical Signaling

Some of the most extensive efforts to understand the evolution of chemical signaling in lizards have taken a macro-evolutionary, interspecific comparative approach. These studies examine traits of the chemical signaling system across species to detect covariation with ecological or social variables, aiming to infer the selective pressures and constraints that shape signal diversity. By identifying patterns of convergence, researchers can infer adaptive responses to similar environmental drivers or social pressures (Losos, 2011).

3.1 Abiotic Correlates and Efficacy-Based Selection

A major focus of these interspecific comparative studies has been to assess how environmental conditions shape chemical signal design, as predicted by efficacy-based selection theory (Endler, 1992, 1993; Guilford & Dawkins, 1991; Hebets & Papaj, 2005). This framework posits that signal traits that maximize transmission and reception are selectively favored, depending on the physical and sensory environment. In the context of chemical signaling, this involves understanding how abiotic factors (e.g., temperature, humidity, wind, and UV radiation) influence signal performance and drive evolutionary shifts across lizard lineages inhabiting different environments.

Harsh signaling environments—such as hot, windy, and high-UV habitats, where scent marks used for social communication degrade rapidly—can impose strong selective pressures on chemical communication systems (Alberts, 1992; Martín et al., 2024; Martín & López, 2013). One proposed response, the "within-channel compensation hypothesis," suggests that species may increase their investment in chemical signaling by producing more secretion or modifying its composition for greater stability or longevity. Alternatively, the "between-channel compensation hypothesis" proposes that species may reduce reliance on chemical signaling and instead shift toward other modalities (e.g., visual signals) that perform more effectively under such environmental constraints (Alberts, 1992; Baeckens et al., 2015).

In lizards, comparative studies have tested these predictions by examining both the epidermal glands responsible for signal production and the chemical composition of the gland secretions. Early work focused on the presence and number of glands as a proxy for signal investment, owing to the relative ease of measuring this trait across museum specimens and field collections (Baeckens, 2017). More recently, researchers have also investigated the lipophilic and proteinic composition of glandular secretions (Mangiacotti, et al., 2023b), seeking to understand how variation in specific compound classes or ratios corresponds to environmental gradients. To date, most of this comparative work has focused on four major lizard groups—the South American *Liolaemus*, North American *Sceloporus*, Australian *Heteronotia* geckos, and, most prominently, the Old-World family Lacertidae—providing broad phylogenetic, geographic, and ecological coverage.

In *Liolaemus*, Escobar et al. (2001) were among the first to report environmental correlates of gland number. They proposed that harsh signaling environments—particularly those with high thermal variability or wind exposure—might favor increased glandular output to compensate for rapid signal degradation. However, later studies incorporating phylogenetic information and broader taxonomic sampling have added nuance. Overall, the phylogenetic signal tends to be strong, with gland number showing evolutionary conservatism within certain clades (García-Roa, Jara, et al., 2017c; Jara et al., 2018). In many cases, climatic factors explain only a small proportion of the variation in gland number, even within *Liolaemus* (Jara et al., 2018; Pincheira-Donoso et al., 2008).

In lacertids, no significant correlation between gland number and climate has been found (Baeckens, et al., 2015). Instead, gland number correlates more strongly with substrate use: species that climb shrubs or vertical structures tend to have fewer glands. Baeckens et al. (2015) argue that three-dimensional, shrubby microhabitats pose particular challenges for chemical signaling, as effective signal coverage across a vertically complex space would require large quantities of secretions. As such, selection may favor reduced investment in chemical signaling, potentially encouraging shifts toward more energetically efficient modalities better suited to these structurally complex habitats. This interpretation contrasts with the views by Escobar et al. (2001), which predicted increased glandular investment under comparable environmental constraints.

With the growing availability of techniques for processing large chemical datasets, researchers have increasingly shifted focus from gland number to the chemical composition of gland secretions. In a comparative study of 64 lacertid species, researchers found a strong association between climatic conditions and secretion composition (Baeckens, et al., 2018a). Lizards inhabiting xeric environments—dry and warm habitats—produced secretions dominated by high-molecular-weight alcohols and fatty acid esters, compounds linked to greater signal longevity under conditions of high evaporation and diffusion. Conversely, lizards from mesic environments—characterized by high humidity, precipitation, wind, and solar radiation—exhibited greater chemical richness, with large proportions of aldehydes and low-molecular-weight alcohols. These volatile, highly aromatic compounds are thought to enhance signal detectability and facilitate airborne transmission, potentially supporting longer-range communication (Alberts, 1992; Apps et al., 2015; Weldon et al., 2008).

However, parallel studies on spiny lizards (*Sceloporus*) and geckos (*Heteronotia*) revealed contrasting patterns (Campos et al., 2020; Zozaya et al., 2022). In *Sceloporus* no relationship was found between chemical richness and climate, but a positive correlation emerged between aldehyde proportions and ambient temperature—a surprising result, given that aldehydes degrade rapidly under hot conditions. This contrasts with patterns observed in lacertids, where species from warmer environments tend to invest less in aldehydes, presumably to reduce signal degradation. Interestingly, *Sceloporus* species also showed very low levels of volatile unsaturated fatty acids—a metric not assessed in the lacertid dataset. In Australian *Heteronotia* geckos, patterns diverged yet again: secretion composition was associated with precipitation levels, but not with temperature—contrasting with both *Sceloporus* and lacertid results. One possible explanation is that these geckos are nocturnal and may not experience the extreme daytime heat affecting diurnal lizards, such as *Sceloporus* and lacertids. These conflicting findings underscore the need for broader, integrative analyses across lineages and environments.

Beyond investigating how species adjust chemical signal investment in response to abiotic constraints, several studies have more directly examined the between-channel compensation hypothesis (i.e., the idea that species may shift investment across sensory modalities when one channel is compromised; Baeckens et al., 2015). This line of inquiry asks whether limitations on one mode of signaling (e.g., chemical) lead to increased reliance on another (e.g., visual), or vice versa. In *Sceloporus* lizards, a series of studies has tested this hypothesis by evaluating trade-offs between chemical and visual signals. Although based on limited species samples and largely non-phylogenetic comparisons, these studies reveal a compelling trend: species lacking prominent abdominal color patches—a key visual display trait—tend to exhibit increased investment in chemical signaling (Hews & Benard, 2001; Hews et al., 2011; Martins et al., 2015, 2018; Pruett et al., 2016; Romero-Diaz et al., 2021). These findings support the idea that sensory investment is flexible, shifting in response to environmental and functional demands—a central tenet of efficacy-based signal evolution. Surprisingly, comparable analyses have yet to be conducted in lacertids, despite their marked interspecific variation in both chemical signaling and conspicuous coloration (Baeckens, 2019; Pérez i de Lanuza & Font, 2016). Examining whether similar modality trade-offs exist in this group—or in other colorful lineages—could offer key insights into the dynamics of multimodal signal evolution.

3.2 Socio-ecological Correlates and Content-Based Selection

A growing body of interspecific comparative research has explored predictions derived from content-based selection theory (Guilford & Dawkins, 1991; Hebets & Papaj, 2005). This framework posits that selection acts on the informational content of signals—favoring traits that reliably convey fitness-relevant characteristics of the sender, such as condition, quality, and social status. In the context of chemical signaling, this involves examining how variation in signal composition reflects individual or species-level traits, including morphology, physiology, reproductive status, and behavior.

Content-based selection has historically been the most studied aspect of chemical signaling research—not only in lizards but across animals more broadly (Martín & López, 2024; Wyatt, 2014). Most of this research has focused on within-population variation, linking differences in chemical signals to individual quality metrics, such as body condition and parasite load (Martín & López, 2015, 2024). While these studies have been invaluable for understanding the information content of chemical signals, they offer limited insight into broader evolutionary forces shaping signal divergence across species. Nonetheless, the few macro-evolutionary studies that have been conducted suggest that interspecific variation in chemical signals may be driven by multiple ecological and social drivers (e.g., mate choice and intrasexual competition, social organization, dietary ecology, and species recognition), offering a promising yet underexplored avenue for understanding the evolution of lizard chemical communication.

One of the most obvious socio-ecological drivers of chemical signal evolution is the intensity and form of sexual selection. Across many lizard species, chemical signals display classic features of sexually selected traits: they are typically male-biased, emerge or intensify at sexual maturity, peak during the breeding season, and are often regulated by testosterone levels (Baeckens et al., 2017b; García-Roa, et al., 2017c; Mangiacotti et al., 2019; Martín et al., 2007). These patterns suggest that glandular secretions function as secondary sexual characters shaped by mate choice, male-male competition, or both. Behavioral evidence supports this view. Numerous studies have shown that chemical signals mediate female mate choice (e.g., Martín & López, 2000, 2006a, b, 2024; Olsson et al., 2003) and play a role in intrasexual competition and territory defense (e.g., López & Martín, 2002, 2011; Martín & López, 2007). Moreover, experimental work demonstrates that signal content is often condition-dependent, encoding information about health, hormonal status, immune competence, or competitive ability (Martín & López, 2015, 2024). Given this, it is plausible that variation in the strength of sexual selection across species drives divergence in chemical signal composition or investment. This hypothesis mirrors trends observed in other signaling modalities, where interspecific differences in sexual selection regimes explain significant variation in visual display, coloration, and song complexity (Chen et al., 2012; Garamszegi et al., 2005; Ord et al., 2001; Soma & Garamszegi, 2011). Recent comparative work supports this idea in lizards. In lacertids, species with greater sexual size dimorphism—a common proxy for sexual selection—showed increased expression of five distinct molecular weight regions in the proteinaceous fraction of their glandular secretions (Mangiacotti, et al., 2023c, 2023b). Interestingly, a parallel analysis of the lipid fraction found no such correlation (Baeckens, et al., 2018b), suggesting that lipids and proteins may convey different types of information.

Social organization represents another key axis along which chemical signaling systems may diversify. Just as sexual selection can shape signaling traits, so too can the demands of social living, which rely heavily on effective communication among conspecifics. It has been proposed that increased reliance on chemical signals for intraspecific communication facilitates more efficient short-range interactions, thereby creating conditions favorable for transitions from solitary to group living. Some of the best support for this idea comes from insect societies, where increasing pheromone complexity has been linked to the evolution of sociality (Fischman et al., 2011; Van Oystaeyen et al., 2014). More recently, macro-evolutionary studies suggest that similar processes may operate in lizards. In this group, phylogenetic comparative studies have shown that the presence of epidermal glands is associated with an elevated evolutionary rate of transitions toward social groupings, including prolonged parent-offspring associations (Baeckens & Whiting, 2021). Such social structures evolved more readily in lineages with chemical signaling glands than in those without. This evidence suggests that chemical signaling may have helped create key selective conditions for the evolution of sociality in lizards—by facilitating kin recognition, supporting prolonged contact, and enhancing group cohesion (Baeckens & Whiting, 2021). While it is probable that efficient chemical communication indirectly fosters conditions conducive to social aggregation, the precise mechanisms linking chemical signaling to the emergence of sociality remain to be fully elucidated.

Diet has also been proposed as a potential driver of variation in chemical signaling, as differences in the intake of specific dietary compounds may influence the cues and signals synthesized or acquired by individuals or species (Symonds & Elgar, 2008). In lizards, dietary effects on chemical signals have been demonstrated experimentally at the individual level: supplementation with carotenoids and vitamin E has been shown to alter glandular secretion composition, increasing the proportion of α-tocopherol in treated individuals (García-Roa, Sáiz, et al., 2017a; Kopena et al., 2011, 2014). However, few studies have tackled this question from a macro-evolutionary perspective, and those that have found limited support for diet as a major driver of signal evolution in lizards. A broad comparative study in lacertids, for example, reported no correlation between diet breadth and chemical diversity or richness, nor between the intake of plants or ants and variation in the major chemical classes present in secretions (Baeckens, et al., 2017d). One exception was that species with an exclusive arthropod diet exhibited lower chemical signal richness, suggesting that incorporating plant material may allow for greater compound diversity, potentially by providing access to plant-derived compounds not found in animal prey (Baeckens, et al., 2017d). Overall, while diet may influence chemical signals within species or contribute to individual-level plasticity, it appears to be a relatively weak predictor of interspecific variation in the broad chemical profiles of lizard secretions.

Finally, species recognition, especially in zones of sympatry between closely related species, may drive signal divergence to avoid costly interspecific interactions. Coexistence in zones of sympatry can impose substantial fitness costs due to interspecific aggression, resource competition, or reproductive interference, including hybridization (Grether et al., 2017). These pressures may favor divergence in signaling traits involved in mate choice and competitive interactions (Ord & Stamps, 2009; Schaefer & Ruxton, 2015). In this context, chemical signals may play a key role in promoting pre-mating isolation and reducing the likelihood of costly heterospecific encounters (Smadja & Butlin, 2009). This appears to be the case in sympatric *Podarcis* lizards—a genus with well-documented instances of both current and historical hybridization (Yang et al., 2021). Recent work shows that sympatric *Podarcis* species exhibit greater divergence in glandular secretions than closely related allopatric species (Mangiacotti, et al., 2021). This pattern is consistent with reproductive character displacement, in which signals diverge in sympatry to improve conspecific recognition during courtship or male-male contests (Grether et al., 2017; Pfennig & Pfennig, 2009). Similar patterns have been documented in Australian geckos, where sympatric species show more pronounced differences in chemical profiles than their allopatric counterparts, suggesting a convergent evolutionary response (Zozaya et al., 2019). By enhancing signal distinctiveness, chemical divergence likely improves species recognition, reinforcing reproductive isolation and contributing to the stable coexistence of closely related taxa.

4 Micro-evolutionary Patterns in Lizard Chemical Signaling

In the last 10 years, large-scale phylogenetic comparative analyses have taken crucial steps toward understanding the evolution of chemical signal diversity in lizards. These analyses have revealed important correlative patterns, linking signal traits to both abiotic and biotic factors across species. Yet to move beyond these broad associations and toward a more mechanistic understanding of signal evolution, it is valuable to examine how such patterns translate to finer evolutionary scales (Reznick & Ricklefs, 2009; Wright, 1982). Micro-evolutionary studies—focused on within-species signal evolution across space and time—allow for investigation into how chemical signal traits respond to selection pressures on ecological timescales. This approach also offers a valuable framework for experimentally testing hypotheses derived from macro-evolutionary analyses, by manipulating specific ecological or social factors thought to influence signal design. So far, most micro-evolutionary work in lizards has relied on among-population comparisons, with only a few studies taking longitudinal or experimental approaches. These early efforts have begun to uncover the role of plasticity, local adaptation, and natural selection in shaping chemical communication at fine evolutionary scales.

4.1 Among-Population Comparisons

Interpopulation comparisons represent the most common micro-evolutionary approach to studying chemical signaling in lizards. Yet, most studies to date have focused purely on reporting differences in gland number or secretion composition between two or more populations, and in some cases, paired these analyses with behavioral assays demonstrating that lizards can discriminate between populations based solely on scent (Escobar et al., 2003; Gabirot et al., 2008, 2010, 2012, 2013; García-Roa, Megía-Palma, et al., 2017b; Ibáñez et al., 2017; Iraeta et al., 2011; Khannoon et al., 2013; Martín et al., 2015; Martín & López, 2006a; Ortega et al., 2019). While such studies are valuable for revealing population-level divergence, their findings are much too often interpreted as evidence of local adaptation, despite lacking formal testing against environmental or socio-ecological gradients.

In a smaller number of cases, however, population-level differences in chemical signaling investment or composition have been rigorously tested against ecological predictors, with findings that echo the complexity observed at macro-evolutionary scales. For instance, Romero-Diaz et al. (2025) analyzed 55 populations of wall lizards (*Podarcis muralis*) across the Pyrenees, integrating detailed climatic and sociosexual variables. They found that investment in chemical signaling (as estimated by epidermal gland number) was elevated in populations with greater sexual size dimorphism, as well as in warmer environments with more vegetation complexity. These patterns suggest that chemical signaling investment may be shaped by both efficacy-based selection favoring greater investment under harsher environmental conditions and the local intensity of sexual selection.

Another important insight from among-population comparisons is that stochastic processes may also contribute to chemical signal divergence. These findings serve as a reminder that variation in chemical signal composition does not necessarily reflect adaptive evolution and caution against the temptation of adaptationist storytelling. In some cases, divergence may instead arise through neutral evolutionary change, especially in isolated populations or secondary contact zones. For instance, in the common wall lizard (*Podarcis muralis*), divergence in epidermal gland secretions between two lineages did not predict hybridization success, and introgression patterns closely resembled those of neutral genetic markers—suggesting that signal divergence may reflect drift rather than selection (MacGregor et al., 2017). In the Skyros wall lizard (*Podarcis gaigeae*), secretion composition varied most strongly among isolated islet populations and was positively correlated with neutral genetic divergence. Notably, mate preferences did not consistently align with chemical differences, pointing to a potential decoupling between signal evolution and sexual selection (Runemark et al., 2011). These cases illustrate that chemical signals and associated preferences can diverge under both stochastic and selective forces, complicating assumptions about their evolutionary drivers (Baeckens et al. 2026).

4.2 Longitudinal and Experimental Perspectives

While most micro-evolutionary studies have relied on spatial comparisons (i.e., allochronic study design), a smaller number have adopted temporal and experimental approaches (i.e., synchronic study design, sensu Hendry & Kinnison, 1999), offering valuable insight into how chemical signals shift over ecological timescales.

One particularly powerful example comes from an island introduction experiment with Aegean wall lizards (*Podarcis erhardii*) in the Greek Cyclades. Individuals from a predator-rich source population were translocated to five small, predator-free islets—environments lacking snakes or other terrestrial vertebrates (Fig. 2a; Donihue, et al., 2020a). This ecological release removed the need to limit conspicuous chemical signaling to avoid eavesdropping predators. Simultaneously, population densities increased fivefold within 4 years, intensifying intraspecific competition (Fig. 2b). Bite scar frequency nearly doubled in males compared to the source population, indicating elevated male-male aggression (Fig. 2c). In response to these rapid ecological changes—occurring over just four generations—males from the experimental islands developed richer

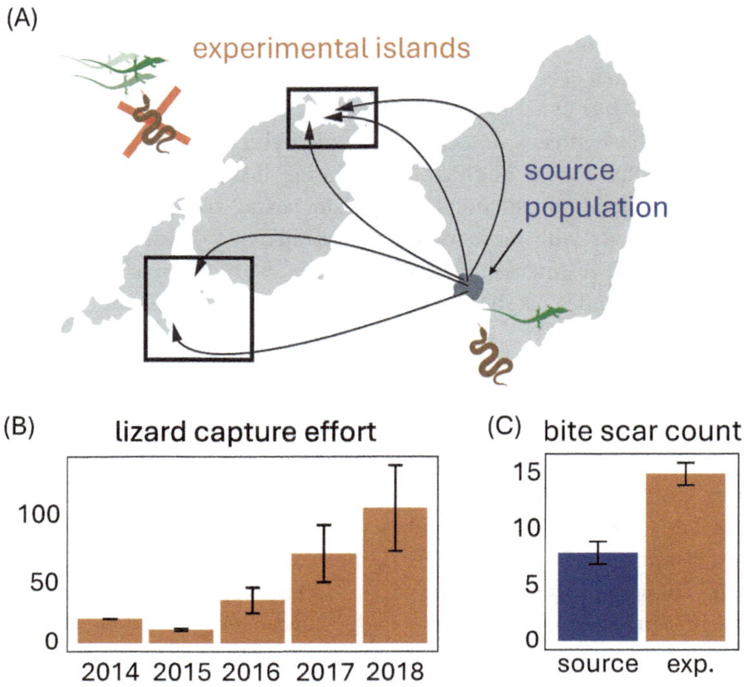

Fig. 2. Studying the speed and predictability of chemical signal divergence using an island introduction experiment. (**a**) Map of the Cycladic islands (Greece), showing the source population of *Podarcis erhardii* on the large island of Naxos and the 5 surrounding predator-free islets to which 20 lizards each were experimentally introduced in 2014 (arrows indicate translocation). (**b**) Mean annual lizard capture effort (± SE) pooled across all experimental islands from 2014 to 2018, illustrating the rapid increase in population densities following introduction. (**c**) Mean number of bite scars (± SE) recorded in 2018 from males in the source population and those on the experimental islands, highlighting elevated levels of intraspecific aggression in the predator-free experimental environments. (Data from Donihue, et al., 2020a; Baeckens, et al., 2025).

chemical secretions, with elevated levels of three lipids potentially linked to territoriality and mate attraction (Donihue, et al., 2020a). These shifts illustrate how signal design may evolve rapidly when selection pressures are relaxed on one front (i.e., predation) but heightened on another (i.e., social competition).

A follow-up longitudinal study on these same populations examined changes in scent marks from individually marked lizards over 2 consecutive years (Baeckens, et al., 2025). Across both individuals and populations, overall chemical composition shifted consistently in direction and magnitude. While the proportions of certain compounds (e.g., α-tocopherol) remained stable, others (e.g., oleic acid and octadecanoic acid) varied either over time or in a body size-dependent manner, highlighting the plasticity and complexity of individual chemical profiles. These findings provide rare empirical evidence that lizard chemical signals are dynamic traits, capable of responding to short-term ecological change. They also underscore the value of experimental and longitudinal approaches for linking specific environmental context to evolutionary shifts in communication systems.

5 Chemical Signal Evolution in the Anthropocene

We are living in an era in which human activities are reshaping ecosystems at unprecedented scales and rates (Barnosky et al., 2012; Ceballos et al., 2015). Understanding how, and how fast, animals respond to this rapidly changing world is essential for predicting population persistence and preventing extinctions (Shine & Baeckens, 2023; Stockwell et al., 2003). At the same time, such rapid anthropogenic changes offer an exceptional "natural laboratory" for studying evolution in action (Sarrazin & Lecomte, 2016).

Major anthropogenic perturbations, such as human-caused extreme climate events, biological invasions, and urbanization, are increasingly recognized as powerful evolutionary drivers (Baeckens & Donihue 2025; Grant et al., 2017; Johnson & Munshi-South, 2017; Moran & Alexander, 2014). Examples from across the animal kingdom show how such disturbances can lead to phenotypic change over short timescales: hurricanes have driven morphological evolution in anole lizards (Donihue, et al., 2020b); the introduction of cane toads in Australia has spurred evolutionary responses in both the invaders and native predators (Phillips et al., 2006; Shine, 2012), and urban environments have altered sexual signals such as frog croaks (Halfwerk et al., 2018), bird song (Slabbekoorn & Peet, 2003), and body coloration (Peneaux et al., 2021).

Yet despite the growing use of these perturbations as focal systems in evolutionary research, they have been rarely applied to the study of vertebrate chemical signaling—and remain almost entirely absent in lizard systems. This gap is striking, given that chemical signal production and efficacy are sensitive to environmental parameters that are commonly altered in human-dominated landscapes, such as temperature, pollution, and substrate structure. There is clear opportunity to explore how the urban heat island effects, soil contamination, or airborne pollutants influence chemical signal production and detectability; how invasive species reshape native signal-receiver dynamics, or how climate extremes (e.g., drought) disrupt the stability and persistence of chemical signals (Boullis et al., 2016; Groot & Zizzari, 2019). A recent study on sand lizards (*Lacerta agilis*) in Poland (Ibáñez et al., 2025) has begun to document shifts in chemical signal composition and diversity across urbanization gradients, but such efforts remain rare. Alarmingly little is known about how urbanization (delBarco-Trillo, 2026, this volume) and other anthropogenic pressures affect vertebrate chemical signaling more broadly. Expanding this research frontier is timely and necessary. Integrating chemical signaling into the broader framework of Anthropocene evolutionary biology will not only improve our understanding of how species respond to abrupt environmental change but may also help anticipate which traits or populations are most vulnerable in our rapidly changing world.

6 Outlook

Over the past decade, lizards have emerged as a promising system for studying the evolution of chemical communication. Macro-evolutionary studies have begun to reveal how climatic, ecological, and social factors shape signal diversity across species. Meanwhile, micro-evolutionary research, though still in its early stages, is uncovering how signals shift on ecological timescales.

Nonetheless, major gaps remain. Comparative work has focused heavily on a few clades, especially Old World lacertids, limiting broader generalizations. Expanding research across more lizard families and continents—while integrating multiple explanatory factors—will be crucial for developing a more comprehensive understanding of chemical signal evolution. Future studies should move beyond documenting simple correlations and instead model how traits and environments have coevolved across phylogenies (Cornwallis & Griffin, 2024).

Bridging broad evolutionary insights with behavioral function remains another critical challenge. While large-scale studies have mapped chemical diversity, the functional relevance of specific compounds remains largely unknown—although important progress is being made (Martín & López, 2024). Carefully designed behavioral assays, informed by both chemical profiling and comparative findings, will be key to understanding the information content and social utility of signals.

Perhaps most urgently, almost no work has addressed the heritability of lizard chemical signals—a prerequisite for understanding their evolutionary potential. Without estimates of genetic variance, we cannot distinguish evolutionary change from environmentally induced plasticity. Incorporating quantitative genetics and expanding genomic approaches should therefore be a major priority moving forward.

Finally, human-driven environmental change offers a powerful but underutilized opportunity to study chemical signal evolution in real time. Urbanization, biological invasions, and climate extremes are already reshaping habitats and social structures, creating novel selection pressures on communication systems.

By integrating macro- and micro-evolutionary approaches, behavioral validation, genetic foundations, and anthropogenic contexts, future work will not only expand our understanding of lizard chemical communication but also help position reptiles more centrally within the broader field of vertebrate signaling evolution. The future is bright for those eager to use lizards as a window to study the ecoevolutionary dynamics of chemical signals in vertebrates.

Acknowledgements. I am grateful to Lisa M. Schulte and Birgit Szabo for their input, which helped to improve the manuscript.

References

Alberts, A. (1992). Constraints on the design of chemical communication systems in terrestrial vertebrates. *American Naturalist, 139*, 62–89.

Andersson, M. (1982). Female choice selects for extreme tail length in a widowbird. *Nature, 299*, 529–534.

Apps, P. J., Weldon, P. J., & Kramer, M. (2015). Chemical signals in terrestrial vertebrates: Search for design features. *Natural Product Reports, 32*, 1131–1153.

Baeckens, S. (2017). Reliable proxies for glandular secretion production in lacertid lizards. *Acta Herpetologica, 12*, 199–204.

Baeckens, S. (2019). Evolution of animal chemical communication: Insights from non-model species and phylogenetic comparative methods. *Belgian Journal of Zoology, 149*, 63–93.

Baeckens, S., & Whiting, M. J. (2021). Investment in chemical signalling glands facilitates the evolution of sociality in lizards. *Proceedings of the Royal Society B: Biological Sciences, 288*, 20202438.

Baeckens, S., Edwards, S., Huyghe, K., & Van Damme, R. (2015). Chemical signalling in lizards: An interspecific comparison of femoral pore numbers in Lacertidae. *Biological Journal of the Linnean Society, 114*, 44–57.

Baeckens, S., Van Damme, R., & Cooper, W. E. (2017a). How phylogeny and foraging ecology drive the level of chemosensory exploration in lizards and snakes. *Journal of Evolutionary Biology, 30*, 627–640.

Baeckens, S., Huyghe, K., Palme, R., & Van Damme, R. (2017b). Chemical communication in the lacertid lizard *Podarcis muralis*: The functional significance of testosterone. *Acta Zoologica, 98*, 94–103.

Baeckens, S., Herrel, A., Broeckhoven, C., Vasilopoulou-Kampitsi, M., Huyghe, K., Goyens, J., & Van Damme, R. (2017c). Evolutionary morphology of the lizard chemosensory system. *Scientific Reports, 7*, 10141.

Baeckens, S., García-Roa, R., Martín, J., & Van Damme, R. (2017d). The role of diet in shaping the chemical signal design of lacertid lizards. *Journal of Chemical Ecology, 43*, 902–910.

Baeckens, S., Martín, J., Garcia-Roa, R., Pafilis, P., Huyghe, K., & Van Damme, R. (2018a). Environmental conditions shape the chemical signal design of lizards. *Functional Ecology, 32*, 566–580.

Baeckens, S., Martín, J., García-Roa, R., & Van Damme, R. (2018b). Sexual selection and the chemical signal design of lacertid lizards. *Zoological Journal of the Linnean Society, 183*, 445–457.

Baeckens, S., Martín, J., Pafilis, P., Herrel, A., & Donihue, C. M. (2025). Temporal dynamics of scent mark composition in field-experimental lizard populations. *Functional Ecology, 39*, 154–164.

Baeckens, S., Donihue, C. M. (2025). *Evolutionary consequences of extreme climate events. Current Biology, 35(17)*, R850–R864. https://doi.org/10.1016/j.cub.2025.07.046

Baeckens, S., Martín, J., Tadić, Z., Van den Camp, E., Sacchi, R., Van Damme, M., & Mangiacotti, M. (2026). *Non-adaptive processes explain chemical signal divergence in island lizards. Oikos*, 2026(2). https://nsojournals.onlinelibrary.wiley.com/doi/10.1002/oik.11743

Barnosky, A. D., Hadly, E. A., Bascompte, J., Berlow, E. L., Brown, J. H., Fortelius, M., et al. (2012). Approaching a state shift in Earth's biosphere. *Nature, 486*, 52–58.

Bissinger, B. E., & Simon, C. A. (1979). Comparison of tongue extrusions in representatives of six families of lizards. *Journal of Herpetology, 13*, 133.

Boullis, A., Detrain, C., Francis, F., & Verheggen, F. J. (2016). Will climate change affect insect pheromonal communication? *Current Opinion in Insect Science, 17*, 87–91.

Burgin, C. J., Colella, J. P., Kahn, P. L., & Upham, N. S. (2018). How many species of mammals are there? *Journal of Mammalogy, 99*, 1–14.

Campos, S. M., Pruett, J. A., Soini, H. A., Zúñiga-Vega, J. J., Goldberg, J. K., Vital-García, C., et al. (2020). Volatile fatty acid and aldehyde abundances evolve with behavior and habitat temperature in *Sceloporus* lizards. *Behavioral Ecology, 31*, 978–991.

Caro, T. M. (1995). Pursuit deterrence revisited. *Trends in Ecology and Evolution, 10*, 500–503.

Ceballos, G., Ehrlich, P. R., Barnosky, A. D., García, A., Pringle, R. M., & Palmer, T. M. (2015). Accelerated modern human-induced species losses: Entering the sixth mass extinction. *Science Advances, 1*, e1400253.

Chen, I.-P., Stuart-Fox, D., Hugall, A. F., & Symonds, M. R. E. (2012). Sexual selection and the evolution of complex color patterns in dragon lizards. *Evolution, 66*, 3605–3614.

Cooper, W. E. (1994). Chemical discrimination by tongue-flicking in lizards: A review with hypotheses on its origin and its ecological and phylogenetic relationships. *Journal of Chemical Ecology, 20*, 439–487.

Cooper, W. E. (1996). Variation and evolution of forked tongues in squamate reptiles. *Herpetological Natural History, 4*, 135–150.

Cooper, W. E., & Vitt, L. J. (1984). Conspecific odor detection by the male broad-headed skink, *Eumes laticeps*: Effects of sex and site of odor source and of male reproductive condition. *The Journal of Experimental Zoology, 209*, 199–209.

Cornwallis, C. K., & Griffin, A. S. (2024). A guided tour of phylogenetic comparative methods for studying trait evolution. *Annual Review of Ecology, Evolution, and Systematics, 55*, 181–204.

delBarco-Trillo, J. (2026). Odors in the city: Intraspecific chemical communication in urban environments. In L. M. Schulte (Ed.), *Chemical signals in vertebrates* (Vol. 16, pp. 283–294). Springer.

Donihue, C. M., Herrel, A., Martín, J., Foufopoulos, J., Pafilis, P., & Baeckens, S. (2020a). Rapid and repeated divergence of animal chemical signals in an Island introduction experiment. *Journal of Animal Ecology, 89*, 1458–1467.

Donihue, C. M., Kowaleski, A. M., Losos, J. B., Algar, A. C., Baeckens, S., Buchkowski, R. W., et al. (2020b). Hurricane effects on Neotropical lizards span geographic and phylogenetic scales. *Proceedings of the National Academy of Sciences of the United States of America, 117*, 10429–10434.

Ekman, P. (2009). Darwin's contributions to our understanding of emotional expressions. *Philosophical Transactions of the Royal Society B: Biological Sciences, 364*, 3449–3451.

Endler, J. A. (1992). Signals, signal conditions, and the direction of evolution. *The American Naturalist, 139*, S125–S153.

Endler, J. A. (1993). Some general comments on the evolution and design of animal communication systems. *Philosophical Transactions of the Royal Society of London. Series B, Biological Sciences, 340*, 215–225.

Escobar, C. A., Labra, A., & Niemeyer, H.M. (2001). Chemical composition of precloacal secretions of Liolaemus lizards. Journal of Chemical Ecology, 27, 1677–1690.

Escobar, C. M., Escobar, C. A., Labra, A., & Niemeyer, H. M. (2003). Chemical composition of precloacal secretions of two *Liolaemus fabiani* populations: Are they different? *Journal of Chemical Ecology, 29*, 629–638.

Filoramo, N. I., & Schwenk, K. (2009). The mechanism of chemical delivery to the vomeronasal organs in squamate reptiles: A comparative morphological approach. *Journal of Experimental Zoology, 34*, 20–34.

Fischman, B. J., Hollis Woodard, S., & Robinson, G. E. (2011). Molecular evolutionary analyses of insect societies. *Proceedings of the National Academy of Sciences of the United States of America, 108*, 10847–10854.

Fleishman, L., & Font, E. (2019). Sensory processing in relation to signaling behavior. In V. L. Bels & A. P. Russell (Eds.), *Behavior of lizards, evolutionary and mechanistic perspectives* (pp. 207–257). CRC Press.

Fögen, T. (2014). Animal communication. In G. L. Campbell (Ed.), *The Oxford handbook of animals in classical thought and life* (pp. 216–232). Oxford University Press.

Font, E., Barbosa, D., Sampedro, C., & Carazo, P. (2012). Social behavior, chemical communication, and adult neurogenesis: Studies of scent mark function in *Podarcis* wall lizards. *General and Comparative Endocrinology, 177*, 9–17.

Gabirot, M., López, P., Martín, J., de Fraipont, M., Heulin, B., Sinervo, B., & Clobert, J. (2008). Chemical composition of femoral secretions of oviparous and viviparous types of male common lizards *Lacerta vivipara*. *Biochemical Systematics and Ecology, 36*, 539–544.

Gabirot, M., Castilla, A. M., López, P., & Martín, J. (2010). Differences in chemical signals may explain species recognition between an Island lizard, *Podarcis atrata*, and related mainland lizards, P. hispanica. *Biochemical Systematics and Ecology, 38*, 521–528.

Gabirot, M., López, P., & Martín, J. (2012). Interpopulational variation in chemosensory responses to selected steroids from femoral secretions of male lizards, *Podarcis hispanica*, mirrors population differences in chemical signals. *Chemoecology, 22*, 65–73.

Gabirot, M., López, P., & Martín, J. (2013). Female mate choice based on pheromone content may inhibit reproductive isolation between distinct populations of Iberian wall lizards. *Current Zoology, 59*, 210–220.

Garamszegi, L. Z., Eens, M., Erritzøe, J., & Møller, A. P. (2005). Sexually size dimorphic brains and song complexity in passerine birds. *Behavioral Ecology, 16*, 335–345.

García-Roa, R., Carreira, S., López, P., & Martín, J. (2016). Genders matters: Sexual differences in chemical signals of *Liolaemus wiegmannii* lizards (Iguania, Liolaemidae). *Biochemical Systematics and Ecology, 69*, 108–114.

García-Roa, R., Sáiz, J., Gómara, B., López, P., & Martín, J. (2017a). Dietary constraints can preclude the expression of an honest chemical sexual signal. *Scientific Reports, 7*, 6073.

García-Roa, R., Megía-Palma, R., Ortega, J., Jara, M., López, P., & Martín, J. (2017b). Interpopulational and seasonal variation in the chemical signals of the lizard *Gallotia galloti*. *PeerJ, 5*, e3992.

García-Roa, R., Jara, M., Baeckens, S., López, P., Van Damme, R., Martín, J., & Pincheira-Donoso, D. (2017c). Macroevolutionary diversification of glands for chemical communication in squamate reptiles. *Scientific Reports, 7*, 9288.

Grant, P. R., Schmitt, J., Grant, B. R., Huey, R. B., Johnson, M. T. J., & Knoll, A. H. (2017). Evolution caused by extreme events. *Philosophical Transactions of the Royal Society B: Biological Sciences, 372*, 20160146.

Graves, B. M. (1993). Chemical delivery to the vomeronasal organs and functional domain of squamate chemoreception. *Brain, Behavior and Evolution, 41*, 198–202.

Grether, G. F., Peiman, K. S., Tobias, J. A., & Robinson, B. W. (2017). Causes and consequences of behavioral interference between species. *Trends in Ecology and Evolution, 32*, 760–772.

Groot, A. T., & Zizzari, Z. V. (2019). Does climate warming influence sexual chemical signaling? *Animal Biology, 69*, 83–93.

Guilford, T. I. M., & Dawkins, M. S. (1991). Receiver psychology and the evolution of animal signals. *Animal Behaviour, 42*, 1–14.

Halfwerk, W., Blaas, M., Kramer, L., Hijner, N., Trillo, P. A., Bernal, X. E., et al. (2018). Adaptive changes in sexual signalling in response to urbanization. *Nature Ecology & Evolution, 3*, 374–380.

Hartmann, T. (2008). The lost origin of chemical ecology in the late 19th century. *Proceedings of the National Academy of Sciences of the United States of America, 105*, 4541–4546.

Hebets, E. A., & Papaj, D. R. (2005). Complex signal function: Developing a framework of testable hypotheses. *Behavioral Ecology and Sociobiology, 57*, 197–214.

Hendry, A. P., & Kinnison, M. T. (1999). The pace of modern life: Measuring rates of contemporary microevolution. *Evolution, 53*, 1637–1653.

Hews, D. K., & Benard, M. F. (2001). Negative association between conspicuous visual display and chemosensory behavior in two phrynosomatid lizards. *Ethology, 107*, 839–850.

Hews, D. K., Date, P., Hara, E., & Castellano, M. J. (2011). Field presentation of male secretions alters social display in *Sceloporus virgatus* but not *S. Undulatus* lizards. *Behavioral Ecology and Sociobiology, 65*, 1403–1410.

Huey, R., Pianka, E., & Schoener, T. (2007). *Lizard ecology, Studies of a model organism*. de Gruyter.

Ibáñez, A., Menke, M., Quezada, G., Jiménez-Uzcátegui, G., Schulz, S., & Steinfartz, S. (2017). Diversity of compounds in femoral secretions of Galápagos iguanas (genera: *Amblyrhynchus* and *Conolophus*), and their potential role in sexual communication in lek-mating marine iguanas (*Amblyrhynchus cristatus*). *PeerJ, 5*, e3689.

Ibáñez, A., Zając, B., Sambak, I., Woźniakiewicz, M., Woźniakiewicz, A., & Pabijan, M. (2025). Chemical signal diversity in male sand lizards (*Lacerta agilis*) along an urbanization gradient. *Scientific Reports, 15*, 6958.

Iraeta, P., Monasterio, C., Salvador, A., & Díaz, J. A. (2011). Sexual dimorphism and interpopulation differences in lizard hind limb length: Locomotor performance or chemical signalling? *Biological Journal of the Linnean Society, 104*, 318–329.

Jara, M., Frias-De-Diego, A., García-Roa, R., Saldarriaga-Córdoba, M., Harvey, L. P., Hickcox, R. P., & Pincheira-Donoso, D. (2018). The macroecology of chemical communication in lizards: Do climatic factors drive the evolution of signalling glands? *Evolutionary Biology, 45*, 259–267.

Johnson, M., & Munshi-South, J. (2017). Evolution of life in urban environments. *Science, 358*, eaam8327.

Keverne, E. B. (1999). The vomeronasal organ. *Science, 286*, 716.

Khannoon, E., Lunt, D., Schulz, S., & Hardege, J. (2013). Divergence of scent pheromones in allopatric populations of *Acanthodactylus boskianus* (Squamata: Lacertidae). *Zoological Science, 30*, 380–385.

Kopena, R., Martín, J., López, P., & Herczeg, G. (2011). Vitamin E supplementation increases the attractiveness of males' scent for female European green lizards. *PLoS One, 6*, e19410.

Kopena, R., López, P., & Martín, J. (2014). Relative contribution of dietary carotenoids and vitamin E to visual and chemical sexual signals of male Iberian green lizards: An experimental test. *Behavioral Ecology and Sociobiology, 68*, 571–581.

Labra, A. (2008). Multi-contextual use of chemosignals by *Liolaemus* lizards. In J. Hurst, R. Beynon, S. Roberts, & T. Wyatt (Eds.), *Chemical signals in vertebrates 11* (pp. 357–365). Springer.

Le Busque, B., & Litchfield, C. (2023). Sharks, spiders, snakes, oh my: A review of creature feature films. *Journal of Environmental Media, 4*, 49–75.

López, P., & Martín, J. (2002). Chemical rival recognition decreases aggression levels in male Iberian wall lizards, *Podarcis hispanica. Behavioral Ecology and Sociobiology, 51*, 461–465.

López, P., & Martín, J. (2011). Male Iberian rock lizards may reduce the costs of fighting by scent matching of the resource holders. *Behavioral Ecology and Sociobiology, 65*, 1891–1898.

Losos, J. B. (2011). Convergence, adaptation, and constraint. *Evolution, 65*, 1827–1840.

Losos, J. B., & Ricklefs, R. E. (2009). Adaptation and diversification on islands. *Nature, 457*, 830–836.

MacGregor, H. E. A., Lewandowsky, R. A. M., D'Ettorre, P., Leroy, C., Davies, N. W., While, G. M., & Uller, T. (2017). Chemical communication, sexual selection, and introgression in wall lizards. *Evolution, 71*, 2327–2343.

Mancini, G., Santini, L., Cazalis, V., Ficetola, G. F., Meiri, S., Roll, U., et al. (2025). Generation length of the world's amphibians and reptiles. *Ecography*, e07527.

Mangiacotti, M., Pezzi, S., Fumagalli, M., Coladonato, A. J., d'Ettorre, P., Leroy, C., et al. (2019). Seasonal variations in femoral gland secretions reveals some unexpected correlations between protein and lipid components in a lacertid lizard. *Journal of Chemical Ecology, 45*, 673–683.

Mangiacotti, M., Baeckens, S., Scali, S., Martín, J., Van Damme, R., & Sacchi, R. (2021). Evolutionary and biogeographical support for species-specific proteins in lizard chemical signals. *Biological Journal of the Linnean Society, 134*, 912–928.

Mangiacotti, M., Fumagalli, M., Casali, C., Biggiogera, M., Forneris, F., & Sacchi, R. (2023a). Carbonic anhydrase IV in lizard chemical signals. *Scientific Reports, 13*, 14164.

Mangiacotti, M., Baeckens, S., Fumagalli, M., Martín, J., Scali, S., & Sacchi, R. (2023b). Protein–lipid association in lizard chemical signals. *Integrative Organismal Biology, 5*, obad016.

Mangiacotti, M., Baeckens, S., Fumagalli, M., Martín, J., Scali, S., & Sacchi, R. (2023c). Sexual selection and proteinaceous diversity in the femoral gland secretions of lacertid lizards. *Diversity, 15*, 777.

Martín, J., & López, P. (2000). Chemoreception, symmetry and mate choice in lizards. *Proceedings of the Royal Society B Biological Sciences, 267*, 1265–1269.

Martín, J., & López, P. (2006a). Interpopulational differences in chemical composition and chemosensory recognition of femoral gland secretions of male lizards *Podarcis hispanica*: Implications for sexual isolation in a species complex. *Chemoecology, 16*, 31–38.

Martín, J., & López, P. (2006b). Links between male quality, male chemical signals, and female mate choice in Iberian rock lizards. *Functional Ecology, 20*, 1087–1096.

Martín, J., & López, P. (2007). Scent may signal fighting ability in male Iberian rock lizards. *Biology Letters, 3*, 125–127.

Martín, J., & López, P. (2013). Effects of global warming on sensory ecology of rock lizards: Increased temperatures alter the efficacy of sexual chemical signals. *Functional Ecology, 27*, 1332–1340.

Martín, J., & López, P. (2015). Condition-dependent chemosignals in reproductive behavior of lizards. *Hormones and Behavior, 68*, 14–24.

Martín, J., & López, P. (2024). Pheromones and reproduction in reptiles. In D. O. Norris & K. H. Lopez (Eds.), *Hormones and reproduction of vertebrates, volume 3, reptiles* (pp. 135–163). Elsevier.

Martín, J., López, P., Gabirot, M., & Pilz, K. M. (2007). Effects of testosterone supplementation on chemical signals of male Iberian wall lizards: Consequences for female mate choice. *Behavioral Ecology and Sociobiology, 61*, 1275–1282.

Martín, J., Ortega, J., & López, P. (2015). Interpopulational variations in sexual chemical signals of Iberian wall lizards may allow maximizing signal efficiency under different climatic conditions. *PLoS One, 10*, e0131492.

Martín, J., Castilla, A. M., López, P., Al-Jaidah, M., Mohannadi, S., & Al-Hamaidi, A. (2016). Chemical signals in desert lizards: Are femoral gland secretions of male and female spiny-tailed lizards, *Uromastyx aegyptia microlepis* adapted to arid conditions ? *Journal of Arid Environments, 127*, 192–198.

Martín, J., Navarro-Castilla, Á., De La Concha, A., Cuervo, J. J., Barja, I., & López, P. (2024). Heat-altered scent marks of males of a fossorial reptile still allow recognition by females but lose information on male quality. *Behavioral Ecology and Sociobiology, 78*, 77.

Martins, E. P., Ossip-Klein, A. G., Zúñiga-Vega, J. J., Vital García, C., Campos, S. M., & Hews, D. K. (2015). Evolving from static to dynamic signals: Evolutionary compensation between two communicative signals. *Animal Behaviour, 102*, 223–229.

Martins, E. P., Ossip-Drahos, A. G., Vital García, C., Zúñiga-Vega, J. J., Campos, S. M., & Hews, D. K. (2018). Trade-offs between visual and chemical behavioral responses. *Behavioral Ecology and Sociobiology, 72*, 189.

Mason, R. T., & Parker, M. R. (2010). Social behavior and pheromonal communication in reptiles. *Journal of Comparative Physiology. A, Neuroethology, Sensory, Neural, and Behavioral Physiology, 196*, 729–749.

Mayerl, C., Van Damme, R., & Baeckens, S. (2015). Evolution and role of the follicular epidermal gland system in non-ophidian squamates. *Amphibia-Reptilia, 36*, 185–206.

Meiri, S. (2018). Traits of lizards of the world: Variation around a successful evolutionary design. *Global Ecology and Biogeography, 27*, 1168–1172.

Moran, E. V., & Alexander, J. M. (2014). Evolutionary responses to global change: Lessons from invasive species. *Ecology Letters, 17*, 637–649.

Moreira, P. L., López, P., & Martín, J. (2008). Discrimination of conspecific faecal chemicals and spatial decisions in juvenile Iberian rock lizards (*Lacerta monticola*). *Acta Ethologica, 11*, 26–33.

Morgan, D. (2008). *Snakes in myth, magic, and history*. Bloomsbury Publishing.

Mouton, P. L. F. N., Van Rensburg, D. a. J., & Van Wyk, J. H. (2010). Epidermal glands in cordylid lizards, with special reference to generation glands. *Zoological Journal of the Linnean Society, 158*, 312–324.

Müller-Schwarze, D. (2006). *Chemical ecology of vertebrates*. Cambridge University Press.

Müller-Schwarze, D., & Mozzel, M. (1977). *Chemical signals in vertebrates*. Plenum Press.

Novosolov, M., Raia, P., & Meiri, S. (2013). The Island syndrome in lizards. *Global Ecology and Biogeography, 22*, 184–191.

Novosolov, M., Rodda, G. H., Feldman, A., Kadison, A. E., Dor, R., & Meiri, S. (2016). Power in numbers: The evolutionary drivers of high population density in insular lizards. *Global Ecology and Biogeography, 25*, 87–95.

Olsson, M., Madsen, T., Nordby, J., Wapstra, E., Ujvari, B., & Wittsell, H. (2003). Major histocompatibility complex and mate choice in sand lizards. *Proceedings of the Royal Society of London. Series B: Biological Sciences, 270*, 254–256.

Ord, T. J., & Stamps, J. A. (2009). Species identity cues in animal communication. *The American Naturalist, 174*, 585–593.

Ord, T. J., Blumstein, D. T., & Evans, C. S. (2001). Intrasexual selection predicts the evolution of signal complexity in lizards. *Proceedings Royal Society of London Series B Biological Sciences, 268*, 737–744.

Ortega, J., Martín, J., Crochet, P.-A., López, P., & Clobert, J. (2019). Seasonal and interpopulational phenotypic variation in morphology and sexual signals of *Podarcis liolepis* lizards. *PLoS One, 14*, e0211686.

Peneaux, C., Grainger, R., Lermite, F., Machovsky-Capuska, G. E., Gaston, T., & Griffin, A. S. (2021). Detrimental effects of urbanization on the diet, health, and signal coloration of an ecologically successful alien bird. *Science of the Total Environment, 796*, 148828.

Pérez i de Lanuza, G., & Font, E. (2016). The evolution of colour pattern complexity: Selection for conspicuousness favours contrasting within-body colour combinations in lizards. *Journal of Evolutionary Biology, 29*, 942–951.

Pfennig, K. S., & Pfennig, D. W. (2009). Character displacement: Ecological and reproductive responses to a common evolutionary problem. *The Quarterly Review of Biology, 84*, 253–276.

Phillips, B., Brown, G. P., Webb, J. K., & Shine, R. (2006). Invasion and the evolution of speed in toads. *Nature, 439*, 803.

Pincheira-Donoso, D., Hodgson, D. J., & Tregenza, T. (2008). Comparative evidence for strong phylogenetic inertia in precloacal signalling glands in a species-rich lizard clade. *Evolutionary Ecology Research, 10*, 11–28.

Pruett, J. A., Zúñiga-Vega, J. J., Campos, S. M., Soini, H. A., Novotny, M. V., Vital-García, C., et al. (2016). Evolutionary interactions between visual and chemical signals: Chemosignals compensate for the loss of a visual signal in male *Sceloporus* lizards. *Journal of Chemical Ecology, 42*, 1164–1174.

Reznick, D. N., & Ricklefs, R. E. (2009). Darwin's bridge between microevolution and macroevolution. *Nature, 457*, 837–842.

Roll, U., Feldman, A., Novosolov, M., Allison, A., Bauer, A. M., Bernard, R., et al. (2017). The global distribution of tetrapods reveals a need for targeted reptile conservation. *Nature Ecology & Evolution, 1*, 1677–1682.

Romero-Diaz, C., Campos, S. M., Herrmann, M. A., Soini, H. A., Novotny, M. V., Hews, D. K., & Martins, E. P. (2021). Composition and compound proportions affect the response to complex chemical signals in a spiny lizard. *Behavioral Ecology and Sociobiology, 75*, 42.

Romero-Diaz, C., Gómez Ramírez, F., Aguilar, P., Marešová, P., Font, E., & Pérez I de Lanuza, G. (2025). Climate and socio-sexual environment predict interpopulation variation in chemical signaling glands in a widespread lizard. *Integrative Zoology*. https://doi.org/10.1111/1749-4877.12923

Rosenthal, M. F., Gertler, M., Hamilton, A. D., Prasad, S., & Andrade, M. C. B. (2017). Taxonomic bias in animal behaviour publications. *Animal Behaviour, 127*, 83–89.

Runemark, A., Gabirot, M., & Svensson, E. I. (2011). Population divergence in chemical signals and the potential for premating isolation between islet- and mainland populations of the Skyros wall lizard (*Podarcis gaigeae*). *Journal of Evolutionary Biology, 24*, 795–809.

Sarrazin, F., & Lecomte, J. (2016). Evolution in the Anthropocene. *Science, 351*, 922–923.

Schaefer, H. M., & Ruxton, G. D. (2015). Signal diversity, sexual selection, and speciation. *Annual Review of Ecology, Evolution, and Systematics, 46*, 573–592.

Schwenk, K. (1993). The evolution of chemoreception in squamate reptiles: A phylogentic approach. *Brain, Behavior and Evolution, 41*, 124–137.

Schwenk, K. (1994). Why snakes have forked tongues. *Science, 263*, 1573–1577.

Schwenk, K. (1995). Of tongues and noses: Chemoreception in lizards and snakes. *Trends in Ecology and Evolution, 10*, 7–12.

Shine, R. (2012). Invasive species as drivers of evolutionary change: Cane toads in tropical Australia. *Evolutionary Applications, 5*, 107–116.

Shine, R., & Baeckens, S. (2023). Rapidly evolved traits enable new conservation tools: Perspectives from the cane toad invasion of Australia. *Evolution, 77*, 1744–1755.

Sinervo, B., & Lively, C. (1996). The rock-paper-scissors game and the evolution of alternative male strategies. *Nature, 380*, 240–243.

Slabbekoorn, H., & Peet, M. (2003). Birds sing at a higher pitch in urban noise. *Nature, 424*, 267–267.

Smadja, C., & Butlin, R. K. (2009). On the scent of speciation: The chemosensory system and its role in premating isolation. *Heredity, 102*, 77–97.

Smith, C. F., Schwenk, K., Earley, R. L., & Schuett, G. W. (2008). Sexual size dimorphism of the tongue in a north American pitviper. *Journal of Zoology, 274*, 367–374.

Soma, M., & Garamszegi, L. Z. (2011). Rethinking birdsong evolution: Meta-analysis of the relationship between song complexity and reproductive success. *Behavioral Ecology, 22*, 363–371.

Stockwell, C. A., Hendry, A. P., & Kinnison, M. T. (2003). Contemporary evolution meets conservation biology. *Trends in Ecology and Evolution, 18*, 94–101.

Symonds, M. R. E., & Elgar, M. A. (2008). The evolution of pheromone diversity. *Trends in Ecology and Evolution, 23*, 220–228.

Tinbergen, N. (1953). *Social behavior in animals, with special reference to vertebrates*. Methuen Press.

Tonini, J. F. R., Beard, K. H., Ferreira, R. B., Jetz, W., & Pyron, R. A. (2016). Fully-sampled phylogenies of squamates reveal evolutionary patterns in threat status. *Biological Conservation, 204*, 23–31.

Troudet, J., Grandcolas, P., Blin, A., Vignes-Lebbe, R., & Legendre, F. (2017). Taxonomic bias in biodiversity data and societal preferences. *Scientific Reports, 7*, 9132.

Van Oystaeyen, A., Oliveira, R. C., Holman, L., van Zweden, J. S., Romero, C., Oi, C. A., & Wenseleers, T. (2014). Conserved class of queen pheromones stops social insect workers from reproducing. *Science, 343*, 287–291.

von Frisch, K. (1967). *The dance language and orientation of bees*. Harvard University Press.

Weldon, P. J., Flachsbarth, B., & Schulz, S. (2008). Natural products from the integument of nonavian reptiles. *Natural Product Reports, 25*, 738–756.

Wilson, E. O. (1975). *Sociobiology, The new synthesis*. Harvard University Press.

Witzgall, P., Kirsch, P., & Cork, A. (2010). Sex pheromones and their impact on pest management. *Journal of Chemical Ecology, 36*, 80–100.

Wright, S. (1982). Character, change, speciation and higher taxa. *Evolution, 36*, 427–443.

Wyatt, T. (2014). *Pheromones and animal behaviour, Chemical signals and signatures*. Cambridge University Press.

Yang, W., Feiner, N., Pinho, C., While, G. M., Kaliontzopoulou, A., Harris, D. J., et al. (2021). Extensive introgression and mosaic genomes of Mediterranean endemic lizards. *Nature Communications, 12*, 2762.

Yew, J. Y., & Chung, H. (2015). Insect pheromones: An overview of function, form, and discovery. *Progress in Lipid Research, 59*, 88–105.

Zozaya, S. M., Higgie, M., Moritz, C., & Hoskin, C. J. (2019). Are pheromones key to unlocking cryptic lizard diversity? *American Naturalist, 194*, 168–182.

Zozaya, S. M., Teasdale, L. C., Moritz, C., Higgie, M., & Hoskin, C. J. (2022). Composition of a chemical signalling trait varies with phylogeny and precipitation across an Australian lizard radiation. *Journal of Evolutionary Biology, 35*, 919–933.

Zuk, M., & Kolluru, G. R. (1998). Exploitation of sexual signals by predators and parasitoids. *The Quarterly Review of Biology, 73*, 415–438.

Chemosensory Self-Inspection

Jan Havlíček[1(✉)], Dagmar Schwambergová[1], and S. Craig Roberts[2]

[1] Department of Zoology, Faculty of Science, Charles University, Prague, Czech Republic
jhavlicek@natur.cuni.cz; dagmar.schwambergova@natur.cuni.cz
[2] Department of Psychology, University of Stirling, Stirling, UK
craig.roberts@stir.ac.uk

Abstract. Self-sniffing appears to be a widespread behaviour across various vertebrate taxa, yet it has rarely been the subject of systematic scientific inquiry. Here, we review the available evidence on self-sniffing behaviour and explore its possible functions. Most research on this topic comes from self-report studies in humans, which indicate that people most frequently smell their hands and armpits. The frequency of this behaviour has been linked to health complaints, suggesting a role in monitoring one's health status. Additionally, evidence from fish to mammals suggests that individuals can distinguish their own chemosensory cues from those of conspecifics, including familiar ones. This ability is likely involved in self-referencing, which in turn is employed in scent-marking and kin recognition. Furthermore, as self-grooming allows individuals to adjust their chemosignals, self-sniffing may serve as a control mechanism for such adjustments. However, little is known about the specific contexts in which self-sniffing occurs or which body parts and excretions are most commonly sniffed. To move beyond speculation, systematic investigations of this overlooked phenomenon are needed across species as animals behave in their natural habitats.

Keywords: Kin recognition · Odour · Scent-marking · Self-grooming · Self-recognition · Sniffing

1 Introduction

Animals use their chemical senses, including olfaction, to gain information about not only their environment and conspecifics but also themselves. Naturalists, pet keepers, and other individuals who are in frequent contact with various vertebrate species are well aware that sniffing its own body parts—such as its genital area or its secretions, such as urine—is a routine part of an animal's behavioural repertoire. Nevertheless, self-sniffing is rarely a focus of systematic scientific inquiry. This lack of attention is remarkable not only because of the apparently high occurrence of self-sniffing in some species but also because chemosensory self-recognition is important in self-grooming

L. M. Schulte (ed.), *Chemical Signals in Vertebrates 16*,
https://doi.org/10.1007/978-3-032-30752-1_5

and monitoring health status, as well as being a central assumption of several influential theories, such as self-referencing and kin recognition.

The main aim of this paper is to provide an overview of the current knowledge on self-sniffing behaviour in vertebrates and to suggest its possible functions. We start with studies that describe the frequency and context of self-sniffing behaviour in primates and humans. Subsequently, we focus on chemosensory self-recognition in species ranging from reptiles to mammals, analysing the possible functions of this ability and providing brief outlines of potential future studies on the underexplored behaviour of self-sniffing. It is likely that in many species, chemosensory self-inspection involves not only olfaction but also the vomeronasal organ (VNO, especially in reptiles) and perhaps other chemical senses, such as taste (especially in fish). Nevertheless, in most species with multiple functional chemical senses, we do not yet know which are involved in self-inspection. Therefore, we use the umbrella term *chemosensory self-inspection*. Only in humans and birds, where the VNO is non-functional (Witt & Hummel, 2006; Silva & Antunes, 2017), do we specifically refer to *olfactory self-inspection*.

2 Self-Sniffing Behaviour in Primates

Observational studies on self-sniffing are rare and have only been conducted in a few primate species, including humans. Unfortunately, no evidence is currently available for other vertebrate taxa; moreover, as existing data are often collected as part of general sniffing behaviour, they often lack details on the specific body parts sniffed and the contexts in which the behaviour occurs.

2.1 Self-Sniffing Behaviour in Non-humans

We first focus on available evidence in non-human primates. One study examined seasonal fluctuations in sniffing behaviour in a zoo-housed group of ring-tailed lemurs (*Lemur catta*; Palagi et al., 2003). Regarding self-sniffing, the authors reported only genital inspection, which occurred more frequently during the breeding season, although it was not restricted to copulation. Males engaged in genital self-sniffing more often than females, and as the authors noted, this behaviour was more closely associated with penis erection rather than copulation itself. Self-sniffing behaviour was also observed in Barbary macaques (*Macaca sylvanus*; Simon et al., 2024) within a fenced forest; while the majority of sniffing was directed at food (83%), social sniffing accounted for 8% and self-sniffing for 4.5% of occurrences, but no further details on the context or potential sex differences were provided. Matsumoto-Oda et al. (2007) also reported self-sniffing behaviour in wild common chimpanzees (*Pan troglodytes*) in the Mahale Mountains, Tanzania; in contrast to Barbary macaques, self-sniffing was more frequent in females than males. Such sex differences in self-sniffing across different primate species might be related to variations in reproductive strategies: mammalian females are generally more susceptible to sexually transmitted diseases, which in promiscuous species may lead to more frequent genital self-sniffing as a form of health monitoring compared to polygynous or monogamous species.

This brief overview demonstrates that our knowledge of self-sniffing is extremely limited. Therefore, first, there is a clear need to study a broader range of species, particularly those that rely primarily on chemical senses for exploration of their environment. Such species would be expected to use their chemical senses to explore their own bodies and excretions. For instance, carnivores that routinely use their sense of smell to track prey—and many of which also mark their territories (Svensson et al., 2014)—are likely to exhibit a relatively high frequency of self-sniffing. Second, we need more detailed data not only on the general occurrence of self-sniffing but also on the body parts or excretions to which it is primarily directed. As we elaborate below, this is important because it can provide insights into the possible functions of this behaviour. Finally, we need to understand the specific contexts in which self-sniffing occurs, as this may further clarify its function.

2.2 Olfactory Self-Inspection in Humans

While studies on non-human species rely on behavioural observation, human studies can take advantage of self-reports. An initial study by Perl et al. (2020), which analysed a global online sample, found that the majority of people sniff themselves at least sometimes; the most frequently reported body parts were the hands and armpits (respectively, 94.3% and 91.6% of respondents reported smelling these). Similarly, Li et al. (2022a, b) reported that 86.9% of Chinese respondents—and 88.2% of those in the USA—sniffed their hands at least rarely, with slightly lower percentages for sniffing one's armpits (China, 79.8%; USA, 85.4%).

These findings were further confirmed in a sample of Czech participants: Schwambergová et al. (2024) found that, on average, they reported sniffing their hands and armpits several times a week. These authors further explored the underlying structure of sniffing individual body parts and clothing, identifying three factors: Social Acceptability Self-Inspection (including items such as a worn T-shirt, armpits, and breath), Intimate Self-Inspection (e.g. sniffing genitals via a hand or the navel via a finger), and Cosmetic Self-Inspection (e.g. the hair, the neck via a hand, and the hand). An analysis of the possible functions of olfactory self-inspection revealed that individuals with higher Social Acceptability Self-Inspection scores reported a greater frequency of health issues. Additionally, Intimate Self-Inspection was associated with lower levels of hygiene. These findings suggest that sniffing different body parts could serve distinct functions. For example, smelling one's armpit might help monitor health status, whereas sniffing intimate areas might enable individuals to take compensatory action to improve their personal hygiene. Interestingly, none of the three factors were associated with disgust sensitivity, as measured by the Three-Domain Disgust Scale (Tybur et al., 2009) or the Body Odour Disgust Scale (Liuzza et al., 2017). The former scale assesses interindividual differences in general disgust sensitivity and its components: pathogen, sexual, and moral disgust. The latter scale specifically focuses on disgust elicited by bodily odours, such as breath or foot odour.

Another study investigated the possible association between self- and other-directed sniffing and sexual desire (Li et al., 2022a). To assess sniffing behaviour, the authors used a questionnaire comprising three factors: Self-Private Body Odour (e.g. armpit, crotch via hand), Other's Body Odour (e.g. friend, family member), and Self-Common Body Odour (e.g. hand, worn clothes). The findings indicate that the frequency of self-sniffing was positively associated with sexual desire, and this pattern was consistently observed across three culturally diverse samples (from China, India, and the USA).

Several studies have reported gender differences in self-sniffing, although the findings vary across individual studies. One found that women scored higher on the Self-Common Body Odour factor among both Chinese and US respondents, but there were no gender differences in the Self-Private Body Odor factor (Li et al., 2022b). In contrast, Schwambergová et al. (2024) found no gender differences in Social Acceptability or Cosmetic Self-Inspection factors but reported a higher frequency of Intimate Self-Inspection in men. Similarly, Perl et al. (2020) found no gender differences in hand or armpit sniffing but reported a higher frequency of crotch sniffing in men. Since all these findings are based on self-reports, it is difficult to determine whether they reflect actual behavioural differences or whether women, on average, are less willing to admit to a behaviour that is considered inappropriate in some cultures, particularly in Europe and North America. The influence of cultural norms on self-sniffing frequency is further supported by a study comparing US, Chinese, and Indian participants: self-sniffing was more frequent among US than Chinese participants, but there were no significant differences between US and Indian participants (Li et al., 2022a).

Overall, these studies suggest that self-sniffing is a widespread behaviour in humans, with its frequency varying depending on the body part and cultural background of the participants. More cross-cultural comparisons are needed, along with a deeper understanding of local cultural restrictions on self-sniffing behaviour. Additionally, we currently lack knowledge about the specific contexts where this behaviour occurs most frequently. For instance, people may self-inspect in anticipation of social interactions.

It is also important to note that all the studies reviewed above rely on self-reports, which are prone to various biases, especially when investigating socially undesirable behaviours (Baumeister et al., 2007). Observational studies are needed to obtain more reliable estimates. However, this approach presents its own challenges, as self-sniffing of certain body parts (e.g. the genitals) likely occurs primarily in private settings, making direct observation ethically unfeasible. Advances in continuous monitoring technology could provide valuable new insights into this area. For instance, participants could wear a specialised suit equipped with transmitters on body parts of interest. A sensor attached to the participant's nose would monitor which body parts the nose approaches within a certain distance and whether this is followed by sniffing activity.

As mentioned above, humans frequently smell their hands. This behaviour may occur either after touching external objects or after touching body parts that cannot be brought directly to the nose, such as the genital area. It has also been suggested that hand sniffing plays a role in social interactions, particularly in assessing other individuals. This behaviour may be especially relevant in the context of handshaking. Frumin et al. (2015) conducted a series of studies to test this idea. They found that after shaking hands with an individual of the same sex, people tended to increase the duration of face-touching with their right hand; in contrast, face-touching with the left hand increased after shaking hands with an individual of the opposite sex. Higher nasal airflow (i.e. sniffing) was observed when the hand touched the face, although it is important to note that nasal airflow measurements were not taken immediately after social interactions. The authors interpreted their findings as strong evidence that handshaking plays a role in social chemosignalling. However, in a follow-up study, Roberts et al. (2023) found that the frequency of face-touching after a handshake at a graduation ceremony was relatively low (7.2%). Furthermore, in a separate study design in which individuals were covertly filmed, face-touching after a handshake was more frequent (41.1%) with the left hand (i.e. the non-shaking hand) than with the right (i.e. shaking) hand (26.5%). Notably, the frequency of left-hand face-touching was consistently higher, regardless of whether the handshake was with an individual of the same or opposite sex.

Overall, the evidence for face-touching as a form of chemical communication remains controversial. Future studies should examine the frequency of face-touching across various social contexts and cultural settings. More importantly, future research should employ direct measures of sniffing activity, such as facial myography, to provide more conclusive insights.

3 Chemosensory Self-Recognition

Although studies that record instances of direct chemosensory self-inspection are rare, it is likely to be a relatively common behaviour with a fundamental influence on various forms of social interaction. This is because any situation in which an individual adjusts its social behaviour based on the social odours of conspecifics must require a working knowledge not only of the conspecific's odour but also of its own. Indeed, cognitive scientists frequently study self-recognition as part of their research on self-awareness. A standard procedure for testing visual self-recognition is the mirror mark test, which assesses whether an animal reacts to a mark on its own body when seeing it in a mirror (Anderson & Gallup, 2015). It has been repeatedly argued that visual self-recognition may not be ecologically relevant for many species that primarily rely on their chemical senses (Cazzolla Gatti, 2016). Therefore, more recently, an analogous chemosensory self-recognition test has been introduced. However, it should be noted that some of the studies reviewed below did not account for the possible effects of familiarity and did not employ a chemosensory analogue of the body mark.

3.1 Chemosensory Self-Recognition in Non-humans

Chemosensory self-recognition in fish has so far been investigated only in the cichlid *Pelvicachromis taeniatus* (Thünken et al., 2009). Males (females were not tested) spent more time in water containing their own scent than water with the scent of an unfamiliar male. The response to an individual's own scent versus the scent of his brother was also compared; the tested individuals spent more time in the water with their own scent, but only during the last minute of the test period. These results suggest that although the cichlids are able to distinguish the two scents, the task is more challenging, perhaps because the scents of relatives are more similar.

Several studies have investigated chemosensory self-recognition in amphibians. In frogs, three studies focused on Hamilton's frog (*Leiopelma hamiltoni*) from New Zealand, a species that does not use acoustic communication. It was found that both males and females preferred the chemical cues of their own faecal pellets over those of conspecifics (Lee & Waldman, 2002). Similarly, both sexes preferred stimuli marked with their own chemical cues to those marked by individuals from distant home ranges, although there was no preference between their own cues and those from individuals within the same home range (Waldman & Bishop, 2004); these findings suggest that familiarity may be a key factor. In a follow-up study, Waldman (2016) compared different sources of chemical cues and found the strongest reaction to cues from the frog's skin, followed by urine, with no difference in response to own versus conspecific faecal samples. This pattern contrasts with the results of Lee and Waldman (2002), and the author attributed this discrepancy to possible contamination of the faecal pellets with skin cues.

The most extensive knowledge of chemosensory self-recognition comes from studies on the red-backed salamander (*Plethodon cinereus*). Individuals of both sexes prefer their own marked substrate over substrate marked by males; in contrast, there was no difference between their own marked substrate and that marked by females (Jaeger & Gergits, 1979). In a follow-up study, males spent more time in burrows containing their own faecal pellets than in those containing male conspecific pellets; similarly, they tended to prefer their own faecal pellets over control samples (Jaeger et al., 1986). Females also spent more time in burrows with their own faecal pellets than those containing pellets from conspecific females, but they showed more frequent threat postures near the conspecific pellets (Horne & Jaeger, 1988), a behaviour rarely observed in males. Females produced faecal pellets more quickly when exposed to their own chemical cues, whereas males did so in response to male conspecific cues (Mathis, 1990). Altogether, these results indicate that both sexes are capable of discriminating their own chemical cues, but their behavioural responses depend on the level of intrasexual competition over territories. The preference for a shelter marked with one's own faecal pellets, as compared to a clean control, was also tested in two species of alpine salamanders (*Salamandra atra* and *S. lanzai*); in both species, males and females systematically selected the shelter containing their own faecal pellets (Gautier & Miaud, 2003). However, it should be noted that no direct comparison between own and conspecific faecal pellets was performed, which prevents a conclusive assessment of chemosensory self-recognition in these species.

Chemosensory self-recognition has also been tested in several species of reptiles. As a behavioural marker of chemosensory investigation, researchers most often use the frequency of tongue-flicking because this behaviour helps transport low-volatile chemicals to the VNO, and its frequency depends on the chemical cues present in the environment. Several studies have found that different lizard species can distinguish between their own chemosensory cues and those of conspecifics. Alberts (1992a) tested the reactions of male desert iguanas (*Dipsosaurus dorsalis*) to either their own or an unfamiliar male's femoral gland secretions; she found that the lizards exhibited more frequent tongue-flicking directed at their own tails and increased marking behaviour when exposed to their own secretions. Similar results were observed in the common blue-tongued skink (*Tiliqua scincoides*), which exhibited fewer tongue flicks when investigating its own chemical cues compared to conspecific cues (Graves & Halpern, 1991); the authors suggested that chemosensory self-recognition may play a role in scent-marking of the home range. Similarly, Aragon et al. (2001) investigated responses to femoral gland secretions in the Iberian rock lizard (*Lacerta monticola*); importantly, they not only tested responses to an individual's own and unfamiliar conspecific secretions but also included secretions from a familiar male to control for potential familiarity effects. They found the lowest frequency of tongue-flicking in response to an individual's own secretions, a higher frequency towards secretions from a familiar male, and the highest frequency towards secretions from an unfamiliar male. More recently, Szabo and Ringler (2023) investigated discrimination between self and unfamiliar conspecific chemosensory cues from the skin and scats of tokay geckos (*Gekko gecko*); their results showed that females exhibited a higher frequency of tongue-flicking in response to both skin and scat stimuli from unfamiliar individuals compared to their own or control stimuli. Interestingly, males showed almost no tongue-flicking behaviour. Furthermore, exploratory behaviours related to chemosensory discrimination appear to be influenced by metabolic activity seasonality, as shown in Fitzgerald's tree iguana (*Liolaemus fitzgeraldi*): during the post-hibernation period, but not in other periods, they exhibited a higher frequency of tongue-flicking when placed in conspecifics' enclosures (regardless of sex) compared to their own enclosures (Aguilar et al., 2009).

In snakes, early studies similarly found differing reactions to self versus unfamiliar conspecific chemosensory cues. Experiments were conducted with two rattlesnake species (*Crotalus viridis* and *C. horridus*), which were exposed to their own soiled bedding, conspecifics' soiled bedding, or a clean control (Chiszar et al., 1991); there was a lower frequency of tongue-flicking in response to own bedding. Similarly, juvenile eastern garter snakes (*Thamnophis sirtalis sirtalis*) exhibited a lower frequency of tongue-flicking when exposed to their own chemosensory cues compared to those of an unfamiliar conspecific (Halpin, 1990). Burghardt et al. (2021) replicated Halpin's (1990) study while controlling for diet and familiarity; they compared the reactions of juvenile eastern garter snakes to chemical cues from their own substrate, a littermate on the same diet, and a littermate on a different diet. They observed a higher frequency of tongue-flicking towards their own substrate compared to that of littermates on the same diet, but there was no significant difference compared to littermates on a different diet. This pattern emerged only in the final segment of the observation period, and no statistically significant differences were found in females. These results suggest either that self-recognition is restricted to males or that the evidence is not very convincing, as only a single segment showed formally significant differences. Freiburger et al. (2024) also studied eastern garter snakes, finding that they could discriminate between their own chemical cues and their own cues mixed with another odour (olive oil). The mixture of its own cue and another odour was used as a chemosensory analogue of the mirror mark test; if an animal recognises itself, it is expected to show greater interest in the mixture. The snakes exhibited a higher proportion of long tongue flicks towards the mixture of their own cue and the olive oil compared to the mixture of the olive oil and a familiar conspecific's cue. Interestingly, no differences in tongue-flicking were observed when only the own and familiar cues were compared. Freiburger et al. also tested ball pythons (*Python regius*) using the same procedure, but these snakes showed similar frequencies of tongue-flicking towards their own cue mixed with another odour and familiar stimuli.

Several studies have also explored olfactory self-recognition in birds, particularly in Procellariiformes, which have a well-developed sense of smell and produce odorous substances from their uropygial and cloacal glands. De León et al. (2013) used a Y-maze to test European storm petrel (*Hydrobates pelagicus*) chicks; they found that the chicks could discriminate their own odour and preferred it over both a no-odour control sample and the odour of a conspecific chick. Such discrimination is a key task for the chicks because the species nests in burrows, and chicks often explore their surrounding environment; they must, therefore, be able to find their way back to their own nest. In adults, the same ability may produce a different outcome, so that self/non-self-discrimination might lead to preferences for non-self odours. For example, although they used similar methods to De León et al., Bonadonna and Nevitt (2004) found that adult Antarctic prions (*Pachyptila desolata*) orient more towards the odour of conspecifics than their own. A similar pattern was observed in blue petrels (*Halobaena caerulea*), where individuals preferred their partner's odour and the odour of unfamiliar conspecifics over their own (Mardon & Bonadonna, 2009). Such behaviours might serve in the formation and maintenance of pairbonds, and perhaps for inbreeding avoidance (Bonadonna, 2009); nevertheless, the specific mechanisms or chemical compounds responsible for the observed effect are not known.

To our knowledge, chemosensory self-recognition in non-human mammals has been tested only in rodents and carnivores. The first such test was done in wild house mice (*Mus domesticus*), in which males (females were not tested) spent more time exploring the scent marks of an unfamiliar male compared to their own (Hurst et al., 2001). They also distinguished between their own scent marks and those of a brother with a different major urinary protein (MUP) type. In contrast, there was no difference in exploration times of their own scent marks and those of a brother with the same MUP type, indicating that MUPs may be associated with the critical contribution to odour individuality.

In a series of experiments, Ferkin et al. (2008) investigated whether meadow voles (*Microtus pennsylvanicus*) can discriminate between their own urine chemosensory cues and the cues of unfamiliar same-sex individuals or siblings. They found that voles of both sexes could discriminate between the cues of both unfamiliar and familiar (i.e. sibling) individuals. Additionally, they discovered that voles could distinguish their own chemosensory cues before and after a change in their reproductive status. More recently, chemosensory self-recognition was tested in giant pouched rats (*Cricetomys ansorgei*) using a habituation paradigm (Freeman & Ophir, 2021); the ability to discriminate between own urine and that of an unfamiliar same-sex conspecific was found only in females.

In an initial study with dogs, Cazzolla Gatti (2016) recorded the amount of time each of four individuals spent sniffing their own urine sample versus samples from three other familiar dogs; dogs sniffed their own urine sample for significantly shorter periods than the other samples. In two complementary studies, Horowitz (2017) found that dogs spent less time sniffing their own urine than either an unfamiliar dog's urine or their own urine mixed with another odour (similar to the visual mark analogy). A similar pattern was observed in wolves, who sniffed their own urine for a shorter time than modified versions of their own urine (Cazzolla Gatti et al., 2021).

It should be noted, however, that the results of many studies described above may not be incontrovertible evidence for chemosensory self-recognition per se; they might instead be a consequence of differences in familiarity. Depending on the context, unfamiliar stimuli may lead to increased avoidance (e.g. chicks of burrowing birds) or investigation compared to one's own odour, rather than constituting true self-recognition. In contrast, studies in humans can explicitly direct experimental participants to the task of selecting their own odour from an olfactory line-up. We turn to such evidence in the following section.

3.2 Olfactory Self-Recognition in Humans

The most extensive evidence for olfactory self-recognition comes from studies in humans. Most of these have focused on torso body odour (collected on T-shirts) or specifically from the armpits (collected on axillary cotton/gauze pads). For instance, Schleidt and colleagues (Hold & Schleidt, 1977; Schleidt, 1980; Schleidt et al., 1981) collected shirts from a sample of men and women worn over seven nights. Participants were then asked to choose their own shirt—based on its smell—from a set of ten stimuli. The success rates were above those expected by chance, ranging from 25% in Japan to 38% in Italy. In contrast, other studies reported considerably higher success rates, such as 75% in the USA (Russell, 1976). Table 1 presents an overview of all human studies on olfactory self-recognition.

One may argue that the discrepancies observed between some of these studies can be attributed to the mode of presentation. For example, Schleidt and colleagues (Hold & Schleidt, 1977; Schleidt, 1980; Schleidt et al., 1981) used a ten-choice test, whereas Russell (1976) used a three-choice test; such differences not only change the likelihood of success by choosing at random but also require a much lower cognitive load. However, this explanation does not entirely account for the extensive variation, as some other studies using a ten-choice test have also reported high success rates, even up to 74% (Lord & Kasprzak, 1989). Moreover, two other studies found success rates that were not different from chance, with 12% in a seven-choice test (Übel et al., 2017) and 5.6% in men with a five-choice test (Platek et al., 2001). It should be noted, however, that the latter two studies collected odours for only 30 minutes, which may not have been sufficient to obtain recognisable samples, at least for some participants.

Several studies have also found higher self-recognition rates in women (Lord & Kasprzak, 1989; Platek et al., 2001), which might be attributed to generally higher olfactory performance in women (Sorokowski et al., 2019). Unfortunately, no sex differences were tested in many other studies (Olsson et al., 2006; Schleidt, 1980; Schleidt et al., 1981; Übel et al., 2017).

Participants also tend to perceive their own body odour as more pleasant compared to the body odours of others (Schleidt et al., 1981); importantly, this applies even in cases where they do not explicitly recognise their own body odour (Übel et al., 2017). These results suggest that self-recognition may not be (or may not primarily be) explicit;

Table 1. An overview of studies on olfactory self-recognition in humans. The table presents the authors of the studies, the sampling media, the length of sampling, the number of alternatives in the forced-choice test, the number of male (M) and female (F) participants, and the results in terms of success rates (overall and separately for male and female participants). All reported results are statistically significant unless otherwise noted. 'NS' denotes that the success rate was not statistically significant at the $p < 0.05$ level.

Authors	Sampling media	Length of sampling	Number of alternatives	Participants	Results
McBurney et al. (1976)	T-shirt	48 h	12	11 M	27%
Russell (1976)	T-shirt	24 h	3	16 M, 13 F	75% (M, 81%; F, 69%)
Hold and Schleidt (1977)	Undershirt	7 nights	10	23 M, 23 F	31% (M, 22%; F, 39%)
Schleidt (1980)	Shirt	7 nights	10	25 M, 25 F	26%
Schleidt et al. (1981)	Shirt	7 nights	10	Germany: 23 M, 23 F Italy: 25 M, 25 F Japan: 22 M, 22 F	Germany: 31% Italy: 38% Japan: 25%
Lord and Kasprzak (1989)	T-shirt	24 h	10	47 M, 53 F	74% (M, 64%; F, 83%)
Platek et al. (2001)	Gauze pad	30-min rest	5	18 M, 32 F	40% (M, 5.6% (NS); F, 59.4%)
Olsson et al. (2006)	T-shirt	3 nights	5	11 M, 26 F	51.6%
Übel et al. (2017)	Cotton pad	30-min exercise	7	35 M, 37 F	12% (NS)

implicit self-recognition could be more efficient. However, this speculation should be tested empirically, potentially using behavioural tests similar to those employed in other species.

4 Possible Functions

In this section, we discuss several adaptive explanations that may account for fluctuations in self-sniffing behaviour. In particular, we focus on self-referencing in the context of the following: (i) scent-marking, (ii) kin recognition, (iii) monitoring one's health status, and (iv) controlling chemosignalling during self-grooming. It should be emphasised that these adaptive explanations are not mutually exclusive, and their use may vary depending on the context and the source of the odour. Moreover, this list is not exhaustive; for instance, the soothing effect of one's own odour is a possible factor that we do not elaborate on here, both for reasons of space and in view of a paucity of data.

4.1 Self-Referencing

Individuals inevitably leave scent traces in the environment as a by-product of their metabolism and activity. Although these traces may be detected by other animals (whether conspecific or heterospecific), the producing individual has little control over their production. In contrast, scent marks are deliberately produced, often by specific glands or in urine/faeces, and deposited in particular locations. They are an intriguing form of indirect communication, as signallers typically produce them in the absence of the perceiver, and perceivers usually detect them without the producer being present (Gosling & Roberts, 2001). Several factors influence how quickly these scent traces or marks degrade in the environment, including the relative composition of volatile and non-volatile compounds in the scent mark, as well as environmental characteristics such as surface adhesiveness, humidity, and temperature (Alberts, 1992b; Apps et al., 2015).

Scent-marking has primarily been linked to marking territory to evaluate potential competitors or deter intruders (e.g. Gosling, 1982; Gosling & McKay, 1990), but it can also be used to attract potential mates (Gosling & Roberts, 2001). A fundamental requirement for the assessment of a scent mark is the ability to determine whether it was produced by someone else or if it is one's own mark. As reviewed in detail above, successful chemosensory self-recognition has been observed in various taxa, amphibians, squamate reptiles, birds, and mammals. However, it has not been detected consistently across all tested species. This is particularly true for reptiles and their socioecology (for details see Szabo, 2026, this volume); for example, chemosensory self-recognition was found in eastern garter snakes but not in ball pythons (Freiburger et al., 2024). The former is highly gregarious, and differentiating one's scent trace from others may be adaptive in various social contexts. In contrast, the ball python is a solitary species with limited contact with other conspecifics, so recognising one's own scent from others may offer little advantage.

Surprisingly, only a few studies have investigated chemosensory self-recognition in mammals. To date, only rodents (Hurst et al., 2001; Ferkin et al., 2008; Freeman & Ophir, 2021), carnivores (Cazzolla Gatti, 2016; Horowitz, 2017; Cazzolla Gatti et al., 2021), and humans (e.g. Schleidt et al., 1981) have been tested. Furthermore, the primary focus of the studies on dogs and wolves has been the cognitive aspect of self-recognition, using a chemosensory analogue of the mirror self-recognition test (Gallup & Anderson, 2018). Passing the mirror self-recognition test is considered a key prerequisite for self-awareness, which has traditionally been restricted to humans and great apes. In this paper, we do not aim to engage further with this complex and heated debate, as it has been extensively discussed elsewhere (see, for instance, De Waal et al., 2005; Gallup & Anderson, 2020). Instead, we focus on the socioecological predictors of chemosensory self-recognition. We predict that species which mark their territories with scent marks or use latrines should be routinely able to differentiate their own odour from that of a stranger or a familiar (e.g. neighbour) individual. Scent-marking is well-known across various mammalian taxa, including carnivores, antelopes, and rodents, to name just a few (Gosling & Roberts, 2001). Various species also tend to counter-mark other individuals' marks (Thom & Hurst, 2004), which also suggests an ability to distinguish between one's own and a conspecific's scent marks. Given the ubiquity of this behaviour and the number of studies on chemosensory communication in rodents (e.g. Roberts, 2007; Ferkin, 2018), we find it surprising that, to the best of our knowledge, only a handful of studies have tested chemosensory self-recognition in rodents (Hurst et al., 2001; Ferkin et al., 2008; Freeman & Ophir, 2021).

4.2 Kin Recognition

There is extensive literature on kin recognition across various vertebrate taxa (Hepper, 1986). Recognising kin can promote cooperation, as relatives share a portion of their genes. It also helps prevent inbreeding, which can reduce offspring viability. Olfaction is a key mechanism for assessing relatedness, as genetically related individuals often share similar odour signatures (Roberts et al., 2005). Kin recognition occurs through two primary mechanisms: (i) familiarity and (ii) phenotype matching (Tang-Martinez, 2001). First, an individual may use the odour of individuals encountered early in ontogeny—such as nestmates—as a template for kin recognition, as these individuals are typically siblings. While interactions with siblings or parents can contribute to the formation of a kin recognition template, this mechanism is not the most reliable, as genetic similarity in an outbred population typically does not exceed 50%. In species like hamsters, where females may mate with multiple males, siblings may share even less than 50% of their genes. A more accurate template would be the mother's odour, but the most precise template of all is an individual's own odour and its similarity to that of another animal, serving as a mechanism for kin recognition (Johnston, 2008). This latter mechanism relies on the ability to recognise one's own odour and use it as a template to compare against others.

Which species use familiarity, phenotype matching, or a combination of both may partly depend on their typical form of parental care (Widdig, 2007). If paternal relatives are not involved in parental care, familiarity cannot be used, and animals must rely on phenotype matching. There is evidence that some rodents use phenotype matching for kin recognition (for a review, see Holmes & Mateo, 2007). Similarly, it has been shown in meerkats (*Suricata suricatta*) that individuals likely use phenotype matching to discriminate the scent of unfamiliar kin from non-kin, as females spent more time investigating the scent of related than unrelated unfamiliar individuals (Leclaire et al., 2013). The mechanism for recognising unfamiliar kin in primates has typically been attributed to social learning; however, it seems that phenotype matching is also present in lemurs (*L. catta*; Charpentier et al., 2010). Notably, storm petrels (*H. pelagicus*) can also distinguish between kin and non-kin based on odours, although whether this is achieved through recognition by familiarity or phenotype matching remains unclear from current evidence (Bonadonna & Sanz-Aguilar, 2012). As noted above, phenotype matching assumes self-recognition, but no studies have explicitly examined this assumption. Clearly, this gap should be considered when testing the phenotype-matching mechanism.

4.3 Monitoring Health Status

Odour cues play a crucial role in conveying information about an individual's health status, particularly in the case of infectious diseases that alter an animal's metabolism and, consequently, its chemical profile. Broadly speaking, infections can affect body odour through four main mechanisms (Schwambergová et al. 2023). First, changes in odour may result directly from the infectious agent's impact on the host's metabolism. The affected odour source depends on the pathogen and the infected tissue; for instance, respiratory infections primarily alter breath exhalations, digestive system infections affect the smell of faeces, and skin infections influence body odour. Urine is particularly informative, as it contains chemical compounds excreted by various tissues. Second, body odour changes may stem from the immune system's response to infection. The exact mechanisms remain largely unknown and likely vary depending on whether the immune response is humoral or cellular, involves inflammation, or engages mucosal immunity, among other factors. Third, infections can influence steroid and stress hormone levels, which in turn affect odour production. Finally, infectious agents can alter the composition of the host's microflora, which plays a key role in producing chemical compounds in many species. These mechanisms are not mutually exclusive and may interact, collectively contributing to infection-induced odour changes.

There is robust evidence, primarily from experiments with rodents, that individuals tend to avoid the scent of infected conspecifics. Most studies have focused on the effects of intestinal parasites, including both protozoans and cestodes, but avoidance has also been observed in response to ectoparasites such as the murine louse (for a review, see Kavaliers et al., 2020). Similarly, in humans, research has shown that individuals rate the body odour of people experimentally injected with lipopolysaccharide (LPS) as less pleasant (Olsson et al., 2014). As a component of the outer membrane of gram-negative bacteria, LPS triggers an innate immune response, potentially altering body odour in a way that indicates infection.

The studies briefly reviewed above indicate that odour cues related to health can be detected in others. It is, therefore, reasonable to expect that animals might also use such cues to monitor their own health status, and this could be important for several reasons. First, an infected animal may engage in self-medication. A growing body of evidence suggests that self-medication is more widespread than previously thought, having been observed not only in primates but also in swallows and some lizards (De Roode et al., 2013; Shurkin, 2014). In primates, analyses suggest that the occurrence of self-medication is positively associated with longevity (Neco et al., 2019). Second, an infected individual may choose to withdraw from social interactions, helping to reduce competition with conspecifics and, importantly, lowering the risk of transmitting infections to kin. Studies in several vertebrate species have shown that sick individuals tend to avoid social contact as part of their sickness behaviours (for recent reviews, see Dantzer, 2021; Lopes et al., 2021).

To our knowledge, only a single study has examined whether the frequency of self-sniffing is associated with health issues in humans. The study found that individuals who reported more frequent health issues also reported higher rates of self-sniffing, which included smelling their own breath or armpits, behaviours that clustered under the 'Social Acceptability' factor (Schwambergová et al., 2024). However, it is important to note that these findings are based on self-reports, and no behavioural data are currently available from either humans or non-human animals. Additionally, these results suggest that odour cues may vary depending on the body part, implying that an overall increase in self-sniffing during periods of high infection risk may not be expected. Instead, specific areas may be more frequently sniffed, potentially varying by the type of infection.

4.4 Self-Grooming

Self-grooming is a widespread behaviour in vertebrates and constitutes a regular part of an animal's daily activities. Traditionally, it has been regarded primarily as a cleaning behaviour, aiding in the removal of ectoparasites (Spruijt et al., 1992), but it has also been proposed that self-grooming plays a role in situation-dependent chemical signalling (Ferkin & Leonard, 2010). Unlike many chemical cues that are passively released into the environment and beyond the direct control of the individual, self-grooming may allow for more targeted and flexible signal enhancement, reducing susceptibility to chemical eavesdropping and rigid signalling constraints.

Studies have repeatedly shown that rodents adjust their level of self-grooming in response to chemical cues from conspecifics. For instance, Ferkin et al. (1996) reported that reproductively active meadow voles increased the frequency of self-grooming in response to the odour of opposite-sex conspecifics. Similarly, females were found to groom more frequently when exposed to the scent of males fed a high-protein diet (Ferkin et al., 1997), suggesting that self-grooming may enhance the dispersion of an individual's own odour, potentially increasing their attractiveness to mates. Additionally, male prairie voles, which are territorial, exhibited increased self-grooming when exposed to the scent of same-sex conspecifics (Ferkin et al., 2001). Interestingly, this pattern was not observed in promiscuous meadow voles (Ferkin et al., 1996), indicating that changes in self-grooming are not solely linked to mating but may also play a role in same-sex interactions, depending on the socioecology of a given species. A recent study showed that orofacial self-grooming in mice enhanced the attractiveness of these individuals, regardless of their sex, through volatile cues from orofacial secretions (Zhang et al., 2022). One may speculate whether a similar function is, for instance, found in bill-wiping behaviour in birds, which appears to be an olfactory display detectable by potential mates and/or rivals (Whittaker et al., 2015).

Since self-grooming and cleaning activities involve consistent exposure to one's own scent and may be an integral part of chemical communication, as suggested above, sniffing one's own odour could serve as a control mechanism for regulating the distribution of chemical compounds in the environment. While it is theoretically possible that the frequency of self-grooming adjusts automatically based on social context, we find this scenario unlikely in cognitively complex animals such as rodents, which engage in highly nuanced chemical communication (e.g. Roberts, 2007; Ferkin, 2018). However, it is important to note that no previous study on self-grooming has specifically investigated self-sniffing, nor have researchers proposing self-grooming as a form of chemical communication explicitly addressed self-sniffing behaviour. Differentiating between self-grooming and self-sniffing may also present methodological challenges, as simple behavioural observations may not be sufficient. A more effective approach could involve the use of facial myography to record active sniffing, though this technique has yet to be applied in this context.

5 Conclusions

In this chapter, we aimed to review evidence on self-sniffing behaviour across vertebrates. However, most of the available knowledge comes from studies on humans and is based on self-reports, which are prone to various biases (Baumeister et al., 2007). Nonetheless, we may tentatively conclude from these findings that chemosensory self-inspection serves as a mechanism for monitoring one's health and hygiene status. In contrast to the scarcity of observational studies, there is ample experimental evidence on chemosensory self-recognition. Interestingly, there are currently twice as many studies on chemosensory self-recognition in reptiles than in mammals; this might simply be because tongue-flicking in reptiles is easier to observe and quantify than the often subtle and difficult-to-distinguish sniffing behaviour in mammals. However, we suspect that this discrepancy may rather reflect the assumption that such an ability is considered unexpected in reptiles but self-evident in mammals. We also reviewed the possible adaptive significance of self-sniffing behaviour, which may play a role in scent-marking, kin recognition, health monitoring, and self-grooming. Given that these activities can have direct fitness consequences—and that self-sniffing is ubiquitous in at least some species—it remains puzzling why chemosensory self-sniffing has been largely overlooked. Although we do not have a definitive answer to this question, we sincerely hope that this paper contributes to changing that.

Acknowledgements. JH and DS are supported by a grant from the Czech Science Foundation (GAČR 21-29772S).

References

Aguilar, P. M., Labra, A., & Niemeyer, H. M. (2009). Chemical self-recognition in the lizard *Liolaemus fitzgeraldi*. *Journal of Ethology, 27*, 181–184.

Alberts, A. C. (1992a). Pheromonal self-recognition in desert iguanas. *Copeia, 1*, 229–232.

Alberts, A. C. (1992b). Constraints on the design of chemical communication systems in terrestrial vertebrates. *American Naturalist, 139*, S62–S89.

Anderson, J. R., & Gallup, G. G. (2015). Mirror self-recognition: A review and critique of attempts to promote and engineer self-recognition in primates. *Primates, 56*, 317–326.

Apps, P. J., Weldon, P. J., & Kramer, M. (2015). Chemical signals in terrestrial vertebrates: Search for design features. *Natural Product Reports, 32*, 1131–1153.

Aragon, P., Lopez, P., & Martín, J. (2001). Discrimination of femoral gland secretions from familiar and unfamiliar conspecifics by male Iberian rock-lizards, *Lacerta monticola*. *Journal of Herpetology, 35*, 346–350.

Baumeister, R. F., Vohs, K. D., & Funder, D. C. (2007). Psychology as the science of self-reports and finger movements: Whatever happened to actual behavior? *Perspectives on Psychological Science, 2*, 396–403.

Bonadonna, F. (2009). Olfaction in petrels: From homing to self-odor avoidance. *Annals of the New York Academy of Sciences, 1170*, 428–433.

Bonadonna, F., & Nevitt, G. A. (2004). Partner-specific odor recognition in an antarctic seabird. *Science, 306*, 835.

Bonadonna, F., & Sanz-Aguilar, A. (2012). Kin recognition and inbreeding avoidance in wild birds: The first evidence for individual kin-related odour recognition. *Animal Behaviour, 84*, 509–513.

Burghardt, G. M., Partin, A. M., Pepper, H. E., Steele, J. M., Liske, S. M., Stokes, A. E., Lathan, A. N., Springer, C. M., & Jenkins, M. S. (2021). Chemically mediated self-recognition in sibling juvenile common gartersnakes (*Thamnophis sirtalis*) reared on same or different diets: Evidence for a chemical mirror? *Behaviour, 158*, 1169–1191.

Cazzolla Gatti, R. (2016). Self-consciousness: Beyond the looking-glass and what dogs found there. *Ethology Ecology and Evolution, 28*, 232–240.

Cazzolla Gatti, R., Velichevskaya, A., Gottesman, B., & Davis, K. (2021). Grey wolf may show signs of self-awareness with the sniff test of self-recognition. *Ethology Ecology and Evolution, 33*, 444–467.

Charpentier, M. J. E., Crawford, J. C., Boulet, M., & Drea, C. M. (2010). Message "scent": Lemurs detect the genetic relatedness and quality of conspecifics via olfactory cues. *Animal Behaviour, 80*, 101–108.

Chiszar, D., Smith, H. M., Bogert, C. M., & Vidaurri, J. (1991). A chemical sense of self in timber and prairie rattlesnakes. *Bulletin of the Psychonomic Society, 29*, 153–154.

Dantzer, R. (2021). Love and fear in the times of sickness. *Comprehensive Psychoneuroendocrinology, 6*, 100032.

de León, A., Mínguez, E., & Belliure, B. (2013). Self-odour recognition in European storm-petrel chicks. *Behaviour, 140*, 925–933.

De Roode, J. C., Lefèvre, T., & Hunter, M. D. (2013). Self-medication in animals. *Science, 340*, 150–151.

De Waal, F. B. M., Dindo, M., Freeman, C. A., & Hall, M. J. (2005). The monkey in the mirror: Hardly a stranger. *Proceedings of the National Academy of Sciences of the United States of America, 102*, 11140–11147.

Ferkin, M. H. (2018). Odor communication and mate choice in rodents. *Biology, 7*, 13.

Ferkin, M. H., & Leonard, S. T. (2010). Self-grooming as a form of olfactory communication in meadow voles and prairie voles (*Microtus* spp.). In A. V. Kalueff, J. L. LaPorte, & C. L. Bergner (Eds.), *Neurobiology of grooming behavior* (pp. 19–45). Cambridge University Press.

Ferkin, M. H., Sorokin, E. S., & Johnston, R. E. (1996). Self-grooming as a sexually dimorphic communicative behaviour in meadow voles, *Microtus pennsylvanicus*. *Animal Behaviour, 51*, 801–810.

Ferkin, M. H., Sorokin, E. S., Johnston, R. E., & Lee, C. J. (1997). Attractiveness of scents varies with protein content of the diet in meadow voles. *Animal Behaviour, 53*, 133–141.

Ferkin, M. H., Leonard, S. T., Heath, L. A., & Paz-y-Miño, C. G. (2001). Self-grooming as a tactic used by prairie voles *Microtus ochrogaster* to enhance sexual communication. *Ethology, 107*, 939–949.

Ferkin, M. H., Pierce, A. A., & Franklin, S. (2008). Self-discrimination in meadow voles, *Microtus pennsylvanicus*. *Ethology, 114*, 863–874.

Freeman, A. R., & Ophir, A. G. (2021). Sex differences in social odor discrimination by southern giant pouched rats (*Cricetomys ansorgei*). *Ethology, 127*, 1019–1032.

Freiburger, T., Miller, N., & Skinner, M. (2024). Olfactory self-recognition in two species of snake. *Proceedings of the Royal Society B: Biological Sciences, 291*, 20240125.

Frumin, I., Perl, O., Endevelt-Shapira, Y., Eisen, A., Eshel, N., Heller, I., Shemesh, M., Ravia, A., Sela, L., Arzi, A., & Sobel, N. (2015). A social chemosignaling function for human handshaking. *eLife, 2015*, e05154.

Gallup, G. G., & Anderson, J. R. (2018). The "olfactory mirror" and other recent attempts to demonstrate self-recognition in non-primate species. *Behavioural Processes, 148*, 16–19.

Gallup, G. G., & Anderson, J. R. (2020). Self-recognition in animals: Where do we stand 50 years later? Lessons from cleaner wrasse and other species. *Psychology of Consciousness: Theory Research, and Practice, 7*, 46–58.

Gautier, P., & Miaud, C. (2003). Faecal pellets used as an economic territorial marker in two terrestrial alpine salamanders. *Ecoscience, 10*, 134–139.

Gosling, L. M. (1982). A reassessment of the function of scent marking in territories. *Zeitschrift für Tierpsychologie, 60*, 89–118.

Gosling, L. M., & McKay, H. V. (1990). Competitor assessment by scent matching: An experimental test. *Behavioral Ecology and Sociobiology, 26*, 415–420.

Gosling, L. M., & Roberts, S. C. (2001). Scent-marking by male mammals: Cheat-proof signals to competitors and mates. *Advances in the Study of Behavior, 30*, 169–217.

Graves, B. M., & Halpern, M. (1991). Discrimination of self from conspecific chemical cues in *Tiliqua scincoides* (Sauria: Scincidae). *Journal of Herpetology, 25*, 125–126.

Halpin, Z. T. (1990). Responses of juvenile Eastern Garter Snakes (*Thamnophis sirtalis sirtalis*) to own, conspecific and clean odors. *Copeia, 1990*, 1157–1160.

Hepper, P. (1986). Kin recognition: Function and mechanisms. *Biology Reviews, 61*, 63–93.

Hold, B., & Schleidt, M. (1977). The importance of human odour in non-verbal communication. *Zeitschrift für Tierpsychologie, 43*, 225–238.

Holmes, W. G., & Mateo, J. M. (2007). Kin recognition in rodents: Issues and evidence. In J. O. Wolff & P. W. Sherman (Eds.), *Rodent societies: An ecological and evolutionary perspective* (pp. 216–228). Chicago University Press.

Horne, E. A., & Jaeger, R. G. (1988). Territorial pheromones of female red-backed salamanders. *Ethology, 78*, 143–152.

Horowitz, A. (2017). Smelling themselves: Dogs investigate their own odours longer when modified in an "olfactory mirror" test. *Behavioural Processes, 143*, 17–24.

Hurst, J. L., Payne, C. E., Nevison, C. M., Marie, A. D., Humphries, R. E., Robertson, D. H. L., Cavaggioni, A., & Beynon, R. J. (2001). Individual recognition in mice mediated by major urinary proteins. *Nature, 414*, 631–634.

Jaeger, R. G., & Gergits, W. F. (1979). Intra- and interspecific communication in salamanders through chemical signals on the substrate. *Animal Behaviour, 27*, 150–156.

Jaeger, R. G., Goy, J. M., Tarver, M., & Márquez, C. E. (1986). Salamander territoriality: Pheromonal markers as advertisement by males. *Animal Behaviour, 34*, 860–864.

Johnston, R. E. (2008). Individual odors and social communication: Individual recognition, kin recognition, and scent over-marking. *Advances in the Study of Behavior, 38*, 439–505.

Kavaliers, M., Ossenkopp, K. P., & Choleris, E. (2020). Pathogens, odors, and disgust in rodents. *Neuroscience and Biobehavioral Reviews, 119*, 281–293.

Leclaire, S., Nielsen, J. F., Thavarajah, N. K., Manser, M., & Clutton-Brock, T. H. (2013). Odour-based kin discrimination in the cooperatively breeding meerkat. *Biology Letters, 9*, 3–6.

Lee, J. S. F., & Waldman, B. (2002). Communication by fecal chemosignals in an archaic frog, *Leiopelma hamiltoni*. *Copeia, 2002*, 679–686.

Li, Z., Hummel, T., & Zou, L. (2022a). Sniffing of body odors and individual significance of olfaction are associated with sexual desire: A cross-cultural study in China, India, and the USA. *Archives of Sexual Behavior, 51*, 3703–3713.

Li, Z., Qi, Y., Mahmut, M. K., & Zou, L. (2022b). Do you often sniff yourself or others? Development of the body odor sniffing questionnaire and a cross-cultural survey in China and the USA. *Physiology & Behavior, 255*, 113934.

Liuzza, M. T., Lindholm, T., Hawley, C., Sendén, M. G., Ekström, I., Olsson, M. J., et al. (2017). The body odor disgust scale (BODS): Development and validation of a novel olfactory disgust assessment. *Chemical Senses, 42*, 499–508.

Lopes, P. C., French, S. S., Woodhams, D. C., & Binning, S. A. (2021). Sickness behaviors across vertebrate taxa: Proximate and ultimate mechanisms. *Journal of Experimental Biology, 224*, jeb225847.

Lord, T., & Kasprzak, M. (1989). Identification of self through olfaction. *Perceptual and Motor Skills, 69*, 219–224.

Mardon, J., & Bonadonna, F. (2009). Atypical homing or self-odour avoidance? Blue petrels (*Halobaena caerulea*) are attracted to their mate's odour but avoid their own. *Behavioral Ecology and Sociobiology, 63*, 537–542.

Mathis, A. (1990). Territorial salamanders assess sexual and competitive information using chemical signals. *Animal Behaviour, 40*, 953–962.

Matsumoto-Oda, A., Kutsukake, N., Hosaka, K., & Matsusaka, T. (2007). Sniffing behaviors in Mahale chimpanzees. *Primates, 48*, 81–85.

McBurney, D. H., Levine, J. M., & Cavanaugh, P. H. (1976). Psychophysical and social ratings of human body odor. Personality and Social Psychology Bulletin, 3, 135–138.

Neco, L. C., Abelson, E. S., Brown, A., Natterson-Horowitz, B., & Blumstein, D. T. (2019). The evolution of self-medication behaviour in mammals. *Biological Journal of the Linnean Society, 128*, 373–378.

Olsson, S. B., Barnard, J., & Turri, L. (2006). Olfaction and identification of unrelated individuals: Examination of the mysteries of human odor recognition. *Journal of Chemical Ecology, 32*, 1635–1645.

Olsson, M. J., Lundström, J. N., Kimball, B. A., Gordon, A. R., Karshikoff, B., Hosseini, N., Sorjonen, K., Olgart Höglund, C., Solares, C., Soop, A., Axelsson, J., & Lekander, M. (2014). The scent of disease: Human body odor contains an early chemosensory cue of sickness. *Psychological Science, 25*, 817–823.

Palagi, E., Telara, S., & Borgognini Tarli, S. M. (2003). Sniffing behavior in *Lemur catta*: Seasonality, sex, and rank. *International Journal of Primatology, 24*, 335–350.

Perl, O., Mishor, E., Ravia, A., Ravreby, I., & Sobel, N. (2020). Are humans constantly but subconsciously smelling themselves? *Philosophical Transactions of the Royal Society B: Biological Sciences, 375*, 20190372.

Platek, S. M., Burch, R. L., & Gallup, G. G. (2001). Sex differences in olfactory self-recognition. *Physiology & Behavior, 73*, 635–640.

Roberts, S. C. (2007). Scent marking. In J. O. Wolff & P. W. Sherman (Eds.), *Rodent societies: An ecological and evolutionary perspective* (pp. 255–266). Chicago University Press.

Roberts, S. C., Gosling, L. M., Spector, T. D., Miller, P., Penn, D. J., & Petrie, M. (2005). Body odor similarity in noncohabiting twins. *Chemical Senses, 30*, 651–656.

Roberts, S. C., Singleton, F. P., Culpepper, P. D., Sorokowski, P., Sorokowska, A., & Havlíček, J. (2023). Handshaking and hand-smelling: On the potential role of handshake greeting in human olfactory communication. In B. Schaal, D. Rekow, M. Keller, & F. Damon (Eds.), *Chemical signals in vertebrates* (Vol. 15, pp. 257–269). Springer Cham.

Russell, M. J. (1976). Human olfactory communication. *Nature, 260*, 520–522.

Schleidt, M. (1980). Personal odor and nonverbal communication. *Ethology and Sociobiology, 1*, 225–231.

Schleidt, M., Hold, B., & Attili, G. (1981). A cross-cultural study on the attitude towards personal odor. *Journal of Chemical Ecology, 7*, 19–31.

Schwambergová, D., Třebická Fialová J., & Havlíček, J. (2023). Olfactory mediated pathogen avoidance in mammals. In B. Schaal, M. Keller, D. Rekow (Eds.), Chemical signals in vertebrates (vol. 15, pp. 207–232). Springer.

Schwambergová, D., Třebická Fialová, J., & Havlíček, J. (2024). Olfactory self-inspection: Own body odour provides cues to one's health and hygiene status. *Physiology & Behavior, 275*, 114449.

Shurkin, J. (2014). Animals that self-medicate. *Proceedings of the National Academy of Sciences of the United States of America, 111*, 17339–17341.

Silva, L., & Antunes, A. (2017). Vomeronasal receptors in vertebrates and the evolution of pheromone detection. *Annual Review of Animal Biosciences, 5*, 353–370

Simon, M., Widdig, A., & Weiß, B. M. (2024). Sniffing behavior of semi free-ranging barbary macaques (*Macaca sylvanus*). *American Journal of Primatology, 86*, e23611.

Sorokowski, P., Karwowski, M., Misiak, M., Marczak, M. K., Dziekan, M., Hummel, T., & Sorokowska, A. (2019). Sex differences in human olfaction: A meta-analysis. *Frontiers in Psychology, 10*, 242.

Spruijt, B. M., Van Hooff, J. A. R. A. M., & Gispen, W. H. (1992). Ethology and neurobiology of grooming behavior. *Physiological Reviews, 72*, 825–852.

Svensson, G. P., Strandh, M., & Löfstedt, C. (2014). Movements in the olfactory landscape. In L.-A. Hansson & S. Åkesson (Eds.), *Animal movement across scales* (pp. 45–66). Oxford University Press.

Szabo, B. (2026). A new perspective on squamate social Cognition – The use of semiochemicals. In L. M. Schulte (Ed.), *Chemical signals in vertebrates* (Vol. 16, pp. 109–125). Springer.

Szabo, B., & Ringler, E. (2023). Geckos differentiate self from other using both skin and faecal chemicals: Evidence towards self-recognition? *Animal Cognition, 26*, 1011–1019.

Tang-Martinez, Z. (2001). The mechanisms of kin discrimination and the evolution of kin recognition in vertebrates: A critical re-evaluation. *Behavioural Processes, 53*, 21–40.

Thom, M. D., & Hurst, J. L. (2004). Individual recognition by scent. *Annales Zoologici Fennici, 41*, 765–787.

Thünken, T., Waltschyk, N., Bakker, T. C. M., & Kullmann, H. (2009). Olfactory self-recognition in a cichlid fish. *Animal Cognition, 12*, 717–724.

Tybur, J. M., Lieberman, D., & Griskevicius, V. (2009). Microbes, mating, and morality: Individual differences in three functional domains of disgust. *Journal of Personality and Social Psychology, 97*, 103–122.

Übel, S., Wabnegger, A., Arendasy, M., Zorjan, S., & Schienle, A. (2017). Affective evaluation of one's own and others' body odor: The role of disgust proneness. *Perception, 46*, 1427–1433.

Waldman, B. (2016). Chemical communication in archaic New Zealand frogs. In B. A. Schulte, T. E. Goodwin, & M. H. Ferkin (Eds.), *Chemical signals in vertebrates* (Vol. 13, pp. 351–360). Springer.

Waldman, B., & Bishop, P. J. (2004). Chemical communication in an archaic anuran amphibian. *Behavioral Ecology, 15*, 88–93.

Whittaker, D. J., Reichard, D. G., Drouilly, M., Battle, K., & Ziegenfus, C. (2015). Avian olfactory displays: A hypothesis for the function of bill-wiping in a social context. *Behavioral Ecology and Sociobiology, 69*, 159–167.

Widdig, A. (2007). Paternal kin discrimination: The evidence and likely mechanisms. *Biological Reviews, 82*, 319–334.

Witt, M., & Hummel, T. (2006). Vomeronasal versus olfactory epithelium: Is there a cellular basis for human vomeronasal perception? *International Review of Cytology, 248*, 209–259.

Zhang, Y. F., Janke, E., Bhattarai, J. P., Wesson, D. W., & Ma, M. (2022). Self-directed orofacial grooming promotes social attraction in mice via chemosensory communication. *iScience, 25*, 104284.

Part II

Intra- and Interspecific Chemical Communication

A New Perspective on Squamate Social Cognition – The Use of Semiochemicals

Birgit Szabo[1(✉)]

[1] Centre for Research on Ecology, Cognition and Behaviour of Birds, University of Gent, Gent, Belgium

Abstract. The Social Intelligence Hypothesis suggests that cognition might be key to enable animals to live in social groups. Especially social cognition is important as it allows animals to respond appropriately to conspecifics and ensure group cohesion. Social cognition is extensively studied in mammals and birds, but to gain a broad understanding of the benefits of social cognitive processes in social interactions, we need a broader phylogenetic approach. In this opinion paper, I suggest squamates (lizards, snakes and worm lizards) as promising models due to their diverse but facultative sociality and reliance on semiochemical communication in social contexts. Squamates possess a highly developed vomeronasal system to detect semiochemicals for social recognition and discrimination. Similar to the well-studied rodents, squamates detect a wide range of information within chemical cues, but research on the associated decision-making processes, individual differences and development of these abilities is still scarce. Comparative approaches leveraging squamates' semiochemical communication and sociobiological diversity could provide important new insights into the evolution of social cognition. Future research should further focus on individual abilities, their link to environmental and social demands and consequences for fitness, advancing our understanding of adaptive social cognitive skills across taxa.

Keywords: Chemical communication · Chemical ecology · Discrimination · Lizard · Recognition · Reptile · Snake · Social evolution · Vomerolfaction · Worm lizards

1 Introduction

Social cognition involves all neural processes by which individuals collect, retain, process and use information that are beneficial in a social context to avoid competition and conflict or aid cooperation and group cohesion (Shettleworth, 2009; Seyfarth & Cheney, 2015). As such, social cognition plays a crucial role in recognition and memory of specific individuals which forms the basis of any social aggregation, may it be mate guarding, pair formation, parent-offspring aggregation and, at its extreme, long-term group

L. M. Schulte (ed.), *Chemical Signals in Vertebrates 16*,
https://doi.org/10.1007/978-3-032-30752-1_6

living (Ward & Webster, 2016; Rubenstein & Abbot, 2017). The role that cognition plays in social group living was first highlighted by Chance and Mead (1953), Humphrey (1976) and Jolly (1966) on the basis of observations in primates which demonstrated that species living in social groups possess better cognitive abilities which led to the formulation of the Social Intelligence Hypothesis (Chance & Mead, 1953; Humphrey, 1976; Jolly, 1966; Byrne & Whiten, 1988). It suggests that having to discriminate, track and remember specific individuals and their relationships poses a challenge that can be overcome by developing enhanced cognitive skills. Consequently, individuals with better cognitive skill fare better in their social environment and produce more offspring (Zuberbühler & Byrne, 2006). The Social Intelligence Hypothesis has been tested widely across mammal and bird species confirming the link between sociality and cognition across taxa (Speechley et al., 2024). However, most studies have focused on more general cognitive skills such as associative learning and flexibility (e.g. Borrego & Gaines, 2016; Ashton et al., 2018; Berhane & Gazes, 2020). One might argue that a stronger focus on social cognition is warranted as we would expect the largest effects of sociality to be found in the social cognitive domain (e.g. MacLean et al., 2013).

Even though the link between sociality and cognition has received much attention, most studies have been conducted in mammals and birds with investigations in other taxa lacking, including reptiles (Speechley et al., 2024). This gap might be linked to the still prevailing, but shown to be incorrect, view of reptiles being asocial and cognitively limited (Szabo et al., 2021; Font et al., 2023). Social cognition involves not just widely studied phenomena such as social learning or highly complex processes such as knowledge attribution (i.e. "theory of mind"), but more subtle phenomena such as recognising and remembering specific individuals, detecting specific social information and responding appropriately to the gathered information (Shettleworth, 2009; Seyfarth & Cheney, 2015; Kavaliers & Choleris, 2017). A growing body of literature has demonstrated that reptiles possess good non-social cognitive abilities (Burghardt, 2013; Szabo et al., 2021) and even (seemingly) "non-social" reptiles are capable of social learning (e.g. Wilkinson et al., 2010; Damas-Moreira et al., 2018) one aspect of social cognition. Furthermore, especially squamates (lizards, snakes and worm lizards) express a large diversity in social complexity, and even closely related species might express large variation from solitary living to long-term family group living (Whiting & While, 2017; Doody et al., 2021). Consequently, squamates provide an exciting opportunity to study the evolution of social cognition in relation to social complexity more generally but especially the early stages in the evolution of social cognition using a comparative approach.

Squamates rely heavily on semiochemicals (one or more chemicals that inform the behaviour of conspecifics) for intra-specific social communication (Mason & Parker, 2010; Martín & López, 2024) although some species can also use visual cues to recognise individuals (e.g. Van Dyk & Evans, 2007). The use of semiochemicals for such communication is beneficial because chemicals can be deposited without the receiver present, last for a long time and can be detected without the sender present (Norris & Lopez, 2011). Furthermore, semiochemicals can be well characterised by several analytical means to understand the mechanistic basis of social cognitive abilities. Squamates have a highly developed vomeronasal system with which they process both volatile and non-volatile compounds collected with their tongue (Norris & Lopez, 2011). As such, recognition, discrimination and interest in a chemical can be easily quantified by recording the frequency of sampling, also called tongue flicking (Cooper, 1994, 1998). This quantification method has been used widely to gain insights into squamate intra-specific social chemical ecology. For example, self-other discrimination based on chemicals has been demonstrated in a range of squamate species (lizards, Aguilar et al., 2009; Szabo & Ringler, 2023; snakes, Chiszar et al., 1991; Burghardt et al., 2021; Freiburger et al., 2024; worm lizards, Martín et al., 2020). Furthermore, different squamate species can discriminate familiar versus unfamiliar individuals in a territorial (e.g. Aragón et al., 2001), and a mating context (e.g. Cooper, 1996; Martín et al., 2020; Verger et al., 2024) as well as in the context of parental care (e.g. Bull et al., 1994; Martín et al., 2021). Moreover, chemical secretions might carry information about sex (e.g. Cooper & Pèrez-Mellado, 2002; Martín et al., 2020), size (e.g. Shine et al., 2003; Labra, 2006; Martín et al., 2024a, b), age (e.g. López et al., 2003; Gabirot et al., 2012), kinship (e.g. Léna & de Fraipont, 1998; Bull et al., 2001; O'Connor & Shine, 2006; Pernetta et al., 2009), group membership (e.g. Bull et al., 2000), reproductive status (e.g. Cooper & Pèrez-Mellado, 2002), health (e.g. Martín et al., 2024a, b), dominance status (e.g. Martín et al., 2007) and even individual identity (e.g. Bull et al., 1999; Carazo et al., 2008; Mangiacotti et al., 2019). Despite this large body of knowledge accumulating in squamates, these recognition and discrimination abilities are not viewed from a social cognitive perspective as part of the social cognitive repertoire of a tested species. Furthermore, despite this wealth of knowledge on what information can be recognised, we have little knowledge of when and in which social contexts this information might be used (for a discussion of the possible adaptive significance of self-sniffing behaviour, see Havlíček et al., 2026, this volume). A recent phylogenetic analysis highlighted the potential important role of this social chemical communication in lizard social evolution linking the presence of signalling glands to the evolution of social aggregations (Baeckens & Whiting, 2021). Given the diversity in information content within the chemical signals of squamates and their diverse social expression, it is surprising that chemical social communication has not yet been considered from a social cognitive perspective.

Therefore, the aim of this paper is to highlight the potential benefits of utilising chemical communication to better understand social cognitive skills and their relationship to social expression in squamates and, hopefully, inspire future work into this fascinating topic. I first performed a systematic literature search in order to highlight some studies which provide excellent foundational work on which future investigation with a more social cognitive focus can build upon. Then, I will shortly provide some knowledge gaps that need to be filled to produce a complete picture of social cognition based on chemicals in squamates. And finally, I will provide some broader future directions that can produce novel insights into the evolution of adaptive social cognitive skills to deal with social challenges.

2 A Solid Foundation: Model Studies in Squamates

I performed a systematic literature search in August 2024 to evaluate the number of studies in squamates that focus on intra-specific chemical communication (for details on the search results, see Szabo, 2024). In total I identified 152 studies focusing on 97 species (30 snakes, 65 lizards and 2 worm lizards; for more details on the search and a full list of the selected literature, see Szabo, 2024). Given the vast diversity of squamates which include about 12,386 extant species (as of January 2025; Uetz et al., 2025), our understanding of their use of chemicals for social communication is still limited. Nonetheless, though small, the literature is diverse. In the following sections, I present some selected studies that provide a solid foundation from which to delve deeper into squamate chemical-based social cognition.

2.1 Selected Studies on Intra-Specific Chemical Communication Based on Tongue Flicks in Squamates

2.1.1 Lizards. An interesting model species is the tokay gecko (*Gekko gecko*), which I have used in my research. This is a social lizard species that forms pairs and shows biparental care and family group living (Grossmann, 2007), with natural variation in pair association (in the lab, Krummenacher, 2025) and family group size (Grossmann, 2007). In our research, we have demonstrated that these lizards can discriminate their own chemicals from those of same-sex unfamiliar conspecifics. They also show self-directed behaviour and increase the sampling rate (tongue flicks) of their own chemicals in response to the chemicals of same-sex unfamiliar conspecifics. Importantly, tokay geckos can use both skin-derived and faecal chemicals to make the discrimination (Fig. 1; Szabo & Ringler, 2023). Self-recognition is an important ability especially in a social context to be able to recognise one's own home range or territory as well as recognise chemicals left by conspecifics such as territory neighbours or intruders to make appropriate decisions regarding territory defence (Freiburger et al., 2024).

After establishing these lizards' ability to discriminate different chemical stimuli, we proceeded to investigate their ability to discriminate familiar from potential new mates and remember familiar mates. Tokay geckos form pairs that perform biparental care in the form of protection of eggs and hatchlings from predators which might be unfamiliar conspecific individuals (Grossmann, 2007). Therefore, recognising a familiar mate's chemicals to make appropriate decisions regarding offspring protection is important in this species. We found that both male and female geckos can discriminate between familiar and potential new mates, but they showed sex-specific responses.

Fig. 1. (continued) directed tongue flicks are defined as the head and tongue tip pointing towards the substrate (e.g. ground, wall) during tongue flicking. The bold line indicates the median, the upper edge of the box the upper quartile, the lower edge the lower quartile and the whisker the maximum and minimum, and dots represent individual data. Top panel shows all data (responses towards skin and faecal chemicals), while the bottom two show data for responses to skin and scat (faecal) chemicals separated. Low average TF due to 0 inflation. * $p < 0.05$, ** $p < 0.01$, *** $p < 0.001$. (Figures were taken and modified from Szabo & Ringler, 2023).

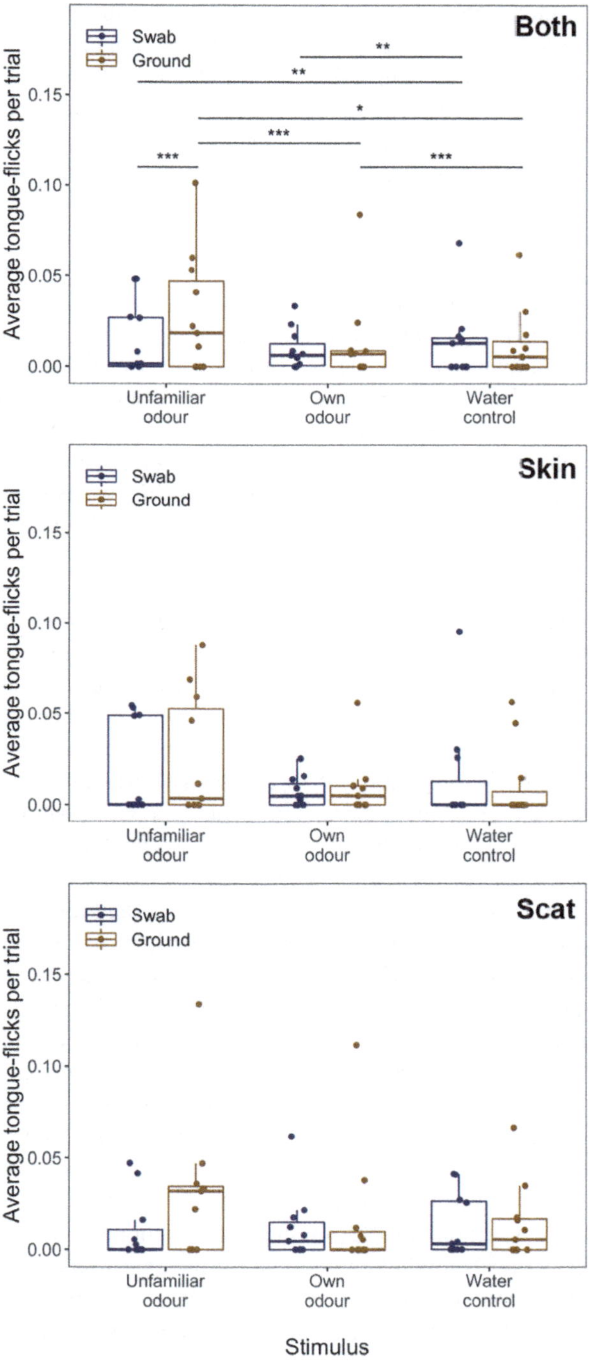

Fig. 1. Boxplots of average tongue flick responses (average across three trials) towards different stimuli presented on swabs within the lizards' home enclosure. Swab-directed tongue flicks are defined as the tip of the tongue pointing towards the swab during tongue flicking. Ground-

Females showed more interest (higher tongue-flick rate) towards the chemical of an unfamiliar male, while males showed more interest in the chemical of a familiar female. Interestingly, males also discriminated their own chemicals from that of their familiar female, showing that they do not just simply label the female with their own chemicals to make the discrimination. Finally, discrimination ability vanished 4–6 weeks after separation from the partner indicating that constant reinforcement is needed for geckos to continue to recognise their mating partner (Verger et al., 2024; Fig. 2).

2.1.2 Snakes. The vast majority of studies in snakes focus on scent trailing behaviour in males which occurs in the mating season and aims at finding mates (Ford, 1986). For example, in gartersnakes (*Thamnophis* sp.), females produce a sexual attractiveness pheromone that communicates female receptivity and size and elicits male courtship (LeMaster & Mason, 2001, 2002; Shine et al., 2003; O'Donnell et al., 2004; Uhrig et al., 2012). However, more recent studies have started to link chemical recognition and discrimination to the sociobiology of different species, especially focusing on differences in self-recognition. These studies show interesting results. For example, social eastern gartersnakes (*T. sirtalis sirtalis*) aggregate frequently across the year with conspecifics (Skinner & Miller, 2020), while more solitary ball pythons (*Python regius*) do not aggregate into groups (Gardner et al., 2016). While gartersnakes show increased interest (tongue flicks) in their marked own scent (a sample of their skin chemicals mixed with olive oil) compared to their own scent, the mark alone (olive oil) and the marked scent of a familiar conspecific (a sample of the skin chemicals of a conspecific mixed with olive oil), ball pythons show no such discrimination (Freiburger et al., 2024). This difference could be attributed to a range of differences in the species ecology, including their feeding ecology, habitat and sociobiology. However, interestingly, both species, but especially ball pythons, show great individual variation in their responses. Furthermore, both gartersnakes and ball pythons show large variation in responses in those conditions including the scent of a familiar individual. Even though the authors state that snakes were familiar, only gartersnakes were housed in groups, and memory of a familiar scent can be limited without constant reinforcement (in geckos; Verger et al., 2024). Therefore, some individuals might have not recognised the scent of familiar individuals as "familiar" which could have increased variation.

As a social species, gartersnakes have been the focus of a large number of studies on the use of intra-specific chemicals (Szabo, 2024). A study in juvenile eastern gartersnakes without previous experience with conspecific chemicals revealed only weak evidence for self-recognition. Females did not differentiate between their own and chemicals of conspecifics but could differentiate individuals based on what diet they were fed, while males could discriminate their own chemicals from those of a sibling on the same diet as well as discriminate individuals based on diet (Burghardt et al., 2021). These results together with the findings of Freiburger et al. (2024) suggest that experience with chemicals might be important for the development of chemical recognition and discrimination; however, research on the development of such skills is almost entirely missing from the literature (but see Léna et al., 2000 in a lizard species).

2.1.3 Worm Lizards. Worm lizards are fossorial animals closely related to lacertids (lizard family; Nisi Cerioni et al., 2024). Our knowledge about their social behaviour is limited due to the difficulty of studying these animals under natural conditions. Nonetheless,

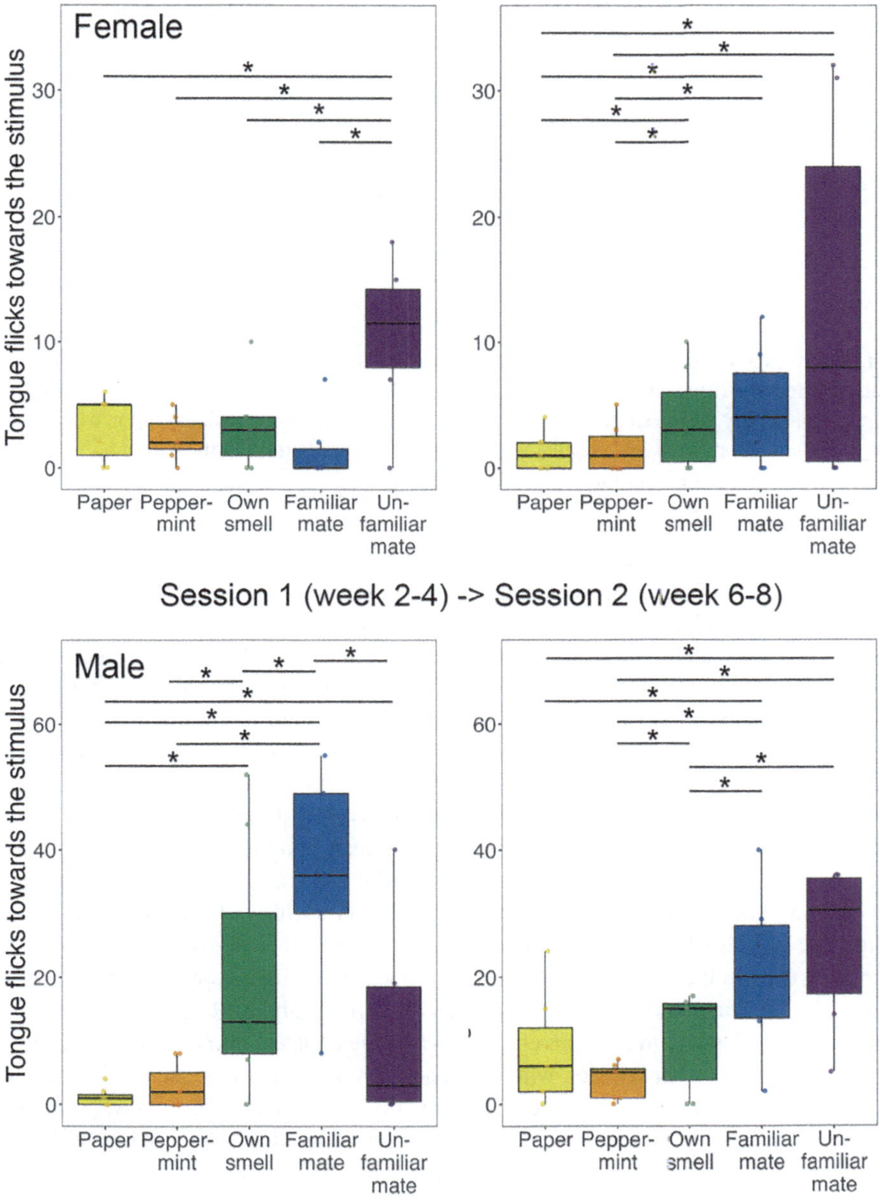

Fig. 2. Boxplots of tongue-flick responses towards different stimuli presented on a piece of filter paper within a glass enclosure. The bold line indicates the median, the upper edge of the box the upper quartile, the lower edge the lower quartile and the whisker the maximum and minimum, and dots represent individual data. The top two figures show female responses in the first and second sessions of the experiment, while the bottom two figures show male responses. * $p < 0.05$. (Figures were taken and modified from Verger et al. (2024)).

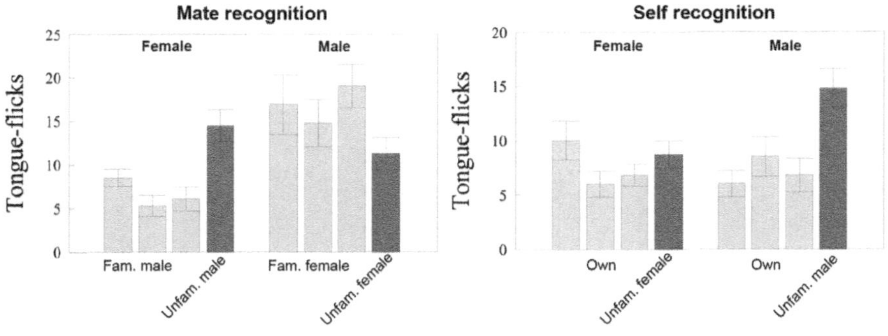

Fig. 3. Average (± SE) number of tongue flicks towards different chemical stimuli emitted by male and female *T. wiegmanni*. A habituation-dishabituation method was used to quantify discrimination ability across stimuli (habituation trials in light grey and dishabituation trials in dark grey). Data for mate- as well as self-recognition are shown. Fam., familiar; unfam., unfamiliar. (Figures were taken and modified from Martín et al. (2020)).

two studies focus on social chemical communication in *Trogonophis wiegmanni*, a species that is frequently observed in social aggregations (Martín et al., 2011). More specifically, they are often found in pairs (more frequently so in the breeding season), and juveniles are often found with adults, most often with a female (Martín et al., 2011). Similar to tokay geckos (Verger et al., 2024), male *T. wiegmanni* respond stronger to the chemicals of a familiar compared to an unfamiliar female mate, while females respond stronger to an unfamiliar male compared to their familiar mate (Fig. 3). Martín and colleagues' hypothesise that chemosensory discrimination of female scent marks by males might facilitate pair bonding and mate guarding but might not be related to parental care as another study showed that only females, but not males, discriminate between familiar and unfamiliar juveniles (Martín et al., 2021). However, about 50% of males did show an ability to discriminate between familiar and unfamiliar juveniles, but what causes this variation is unclear (Martín et al., 2021). Male *T. wiegmanni* also respond stronger to the chemicals of an unfamiliar same-sex conspecific compared to their own odour, while females do not (Martín et al., 2020; Fig. 3). This finding is contrary to tokay geckos (Szabo & Ringler, 2023) in which females showed the same responses as *T. wiegmanni* males. Female tokay geckos are aggressive towards other females, while it is unclear if this is also the case for *T. wiegmanni* females, which could potentially explain the difference in ability. Finally, juvenile *T. wiegmanni* tongue flick more towards the chemicals of familiar adults (male and female) compared to an unfamiliar male which points towards an influence of experience on discrimination ability (Martín et al., 2021) similar to gartersnakes described above. However, how genetic similarity (chemicals of the parents) as compared to familiarity affects responses is unclear.

All these examples take the sociobiology of the tested species into account to better understand their chemical recognition and discrimination abilities. However, so far, individual differences and their consequences or how decision-making is related to recognition and discrimination ability has not been well studied. In the next section, I will describe research methods to study social cognition in very well-studied taxa that also rely heavily on chemicals to communicate in a social context: rodents. This knowledge will help to delve deeper into squamate social semiochemical cognition.

3 Linking Well-Established Social Cognitive Research on Rodents to Squamates

Mate choice in rodents has been considered the outcome of a social cognitive process. It involves decisions regarding who to mate with and when, a process in which the recognition and discrimination of olfactory cues plays a crucial role (Beach, 1942; Kavaliers & Choleris, 2017). Similar to squamates (see above), rodents detect information regarding age, sex, kinship, familiarity, dominance status, reproductive state and body condition as well as individual identity based on chemicals (reviewed in Johnston, 2003). Mate choice is reliant on the detection and processing of this information leading to social decision-making and consequently appropriate social behaviour and mating (Kavaliers & Choleris, 2017). Considering the parallels between rodent and squamate social recognition ability, researchers can utilise similar techniques to better understand social cognition in squamates. The habituation/dishabituation paradigm is one method used to understand social recognition in rodents. First, an animal is repeatedly presented with a social stimulus (an animal or their odour) to which it habituates (shown in a gradual decrease in responses). Thereafter, a new social stimulus is presented. If the test animal can recognise the new stimulus as different from the stimulus it was habituated to, then it will show increased responses. This paradigm can be used to address a broad range of questions from category discrimination to individual recognition with the possibility to take different environmental context into account (Paletta et al., 2023). For example, a recent study in male *Psammodromus algirus* lizards used the habituation/dishabituation paradigm to link age-dependent reproductive strategies (territorial, dominant older males and younger sneaker males) to the ability to discriminate individuals. Older males could discriminate between individual older males but not younger males, while younger males could not discriminate individual males of any age class (Martín et al., 2024a, b). This makes sense as territory holders need to defend against all young sneaker males (no discrimination required) but only against unfamiliar dominant males. Young males, however, need to avoid any other male to be successful (Martín et al., 2024a, b).

Another paradigm used in rodents is the social recognition test. Here, two social stimuli are repeatedly presented at the same time. Test animals are allowed to investigate both across trials. In the test phase, one stimulus is replaced with a new stimulus, and if the new stimulus is investigated more, then it shows the test animal can recognise the familiar stimulus and discriminate it from the new stimulus (Paletta et al., 2023). Furthermore, the duration between training and test can be varied to study recognition memory. A memory of familiar individuals is important in the establishment of social hierarchies, mate choice decisions and parental care (Jacobs et al., 2016). Together, the habituation/dishabituation and social recognition paradigm are excellent methods to answer questions regarding what information animals can detect and discriminate.

For a comprehensive understanding, it is important to also consider the subsequent use of the information gathered from recognition and discrimination in decision-making, but this is far less well studied in squamates (Mason & Parker, 2010). Nonetheless, social information is used during mate choice (e.g. Bruinjé et al., 2022), settlement (e.g. Léna et al., 2000), retreat site (e.g. Scott et al., 2013; Thompson et al., 2020) and foraging decisions (e.g. Clark, 2007) as well as in agonistic encounters (e.g. López & Martín, 2002). However, much research still needs to be done to understand the chemically mediated social decisions in these animals (e.g. utilising choice tests). For example, studies focus on group average ability rather than individual differences even though results can show considerable individual variation (e.g. Martín et al., 2021). Furthermore, the causes (e.g. genetic- or environmental-based developmental plasticity) and consequences of this variation especially under natural conditions in the wild (relationship between decision-making and fitness; Thornton & Lukas, 2012) are poorly understood. For example, dispersal in juvenile common lizards (*Zootoca vivipara*) is associated with attraction and aversion to maternal chemical cues. These differences are already present at birth (common lizards are a viviparous species) and are not influenced by early experience of being raised with or without their mother (Léna et al., 2000). Unfortunately, this study did not link individual ability to choice and dispersal decisions. For future research, it will be important to move beyond studies testing if a species can recognise or discriminate conspecific chemicals or not, towards quantifying individual ability and the source of individual variation such as past experiences, the demands of the social environment and, importantly, if individual variation has consequences for social interactions, decision-making and consequently individual fitness under natural conditions. Only then will we be able to grasp the full extent of social cognition in squamates.

4 A Bright Future: Semiochemical-Based Social Cognition in Squamates

Heritable individual phenotypic variation is the basis on which selections acts upon (Darwin, 1859; Thornton & Lukas, 2012). Therefore, understanding the full extent and variation of chemical-based social cognitive abilities within species is of great interest to link species-specific environmental and sociobiological characteristics to the information content of the chemical signals, detection ability of this information and the decision outcomes and fitness consequences based on the collected information. Such detailed information provides the substrate for comparative studies that focus on broader questions regarding the selective pressures driving the evolution of social cognitive abilities (Völter et al., 2018). I believe that squamates are a powerful comparative model system in this regard, because:

1. A wide range of information is encoded reliably in the chemical signals of squamates.
2. The detection of and preference for this information can be measured through a combination of test on tongue-flick rates and choice tests across species (Szabo, 2024).
3. Squamates express a large diversity in sociality including parthenogenetic species, parental care level (no care to short-term care to long-term care until offspring reach sexual maturity) and facultative sociality (from no group living to long-term stable family groups) (Somma, 2003; Rheubert et al., 2014; Gardner et al., 2016; Whiting & While, 2017; Doody et al., 2021).

Some lizard species have already been successfully used to understand potential environmental factors driving the evolution of cognition. For example, a study on 13 lacertid species showed a link between behavioural flexibility (reversal learning) and environmental variability (De Meester et al., 2022). Similarly, by testing semiochemical social cognition across species, we can answer broad evolutionary questions about what information might be relevant, and therefore detected, and how this information is used for decision-making under different social conditions. For example, depending on the mating system, species should express appropriate social semiochemical cognitive abilities that will help them select the most appropriate mating partner. If females choose, they need to be able to reliably detect male quality and be able to discriminate across males. If females mate multiple times, they should be able to discriminate and remember specific males to avoid remating. On the other hand, if females do not choose, then they would not need to discriminate and remember males. The facultative social nature of squamates also provides a new perspective on the importance of social cognition when species naturally experience variation in sociality that exceeds what is possible in more obligate social species. For instance, even in species that show parental care, not all offspring might receive care (e.g. some siblings disperse while others stay; While et al., 2009) which can be related to semiochemical social cognitive abilities (e.g. Léna et al., 2000). A broader phylogenetic approach to the study of the evolution of cognition in relation to sociality (Social Intelligence Hypothesis), even beyond squamates (e.g. turtles and tortoises; Ibáñez et al., 2012; Ibáñez & Vogt, 2015), will provide a novel perspective on what types of social aggregations exert selective pressure on which social cognitive abilities.

5 Conclusion

Social behaviour and cognition might not be what we readily associate with lizards, snakes and worm lizards. Their social interactions can be inconspicuous, especially when strongly relying on channels other than visual communication with its obvious colours and elaborate display behaviour. However, those who dare to venture into the unknown with a keen eye and an open mind will discover a new world, not as flashy but surely as captivating. I believe that there is much to learn about squamate sociality and the evolution of social behaviour through the study of squamate semiochemical-based social cognition.

Acknowledgements. This work was supported by the University of Bern and Gent University Methusalem Project (01 M00221). I would like to thank Charlène Gémard and Isabel Damas-Moreira for their comments on an early draft of this paper.

References

Aguilar, P. M., Labra, A., & Niemeyer, H. M. (2009). Chemical self-recognition in the lizard *Liolaemus fitzgeraldi*. *Journal of Ethology, 27*(1), 181–184. https://doi.org/10.1007/s10164-008-0088-x

Aragón, P., López, P., & Martín, J. (2001). Discrimination of femoral gland secretions from familiar and unfamiliar conspecifics by male Iberian rock-lizards, *Lacerta monticola*. *Journal of Herpetology, 35*(2), 346–350. https://www.jstor.org/stable/1566131

Ashton, B. J., Ridley, A. R., Edwards, E. K., & Thornton, A. (2018). Cognitive performance is linked to group size and affects fitness in Australian magpies. *Nature, 554*(7692), 364–367. https://doi.org/10.1038/nature25503

Baeckens, S., & Whiting, M. J. (2021). Investment in chemical signalling glands facilitates the evolution of sociality in lizards. *Proceedings of the Royal Society B - Biological Sciences, 288*(1945), 20202438. https://doi.org/10.1098/rspb.2020.2438

Beach, F. A. (1942). Analysis of the stimuli adequate to elicit mating behavior in the sexually inexperienced rat. *Journal of Comparative Psychology, 33*(2), 163–207. https://doi.org/10.1037/h0061692

Berhane, J. F., & Gazes, R. P. (2020). Social monkeys learn more slowly: Social network centrality and age are positively related to learning errors by capuchin monkeys (*Cebus [Sapajus] apella*). *Canadian Journal of Experimental Psychology/Revue canadienne de psychologie expérimentale, 74*(3), 228–234. https://doi.org/10.1037/cep0000202

Borrego, N., & Gaines, M. (2016). Social carnivores outperform asocial carnivores on an innovative problem. *Animal Behaviour, 114*, 21–26. https://doi.org/10.1016/j.anbehav.2016.01.013

Bruinjé, A. C., Paiva, T. M. D., & Costa, G. C. (2022). Multimodal female mate choice in a polymorphic flat rock lizard. *Behavioral Ecology and Sociobiology, 76*(6), 77. https://doi.org/10.1007/s00265-022-03181-x

Bull, C. M., Doherty, M., Schulze, L. R., & Pamula, Y. (1994). Recognition of offspring by females of the Australian skink, *Tiliqua rugosa*. *Journal of Herpetology, 28*(1), 117–120. https://doi.org/10.2307/1564693

Bull, C. M., Griffin, C. L., & Perkins, M. V. (1999). Some properties of a pheromone allowing individual recognition, from the scats of an Australian lizard, Egernia striolata. *Acta Ethologica, 2*(1), 35–42. https://doi.org/10.1007/PL00012230

Bull, C. M., Griffin, C. L., Lanham, E. J., & Johnston, G. R. (2000). Recognition of pheromones from group members in a gregarious lizard, *Egernia stokesii*. *Journal of Herpetology, 34*(1), 92–99. https://doi.org/10.2307/1565244

Bull, C. M., Griffin, C. L., Bonnett, M., Gardner, M. G., & Cooper, S. J. B. (2001). Discrimination between related and unrelated individuals in the Australian lizard *Egernia striolata*. *Behavioral Ecology and Sociobiology, 50*(2), 173–179. https://doi.org/10.1007/s002650100348

Burghardt, G. M. (2013). Environmental enrichment and cognitive complexity in reptiles and amphibians: Concepts, review, and implications for captive populations. *Applied Animal Behaviour Science, 147*(3–4), 286–298. https://doi.org/10.1016/j.applanim.2013.04.013

Burghardt, G. M., Partin, A. M., Pepper, H. E., Steele, J. M., Liske, S. M., Stokes, A. E., Lathan, A. N., Springer, C. M., & Jenkins, M. S. (2021). Chemically mediated self-recognition in sibling juvenile common gartersnakes (*Thamnophis sirtalis*) reared on same or different diets: Evidence for a chemical mirror? *Behaviour, 158*(12–13), 1169–1191. https://doi.org/10.1163/1568539X-bja10131

Byrne, R. W., & Whiten, A. (1988). Machiavellian intelligence: Social expertise and the evolution of intellect in monkeys, apes, and humans. *Behavior and Philosophy, 18*(1), 73–75.

Carazo, P., Font, E., & Desfilis, E. (2008). Beyond 'nasty neighbours' and 'dear enemies'? Individual recognition by scent marks in a lizard (*Podarcis hispanica*). *Animal Behaviour, 76*, 1953–1963. https://doi.org/10.1016/j.anbehav.2008.08.018

Chance, M. R. A., & Mead, A. P. (1953). Social behaviour and primate evolution. *Symposium of the Society for Experimental Biology and Evolution, 7*, 395–439.

Chiszar, D., Smith, H. M., Bogert, C. M., & Vidaurri, J. (1991). A chemical sense of self in timber and prairie rattlesnakes. *Bulletin of the Psychonomic Society, 29*(2), 153–154. https://doi.org/10.3758/BF03335221

Clark, R. W. (2007). Public information for solitary foragers: timber rattlesnakes use conspecific chemical cues to select ambush sites. *Behavioral Ecology, 18*(2), 487–490.

Cooper, W. E. (1994). Chemical discrimination by tongue-flicking in lizards: A review with hypotheses on its origin and its ecological and phylogenetic relationships. *Journal of Chemical Ecology, 20*, 439–487. https://doi.org/10.1007/BF02064449

Cooper, W. E. (1996). Chemosensory recognition of familiar and unfamiliar conspecifics by the scincid lizard *Eumeces laticeps*. *Ethology, 102*(6), 454–464. https://doi.org/10.1111/j.1439-0310.1996.tb01139.x

Cooper, W. E. (1998). Evaluation of swab and related tests as a bioassay for assessing responses by squamate reptiles to chemical stimuli. *Journal of Chemical Ecology, 24*, 841–866. https://doi.org/10.1023/A:1022373517653

Cooper, W. E., & Pèrez-Mellado, V. (2002). Pheromonal discriminations of sex, reproductive condition, and species by the lacertid lizard *Podarcis hispanica*. *Journal of Experimental Zoology, 292*(6), 523–527. https://doi.org/10.1002/jez.10089

Damas-Moreira, I., Oliveira, D., Santos, J. L., Riley, J. L., Harris, D. J., & Whiting, M. J. (2018). Learning from others: An invasive lizard uses social information from both conspecifics and heterospecifics. *Biology Letters, 14*(10), 20180532. https://doi.org/10.1098/rsbl.2018.0532

Darwin, C. (1859). *On the origin of species by means of natural selection* (Vol. 167). John Murray.

De Meester, G., Van Linden, L., Torfs, J., Pafilis, P., Šunje, E., Steenssens, D., Zulčić, T., Sassalos, A., & Van Damme, R. (2022). Learning with lacertids: Studying the link between ecology and cognition within a comparative framework. *Evolution, 76*(11), 2531–2552. https://doi.org/10.1111/evo.14618

Doody, J. S., Dinets, V., & Burghardt, G. M. (2021). *The secret social lives of reptiles*. JHU Press.

Font, E., Burghardt, G. M., & Leal, M. (2023). Brains, behaviour, and cognition: Multiple misconceptions. In C. Warwick, P. C. Arena, & G. M. Burghardt (Eds.), *Health and welfare of captive reptiles* (pp. 211–238). Springer International Publishing.

Ford, N. B. (1986). The role of pheromone trails in the sociobiology of snakes. In D. Duvall, D. Müller-Schwarze, & R. M. Silverstein (Eds.), *Chemical signals in vertebrates 4: Ecology, evolution, and comparative biology* (pp. 261–278). Springer.

Freiburger, T., Miller, N., & Skinner, M. (2024). Olfactory self-recognition in two species of snake. *Proceedings of the Royal Society B-Biological Sciences, 291*(2020), 20240125. https://doi.org/10.1098/rspb.2024.0125

Gabirot, M., Picerno, P., Valencia, J., Lopez, P., & Martin, J. (2012). Chemosensory age discrimination in the snake *Boa constrictor* (Serpentes: Boidae). *Revista De Biologia Tropical, 60*(4), 1603–1611.

Gardner, M. G., Pearson, S. K., Johnston, G. R., & Schwarz, M. P. (2016). Group living in squamate reptiles: A review of evidence for stable aggregations. *Biological Reviews, 91*(4), 925–936. https://doi.org/10.1111/brv.12201

Grossmann W (2007). Der Tokeh: Gekko gecko. Natur und Tier Verlag, Münster.

Havlíček, J., Schwambergová, D., & Roberts, S. C. (2026). Chemosensory self-inspection. In L. M. Schulte (Ed.) *Chemical signals in vertebrates* (Vol 16, pp. 87–105). Springer.

Humphrey, N. K. (1976). The social function of intellect. In P. P. G. Bateson & R. A. Hinde (Eds.), *Growing points in ethology* (pp. 303–317). Cambridge University Press.

Ibáñez, A., & Vogt, R. C. (2015). Chemosensory discrimination of conspecifics in the juvenile yellow-spotted river turtle *Podocnemis unifilis*. *Behaviour, 152*(2), 219–230. https://doi.org/10.1163/1568539X-00003240

Ibáñez, A., López, P., & Martín, J. (2012). Discrimination of conspecifics' chemicals may allow Spanish terrapins to find better partners and avoid competitors. *Animal Behaviour, 83*(4), 1107–1113. https://doi.org/10.1016/j.anbehav.2012.02.001

Jacobs, S. A., Huang, F., Tsien, J. Z., & Wei, W. (2016). Social recognition memory test in rodents. *Bio-Protocol, 6*(9), e1804–e1804. https://doi.org/10.21769/BioProtoc.1804

Johnston, R. E. (2003). Chemical communication in rodents: from pheromones to individual recognition. *Journal ofmammalogy, 84*(4), 1141–1162.

Jolly, A. (1966). Lemur social behavior and primate intelligence: The step from prosimian to monkey intelligence probably took place in a social context. *Science, 153*(3735), 501–506. https://doi.org/10.1126/science.153.3735.501

Kavaliers, M., & Choleris, E. (2017). Social cognition and the neurobiology of rodent mate choice. *Integrative and Comparative Biology, 57*(4), 846–856. https://doi.org/10.1093/icb/icx042

Krummenacher, A. L., & Szabo, B. (2025). Can physical closeness measure variation and change in pair associationstrength in captive geckos?.

Labra, A. (2006). Chemoreception and the assessment of fighting abilities in the lizard *Liolaemus monticola*. *Ethology, 112*(10), 993–999. https://doi.org/10.1111/j.1439-0310.2006.01256.x

LeMaster, M. P., & Mason, R. T. (2001). Evidence for a female sex pheromone mediating male trailing behavior in the red-sided garter snake, *Thamnophis sirtalis parietalis*. *Chemoecology, 11*, 149–152. https://doi.org/10.1007/PL00001845

LeMaster, M. P., & Mason, R. T. (2002). Variation in a female sexual attractiveness pheromone controls male mate choice in garter snakes. *Journal of Chemical Ecology, 28*, 1269–1285. https://doi.org/10.1023/A:1016294003641

Léna, J. P., & de Fraipont, M. (1998). Kin recognition in the common lizard. *Behavioral Ecology and Sociobiology, 42*(5), 341–347. https://doi.org/10.1007/s002650050446

Léna, J. P., De Fraipont, M., & Clobert, J. (2000). Affinity towards maternal odour and offspring dispersal in the common lizard. *Ecology Letters, 3*(4), 300–308. https://doi.org/10.1046/j.1461-0248.2000.00155.x

López, P., & Martín, J. (2002). Chemical rival recognition decreases aggression levels in male Iberian wall lizards, Podarcis hispanica. *Behavioral Ecology and Sociobiology, 51*(5), 461–465. https://doi.org/10.1007/s00265-001-0447-x

López, P., Aragón, P., & Martín, J. (2003). Responses of female lizards, *Lacerta monticola*, to males' chemical cues reflect their mating preference for older males. *Behavioral Ecology and Sociobiology, 55*, 73–79. https://doi.org/10.1007/s00265-003-0675-3

MacLean, E. L., Sandel, A. A., Bray, J., Oldenkamp, R. E., Reddy, R. B., & Hare, B. A. (2013). Group size predicts social but not nonsocial cognition in lemurs. *PLoS One, 8*(6), e66359. https://doi.org/10.1371/journal.pone.0066359

Mangiacotti, M., Gaggiani, S., Coladonato, A. J., Scali, S., Zuffi, M. A. L., & Sacchi, R. (2019). First experimental evidence that proteins from femoral glands convey identity-related information in a lizard. *Acta Ethologica, 22*, 57–65. https://doi.org/10.1007/s10211-018-00307-1

Martín, J., & López, P. (2024). Pheromones and reproduction in reptiles. In D. O. Norris & K. H. Lopez (Eds.), *Hormones and reproduction of vertebrates* (Vol. 3, pp. 135–163). Academic Press.

Martín, J., Moreira, P. L., & López, P. (2007). Status-signalling chemical badges in male Iberian rock lizards. *Functional Ecology, 21*(3), 568–576. https://doi.org/10.1111/j.1365-2435.2007.01262.x

Martín, J., Polo-Cavia, N., Gonzalo, A., López, P., & Civantos, E. (2011). Social aggregation behaviour in the North African amphisbaenian *Trogonophis wiegmanni*. *African Journal of Herpetology, 60*(2), 171–176. https://doi.org/10.1080/21564574.2011.566285

Martín, J., García, E. R., Ortega, J., & López, P. (2020). How to maintain underground social relationships? Chemosensory sex, partner and self recognition in a fossorial amphisbaenian. *PLoS One, 15*(8). https://doi.org/10.1371/journal.pone.0237188

Martín, J., Raya-García, E., Ortega, J., & López, P. (2021). Offspring and adult chemosensory recognition by an amphisbaenian reptile may allow maintaining familiar links in the fossorial environment. *PeerJ, 9*, e10780. https://doi.org/10.7717/peerj.10780

Martín, J., López, P., Rodríguez-Ruiz, G., Cuervo, J. J., Sacchi, R., & Mangiacotti, M. (2024a). Age-dependent self-and individual chemosensory discrimination in male *Psammodromus algirus* lizards is based on lipids and proteins from their scent. *Behavioral Ecology and Sociobiology, 78*(12), 125. https://doi.org/10.1007/s00265-024-03543-7

Martín, J., Rodríguez-Ruiz, G., Navarro-Castilla, A., Barja, I., & López, P. (2024b). Blind date: Female fossorial amphisbaenians prefer scent marks of large and healthy males. *Integrative Zoology, 19*(6), 1018–1033. https://doi.org/10.1111/1749-4877.12802

Mason, R. T., & Parker, M. R. (2010). Social behavior and pheromonal communication in reptiles. *Journal of Comparative Physiology A, 196*, 729–749. https://doi.org/10.1007/s00359-010-0551-3

Nisi Cerioni, P., Giovannotti, M., Slimani, T., S'khifa, A., Splendiani, A., Fioravanti, T., et al. (2024). Satellite DNA supports the monophyly of Lacertibaenia (Amphisbaenia and Lacertidae) in squamate phylogeny. *The European Zoological Journal, 91*(2), 804–816. https://doi.org/10.1080/24750263.2024.2376594

Norris, D. O., & Lopez, K. H. (2011). *Hormones and reproduction of vertebrates, Volume 3: Reptiles*. Elsevier.

O'Donnell, R. P., Ford, N. B., Shine, R., & Mason, R. T. (2004). Male red-sided garter snakes, Thamnophis sirtalis parietalis, determine female mating status from pheromone trails. *Animal Behaviour, 68*(4), 677–683. https://doi.org/10.1016/j.anbehav.2003.09.020

O'Connor, D. E., & Shine, R. (2006). Kin discrimination in the social lizard *Egernia saxatilis* (Scincidae). *Behavioral Ecology, 17*(2), 206–211. https://doi.org/10.1093/beheco/arj019

Paletta, P., Aspesi, D., Bass, N., & Choleris, E. (2023). The study of social cognition: Social recognition and social learning in laboratory rats and mice. In R. G. Paredes, W. Portillo, & M. Bedos (Eds.), *Animal models of reproductive behavior* (pp. 1–31). Springer.

Pernetta, A. P., Reading, C. J., & Allen, J. A. (2009). Chemoreception and kin discrimination by neonate smooth snakes, Coronella austriaca. *Animal Behaviour, 77*(2), 363–368. https://doi.org/10.1016/j.anbehav.2008.10.008

Rheubert, J. L., Siegel, D. S., & Trauth, S. E. (2014). *Reproductive biology and phylogeny of lizards and tuatara*. CRC Press.

Rubenstein, D. R., & Abbot, P. (2017). *Comparative Social Evolution*. Cambridge University Press.

Scott, M. L., Whiting, M. J., Webb, J. K., & Shine, R. (2013). Chemosensory discrimination of social cues mediates space use in snakes, *Cryptophis nigrescens* (Elapidae). *Animal Behaviour, 85*(6), 1493–1500. https://doi.org/10.1016/j.anbehav.2013.04.003

Seyfarth, R. M., & Cheney, D. L. (2015). Social cognition. *Animal Behaviour, 103*, 191–202. https://doi.org/10.1016/j.anbehav.2015.01.030

Shettleworth, S. J. (2009). *Cognition, evolution, and behavior*. Oxford University Press.

Shine, R., Phillips, B., Waye, H., LeMaster, M., & Mason, R. T. (2003). Chemosensory cues allow courting male garter snakes to assess body length and body condition of potential mates. *Behavioral Ecology and Sociobiology, 54*(2), 162–166. https://doi.org/10.1007/s00265-003-0620-5

Skinner, M., & Miller, N. (2020). Aggregation and social interaction in garter snakes (*Thamnophis sirtalis sirtalis*). *Behavioral Ecology and Sociobiology, 74*(5), 51. https://doi.org/10.1007/s00265-020-2827-0

Somma, L. A. (2003). *Parental behavior in Lepidosaurian and Testudinian reptiles: A literature survey*. Krieger Publishing Company.

Speechley, E. M., Ashton, B. J., Foo, Y. Z., Simmons, L. W., & Ridley, A. R. (2024). Meta-analyses reveal support for the social intelligence hypothesis. *Biological Reviews, 99*(5), 1889–1908. https://doi.org/10.1111/brv.13103

Szabo, B. (2024). *Literature list on intraspecific chemical communication in Squalates*. Retrieved from https://doi.org/10.17605/OSF.IO/J9RXN

Szabo, B., & Ringler, E. (2023). Geckos differentiate self from other using both skin and faecal chemicals: Evidence towards self-recognition? *Animal Cognition, 26*(3), 1011–1019. https://doi.org/10.1007/s10071-023-01751-8

Szabo, B., Noble, D. W., & Whiting, M. J. (2021). Learning in non-avian reptiles 40 years on: Advances and promising new directions. *Biological Reviews, 96*(2), 331–356. https://doi.org/10.1111/brv.12658

Thompson, S. A., Pearson, S. K., While, G. M., Chapple, D. G., & Gardner, M. G. (2020). Scat on the doorstep: Refuge choice in a group-living lizard is influenced by the presence of scat piles. *Austral Ecology, 45*(4), 426–434. https://doi.org/10.1111/aec.12864

Thornton, A., & Lukas, D. (2012). Individual variation in cognitive performance: Developmental and evolutionary perspectives. *Philosophical Transactions of the Royal Society B: Biological Sciences, 367*(1603), 2773–2783. https://doi.org/10.1098/rstb.2012.0214

Uetz, P., Freed, P, Aguilar, R., Reyes, F., Kudera, J. & Hošek, J. (2025). *The reptile database*. http://www.reptile-database.org. Accessed 6 Feb 2025.

Uhrig, E. J., Lutterschmidt, D. I., Mason, R. T., & LeMaster, M. P. (2012). Pheromonal mediation of intraseasonal declines in the attractivity of female red-sided garter snakes, *Thamnophis sirtalis parietalis*. *Journal of Chemical Ecology, 38*, 71–80. https://doi.org/10.1007/s10886-011-0054-x

Van Dyk, D. A., & Evans, C. S. (2007). Familiar–unfamiliar discrimination based on visual cues in the Jacky dragon, Amphibolurus muricatus. *Animal Behaviour, 74*(1), 33–44. https://doi.org/10.1016/j.anbehav.2006.06.018

Verger, M. O., Devillebichot, M., Ringler, E., & Szabo, B. (2024). Sex-specific discrimination of familiar and unfamiliar mates in the tokay gecko. *Animal Cognition, 27*(1), 55. https://doi.org/10.1007/s10071-024-01896-0

Völter, C. J., Tinklenberg, B., Call, J., & Seed, A. M. (2018). Comparative psychometrics: Establishing what differs is central to understanding what evolves. *Philosophical Transactions of the Royal Society B: Biological Sciences, 373*(1756), 20170283. https://doi.org/10.1098/rstb.2017.0283

Ward, A., & Webster, M. (2016). *Sociality: The behaviour of group-living animals* (Vol. 407). Springer.

While, G. M., Uller, T., & Wapstra, E. (2009). Within-population variation in social strategies characterize the social and mating system of an Australian lizard, Egernia whitii. *Austral Ecology, 34*(8), 938–949. https://doi.org/10.1111/j.1442-9993.2009.02002.x

Whiting, M. J., & While, G. M. (2017). Sociality in lizards. In D. R. Rubenstein & P. Abbot (Eds.), *Comparative Social Evolution* (pp. 390–426). Cambridge University Press.

Wilkinson, A., Kuenstner, K., Mueller, J., & Huber, L. (2010). Social learning in a non-social reptile (*Geochelone carbonaria*). *Biology Letters, 6*(5), 614–616. https://doi.org/10.1098/rsbl.2010.0092

Zuberbühler, K., & Byrne, R. W. (2006). Social cognition. *Current Biology, 16*(18), R786–R790.

Scent Marking in Wild-Derived Black Rats (*Rattus rattus*): Observations from Different Experimental Procedures

Veronika Rudolfová[1,2] (✉), Iveta Štolhoferová[1], Eliška Rychtecká[1], Marie Berčíková[1], Hana Skalíková[1], and Daniel Frynta[1]

[1] Faculty of Science, Charles University, Department of Zoology, Prague, Czech Republic
[2] National Institute of Mental Health, Klecany, Czech Republic
veronika.rudolfova@natur.cuni.cz

Abstract. Scent marking in animals is a frequently studied phenomenon. Mice and laboratory rats are the most studied species in this respect, and much is known about scent marking and its function in these species. In our work with black rats (*Rattus rattus*), we have seen scent marking only sporadically and the way this species marks seemed to differ substantially from what was previously reported in laboratory rats. In this chapter, we present descriptive data on scent marking in black rats in several experimental designs collected over several years. We report that their scent marking behaviour is indeed different to scent marking of laboratory rats, and we suggest that the ecology of the black rats, or their lower propensity for territoriality, may be the reason why they rarely use scent marking.

Keywords: Urine · Anogenital dragging · Pheromone · Roof rat · Ship rat

1 Introduction

In the last several decades, there has been a substantial research effort put into describing and understanding scent marking in animals (see, e.g. these reviews: Eisenberg & Kleiman, 1972; Gosling, 1982; Hurst, 1987; Gosling & Roberts, 2001a, b; Roberts & Gosling, 2001; Johnston, 2003; Beynon & Hurst, 2004; Roberts, 2007; Arakawa et al., 2008; Johnston, 2008; Bakshi, 2010; Ferkin, 2015, 2018, 2019 and references therein).

Originally, the functional aspect of this behaviour has attracted attention. Most studies reviewing the existing data suggest that it is important in intraspecific communication and that scent marking may facilitate social interactions and communicate information about the individual. It may serve as a cue of a health condition of the animal (Arakawa et al., 2008; Gosling & Roberts, 2001a; Ferkin, 2015) and allow for an easier individual recognition (Johnston, 2003; Hurst & Beynon, 2004; Thom & Hurst, 2004; Hurst et al., 2005; Arakawa et al., 2008; Ferkin, 2015; Gromov, 2015). It may communicate dominance and social status (Gosling, 1982; Gosling & Roberts, 2001a; Hurst et al., 2001; Hurst & Beynon, 2004; Arakawa et al., 2008; Bakshi, 2010;

L. M. Schulte (ed.), *Chemical Signals in Vertebrates 16*,
https://doi.org/10.1007/978-3-032-30752-1_7

Gromov, 2015) and advertise the reproductive status of the animal (Eisenberg & Kleiman, 1972; Roberts, 2007; Arakawa et al., 2008; Bakshi, 2010; Ferkin, 2018; Gómez-Baena et al., 2014; Gromov, 2015), and it may be used in mate choice (Gosling, 1982; Gosling & Roberts, 2001a; Hurst et al., 2001; Thom & Hurst, 2004; Hurst, 2009; Ferkin, 2018; Gromov, 2015). Scent marks probably serve to indicate territorial ownership (Gosling, 1982; Gosling & Roberts, 2001a; Hurst et al., 2001; Hurst & Beynon, 2004; Roberts, 2007; Arakawa et al., 2008; Gromov, 2015), where it may serve as a scent of familiarity indicating a safe place (Bakshi, 2010), warn conspecifics that the site is occupied (Bakshi, 2010) and help the resident orient in the environment (Hurst, 1987; Bakshi, 2010). Scent marking may also allow the assessment of competitive ability of the individual (Gosling & Roberts, 2001a, b; Roberts, 2007). More recently, the chemical profile of scent signals has been intensively studied (e.g. Hurst et al., 2001; Beynon & Hurst, 2004; Hurst & Beynon, 2004; Thom & Hurst, 2004; Hurst et al., 2005; Hurst, 2009; Vettorazzi et al., 2013; Gómez-Baena et al., 2014, 2019).

Despite this effort, we still have only fragmented information on the exact function of scent marking and its role in the life of certain animals. Among rodents, most of the studies focus on laboratory rats and house mice (laboratory, wild-derived and wild). Nevertheless, several other species were studied as well—e.g. gerbils (Thiessen et al., 1970, 1971; Swanson, 1980; Yoshimura, 1981; Gromov, 2015), the giant pouched rat (e.g. Docherty, 2009; Freeman & Ophir, 2018; Freeman et al., 2020; Freeman & Ophir, 2021), hamsters (e.g. Johnston & Lee, 1976; Huck et al., 1985; Ferris et al., 1987), the harvest mouse (e.g. Roberts & Gosling, 2004), voles (e.g. Johnston et al., 1997; Ferkin, 2007, 2010) and scarcely even other species such as ground squirrels (e.g. Steiner, 1974) and deermouse (e.g. Becker et al., 2018). It is universally accepted that some rodents are able to detect age, sex, strain, relatedness, individual identity, health status or reproductive condition of their conspecifics from their urine (e.g. Eisenberg & Kleiman, 1972; Ferkin & Johnston, 1995; Hurst et al., 2001; Johnston, 2003; Beynon & Hurst, 2004). Urine is, therefore, an important means for individual recognition when the animals encounter each other or when they encounter any scent mark left behind.

It may play an important role in both intra- and inter-sexual communication as it may facilitate the determination of dominance status and signal the quality of a potential partner (Eisenberg & Kleiman, 1972; Brown, 1992). Given the role of scent marking in intra-sexual competition and mate choice, age seems to play a major role with adult animals depositing more marks than subadult and juvenile animals (Hurst, 1990; Roberts, 2007). Many studies have shown that the sex of the animal may also be of some importance (Brown, 1977; Drickamer, 1995) and that in some cases males scent mark more frequently than females (Brown, 1977; Gómez-Baena et al., 2023). Other rather important factors are sexual experience (Brown, 1985, 1991) and preparedness for breeding, specifically breeding season in males and oestrus in females (Brown, 1977; Birke, 1978; Birke & Sadler, 1984; Ferkin & Johnston, 1995; Gromov, 2015; Gómez-Baena et al., 2023). Nevertheless, there may also be other, less obvious factors, which may influence the frequency of scent marking in an animal. For example, in deermice (*Peromyscus*), scent marking behaviour and frequency depended on the mating system, with monogamous California deermouse (*P. californicus*) marking more frequently than promiscuous white-footed mouse (*P. leucopus*; Becker et al.,

2018). In laboratory rats, rearing conditions also play a role in the propensity to inspect odours and urine mark (Brown, 1991; Becker et al., 2018). Social structure may, therefore, also play a significant role in the propensity to scent mark the environment.

In rodents, scent marking may be associated with a specific form of application of the scent, with urination and anogenital dragging (accompanied by preputial gland marking) being the most common forms. The ways in which laboratory rats scent mark with urine were described in detail in Brown (1975), where he states that the rats urinate on an object in a lift-leg position (either lifting a leg and urinating on the object while standing or putting one leg on the object and urinating on the object while walking with the leg still up on it). He described that urine was also deposited using an anogenital drag, when the rat rubs its anogenital region on the object, leaving a trail of urine (Brown, 1975). Such anogenital dragging seems to be a common response to a new object (Brown, 1975; Birke & Sadler, 1984). In some species, it has been documented that the animals intentionally direct urine towards objects (laboratory rats, Brown, 1975) or other individuals in the environment (maras, Macdonald et al., 2007).

A specific form of scent marking is countermarking, a behaviour in which a scent mark is placed near or directly on top of another scent mark (the term overmarking is sometimes also used). It is often used in territorial species as a means of territorial display and defence, or in mate choice (e.g. Hurst, 1990; Johnston et al., 1997; Ferkin, 2007, 2010; Becker et al., 2018). In territorial defence it may be viewed as "reactive scent marking" (Roberts, 2007), a form of competition where potential mates or intruders are evaluated (Johnston, 2003).

Among the vast research on scent marking, only a few papers describe what do the marks actually look like or present images from the experiments. It is therefore not an easy task to describe scent marking in a previously neglected species and compare and contrast it to the well-studied species. Studies with laboratory rats typically present data on areas covered in urine (e.g. Brown, 1975, 1977, 1985, 1991; Roberts et al., 2001; Hurst et al., 2005) or the number of marks (Swanson, 1980; Yoshimura, 1981; Matochik & Barfield, 1991; Gromov, 2015) but not the marks themselves. Fortunately, some studies provide images of urine marks (mice, e.g. Desjardins et al., 1973; Beynon & Hurst, 2004; Arakawa et al., 2008; rats, Richards & Stevens, 1974; giant pouched rats, Freeman & Ophir, 2018; harvest mice, Docherty, 2009), and studies with meadow voles (*Microtus pennsylvanicus*) provide quite detailed descriptions of the scent marks. In this species, the "typical scent marks" range from 0.4 to 5 cm in length and 0.1 to 0.3 cm in width (Thomas & Wolff, 2002; Ferkin, 2007). Voles also produce quite a large amount of scent marks—in an empty arena without scents, they produce approximately 16 marks in the first 10 minutes of experiment; when the arena contains scent marks from conspecifics, the number of scent marks increase to 25–30 (Ferkin, 2007). In a 30-minute trial, two species of voles (*M. pennsylvanicus* and *M. ochrogaster)* leave more than a hundred urine marks in the arena (Thomas & Wolff, 2002).

We are aware of only one research area dedicated to the scent marking behaviour of the black rat, *Rattus rattus*, which was, moreover, conducted for the purpose of a more efficient pest management. Two research groups (from India and Slovakia) studied the proteins and potential pheromones used in intra- and inter-sexual communication of this rodent in order to target compounds which would increase the probability of black rat trapping. Interestingly, their research has shown a similar trend previously described in laboratory rats and mice (Beynon & Hurst, 2004) showing that males produce higher levels of urinary proteins (Archunan et al., 2004), and the authors have detected several compounds which may represent pheromones in this species (Archunan et al., 2004; Kamalakkannan et al., 2006; Bala & Babbar, 2024). Nevertheless, these studies did not focus on scent marking as a behaviour, and therefore we are not able to determine how frequent and important is scent marking (both with urine and anogenital dragging) in this species.

The aim of this chapter is to present observations on urine and preputial gland marking of black rats (*Rattus rattus*). In several previous behavioural experiments, we noticed scent marking behaviour which sharply contrasted with what is reported for laboratory rats. To further examine this discrepancy, we conducted a specific experiment in order to enhance our knowledge of scent marking behaviour of the black rat. Here, we present all relevant data gained in various experimental designs with the aim to show the overall trends in scent marking in this species.

2 Methods

2.1 Animals

In this study, we tested wild-derived black rats (*Rattus rattus*) from a laboratory-kept population. The population was founded by wild-caught animals (14 males and 14 females) and is currently kept in our laboratory for approx. 10 years. Once the initial breeding groups were stabilized, the number of individuals never decreased below 100, and on 3 separate occasions, one wild-caught individual was added to the colony to increase its genetic variability. For more details on the origin of founders, see Žampachová et al. (2017). The animals were kept in 12:12 light-dark conditions, on hemp bedding, with standard food pellets for rats and mice (Krysík, VKS Podhledští Dvořáci) and water ad libitum. The cages also contained shredded paper and occasionally hay as nesting material, and they were equipped with flowerpots and cardboard boxes as shelters and enrichment and tree branches for climbing.

The rats were bred in large social groups of 10–30 animals. Once they reached full maturity (13 weeks or older), they were separated from their natal group, and new same-sex groups of 5–9 individuals were established. The cagemates originated from the same breeding group and the same cohort. Since we report results of experiments conducted over the last 5 years, the animals in question represent approximately the 10th–12th generation derived from the founders of the colony depending on the experiment.

2.2 Behavioural Testing

Apart from Experiment 3, all below-described experiments were originally conducted with an aim unrelated to the focus of this chapter. The full experimental designs were complex, and we will, therefore, limit ourselves here to information relevant for the purpose of this chapter. All behavioural testing was carried out in the dark phase of the day, at the peak of activity of the animals. The animals were caught into small plastic containers where they were kept at least 15 minutes to allow them a chance to calm down before the testing. The behavioural testing took place in an otherwise empty experimental room under red light of 5–10 lux. After each trial, the area was cleaned with 96% ethanol. Animal behaviour was scored from video recordings in Boris software (Friard & Gamba, 2016). The specifics of each experiment will be shortly described in the respective sections. All procedures were performed in accordance with the Czech law implementing all corresponding EU regulations and were approved by the Institutional Animal Care and Use Committee (permit nos. MSMT-33802/2021–4 and MSMT-7907/2023–4).

2.2.1 Experiment 1: Odour Discrimination. In this experiment we tested odour discrimination in the black rat (*Rattus rattus*) in an olfactory hole-board test. For this purpose, we used 35 females (housed in same-sex groups) in 15-minute trials. In this experiment we used an 80 x 80 x 80 cm open-field arena with a floor covered by a hole-board desk with 16 holes, each 2.5 cm in diameter. In the four corner holes, we presented a glass container with clean bedding (control), soiled bedding with an unfamiliar black rat odour (same sex as the tested animal) or soiled bedding with a familiar black rat odour (bedding from home cage). Each rat was presented with two control samples, one familiar odour sample and one unfamiliar odour sample in a pseudorandom arrangement; the animals were able to gain direct contact with the odour sample. We noted exploratory activity, time spent sniffing the holes/inspecting the sample (for each of the 16 holes separately), sitting and grooming. Urination, defaecation and marking with anogenital dragging in the arena were recorded as well.

2.2.2 Experiment 2: Observations from Exploration Experiments. In two previously conducted experiments intended to describe explorative behaviour and behavioural syndrome in the black rat, we noticed anogenital dragging in some individuals.

Experiment 2A followed a simple design of free exploration in an open field. In total, we tested 69 adult individuals, 35 males and 34 females. Most animals were formerly (approximately 5 months prior to the experiment) used as founders of new breeding groups (two males and two females) and thus had an opportunity to breed, although a few were housed in same-sex groups. We avoided testing visibly pregnant females or lactating mothers with pups under 2 weeks of age. The tested animal was caught in its home cage, put in a spacious plastic transport box (30 × 17.5 × 15 cm) and left at least 15 minutes unmanipulated. Afterwards, still in its transport box, it was placed in an experimental arena (80 × 80 × 80 cm), and part of the transport box's metal lid was lifted. This created a gap in the lid (11.5 × 7 cm) that allowed the rat to leave the box if it chose to. The experiment lasted for 10 minutes once the rat left its transport box (when the animal had at least three feet on the floor), and during those 10 minutes, it was allowed to return on or inside the box. Observed behaviour included various exploratory activities, sitting and grooming. Moreover, we also noted duration and frequency of anogenital dragging on the transport box (preputial gland marking on the metal lid) and on the floor. We did not observe anogenital dragging on the walls of the arena nor any other form of scent marking.

Experiment 2B followed the same design with a few minor changes. Firstly, the arena's floor was overlaid with a hole-board desk (16 holes, 4 cm in diameter) albeit the holes were empty. Secondly, each individual was tested twice with a 1-day break between the two trials. Lastly, different individuals were tested—in this experiment we tested 148 adults in total, 64 males and 84 females. All individuals were housed in same-sex groups 1–3 months prior to the experiment. In this chapter, we will once again report duration and frequency of anogenital dragging on the transport box and on the floor. No anogenital dragging on the walls was observed and any other form of scent marking was noted.

Finally, in Experiments 2C and 2D, we report incidences of urination across repeated trials of forced exploration in an open-field test and as a factor of trial order. The relevant experiments were once again conducted to investigate black rats' exploratory behaviour, and detailed methodology is available in Štolhoferová et al. (2024). In short, in Experiment 2C, 12 adult males were tested in 10-minute trials of forced exploration in an open-field task in 12 consecutive trials. In experiment 2D, different 12 adult males and 11 adult females were tested twice in 40-minute trials of a forced exploration in an open-field task. We did not observe any anogenital dragging in these experiments, but once the animal was caught from the experimental arena, we counted the number of urine pools or marks visible to the naked eye.

2.2.3 Experiment 3: Odour Discrimination and Countermarking. In this experiment, we intended to test odour discrimination and countermarking in nine males of black rats (*Rattus rattus*) using an olfactory hole-board test. The test arena was an 80 × 80 × 80 cm open-field arena with a wire-mesh back wall for climbing. The floor contained a hole board with 16 regularly spaced holes (4 × 4 grid) with a diameter of 6 cm. Under each hole, there was a metal grid covered with filter paper (with a cut-out hole with a diameter of approx. 2 cm^2 to enable easier passing of the odour through the filter paper). Three types of odour samples were used in this experiment: own odour (soiled bedding from the rat's current home cage), familiar odour (soiled bedding from a cage where animals from the same breeding group and cohort were housed) and unfamiliar odour (soiled bedding from unrelated unknown individuals).

Two odours were presented simultaneously, with three possible pairings (own vs. familiar, own vs. unfamiliar, familiar vs. unfamiliar) rotated in pseudorandom order. Odour samples were always placed below the hole board in two of the corner holes. Each experimental session lasted 23 minutes, with 5 minutes of habituation to the arena with no odour, followed by three 6-minute trials with different odour pairings. Each rat was tested four times across several days (always with 2 or 3 days between the sessions). The experimenters were not present in the room during testing and came in only to change the odour samples.

Behaviour was recorded on a camera and filter papers were inspected under UV light. Measured behavioural parameters included head dipping, sniffing to the odour samples, locomotion, climbing, rearing, grooming, urination and defaecation. The arena was cleaned with ethanol (96%) between sessions.

3 Results

3.1 Experiment 1

In this experiment we did not detect any clear urine trails, which would indicate that black rats use urine to scent mark the arena. Of the 35 females tested in this experiment, only 23 left any urine drops in the arena. We did not detect any large pools of urine or urine trails, which would suggest deliberate scent marking of the arena. In the majority of the cases, only tiny droplets (in some cases even one drop per trial) were left in the arena. Only 2 of the 35 females marked the arena with anogenital dragging. One of the females scent -marked in 67 marking bouts for the total duration of 101 seconds; the other one marked in 13 marking bouts for the total duration of 46 seconds. Both of these females were from the same same-sex group.

3.2 Experiment 2

In experiments 2A and 2B, we observed duration and frequency of anogenital dragging. All observed cases are shown in Table 1. Despite testing about 200 fully adult animals, its incidence was very low (we have seen it only in 18 animals), preventing us from employing proper statistical analyses to test, e.g. the effect of sex or animal weight. Thus, we follow with a mostly descriptive approach.

In experiment 2A, we tested 69 individuals over 9 months old (the precise age of the animals is unknown). There was an evident sex difference with only one female (3%) who performed anogenital dragging contrasted with ten males (29%) who performed this behaviour. The female was housed in a breeding group; her sister (also kept in the same breeding group) did not scent mark. Social hierarchy in black rats is not very straightforward, and it appears to be highly flexible and almost completely dependent on personalities of specific individuals in the group. Therefore, we cannot say whether the scent marking female was dominant over her sister (although it was 9 g heavier). Regarding the ten scent marking males, seven were from breeding groups and three from same-sex groups. In contrast, 11 males from breeding groups and 6 males from same-sex groups did not scent mark, leading to similar odds ratios for both types of groups (7:11 and 3:6, respectively). In five breeding groups where two males were present and tested, neither male scent marked in two of the groups, one male (the heavier one) scent marked in other two groups, and both males scent marked in the last group. All individuals scent marked on the lid of the transport box, with several spending over a minute with this behaviour, i.e. a significant portion of the 10-minute trial. Three individuals additionally scent marked on the floor, but these events were both rarer and shorter.

In experiment 2B, the anogenital dragging was even more sporadic. Nonetheless, the sex difference was still visible: only 2 out of 84 (2%) females performed anogenital dragging, while 5 out 64 (8%) males were observed engaging in this behaviour. Moreover, two males scent marked in both trials. Even though all individuals were kept in same-sex groups at least 1 month before the experiment, it might be worth noting that the seven scent marking individuals originated from four litters only. Two males were

Table 1. Observed cases of anogenital dragging in free exploration of the novel environment, experiments 2A and 2B. In experiment 2A, there were 69 individuals aged 9–15 months and of 162–260 g of body weight. In experiment 2B, there were 148 individuals (each tested twice) aged 6 months and of 121–222 g of body weight. Object scent marking refers to anogenital dragging on the transport box; floor scent marking refers to anogenital dragging along the arena walls, in the middle of the floor or around the empty hole (if present); duration is in seconds.

		Object scent marking		Floor scent marking	
	Sex	Duration	Count	Duration	Count
Experiment 2A (69 individuals)	M	19.3	5	0	0
	F	10.9	1	0	0
	M	5.3	1	0	0
	M	91.2	12	0	0
	M	94.6	17	0	0
	M	89.4	12	0	0
	M	151.9	16	7.4	1
	M	3.2	2	0	0
	M	11.1	4	6.4	2
	M	75.3	24	3.6	1
	M	17.1	5	0	0
Experiment 2B (148 individuals in 2 trials)	M	20.3	2	0	0
	M	17.4	4	1.2	1
	F	1.2	1	51.9	31
	F	0	0	1.0	1
	M	0	0	0.7	1
	M	2.9	1	0	0
	M	6.6	2	13.8	8
	M	3.4	1	0	0
	M	0	0	1.0	1

at least half-brothers, while two males and one female were at least litter-mates. Unlike in the previous experiment, the rats performed anogenital dragging on both the transport box and the floor (both along the walls and around the holes) quite equally.

In experiment 2C, we tested 12 adult males 12 times (1 test per day) and examined how the repeated testing affected the number of deposited urine marks. The number of urine marks was generally low: during most trials, the majority of individuals did not leave any visible urine mark, and the maximum per individual and trial was three urine marks which was, moreover, observed only twice. Despite the low numbers, there appears to be a clear trend of increased deposition of urine marks during later trials (Fig. 1).

In experiment 2D, 12 different males and 11 females were tested in two 40-minute trials. Despite the longer testing period, the number of deposited pools was similar to the 10-minute experiment (with the mean = 0.82, $SD = 1.43$, median = 0 and range = 0–7 marks per trial). It is worth noting that the number of urine marks tended to be slightly larger during the second trial ($\text{mean}_1 = 0.59$, $\text{mean}_2 = 1.04$, $\text{median}_1 = 0$, $\text{median}_2 = 0$), and females tended to left more urine marks than males ($\text{mean}_M = 0.49$, $\text{mean}_F = 1.18$, $\text{median}_M = 0$, $\text{median}_F = 1$).

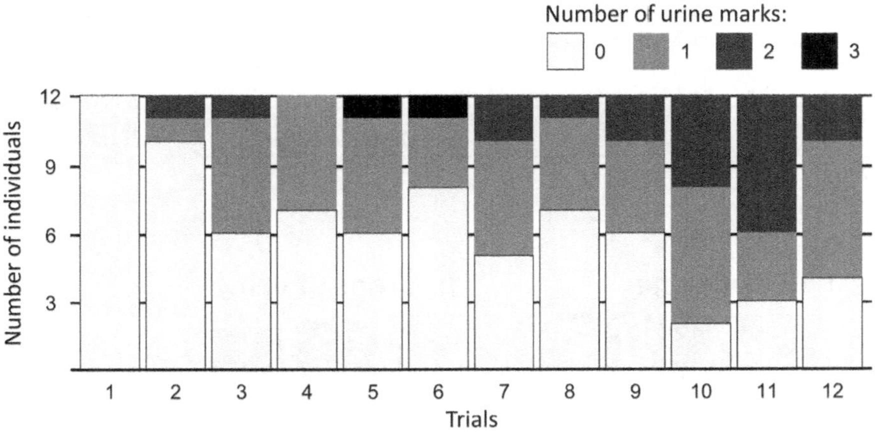

Fig. 1. The effect of trial on the number of deposited urine marks. The respective shades of grey (white, light grey, dark grey and black) indicate how many individuals left 0–3 urine marks during a particular trial of experiment 2C.

3.3 Experiment 3

In this experiment we did not detect any urine marks that would indicate scent marking or countermarking in the sense that was previously described in studies using laboratory rats or mice. While six out of nine males (67%) left a urine mark on one of the filter papers in at least one of the four trials, the actual "countermarking" frequency was very low. Of the 216 filter papers examined under a UV light for urine marks, only 10 (5%) showed any sign of urine (we detected 3 pools of urine and 7 drops of urine in total—for details see Table 2).

Nevertheless, the majority of urine scent marks were quite small and were not present in any shape indicating intentional marking (urine spray or trail of droplets). The size of the urine droplets ranged from 0.5 mm to 5 mm in length and from 0.5 to 3 mm in width; the size of urine pools was at least 1 cm in diameter (for images of the respective scent marks, see Fig. 2). The rest of the arena (the parts that were not covered by filter paper) was often marked with urine. In 29 out of the 36 trials, the animals left tiny droplets of urine somewhere else in the arena suggesting that marking of filter papers may have been random. In conclusion, we have not detected any strong indications of countermarking in black rats since (1) the rats rarely marked on the filter papers and (2) even if they did, the marks were deposited on their own scent in half of the observed cases.

4 Discussion

Scent marking in wild-derived black rats (*Rattus rattus*) appears to be quite different to that of laboratory rats, a closely related species. In this chapter, we aimed to describe overall trends in scent marking with two types of marks: urine scent marking and

Table 2. Observed cases of countermarking in black rats. The rows depict whether the mark was in the form of a urine pool or a single urine drop. The columns indicate whether the rat marked on odour sample from familiar individuals, odour sample from unfamiliar individuals or if it marked on the home-cage odour sample (indicated here as self).

	Unfamiliar odour	Familiar odour	Self-odour
Urine drop	1	3	3
Urine pool	–	1	2

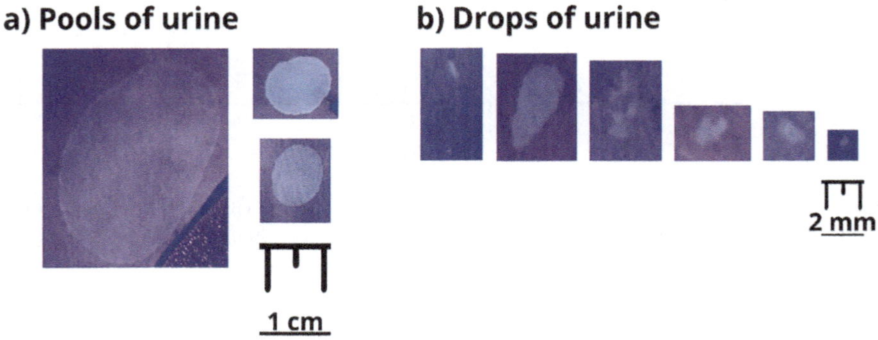

Fig. 2. Images of pools of urine and drops of urine under UV light left on the filter paper in Experiment 3.

preputial gland marking. Regarding the former, we have not detected any specific intentional urine marking or countermarking in a form previously described in other species of rodents (e.g. urine spraying, leg-lift position, urine trail marks). On the contrary, not all animals urinate in the experiments, and when some urine is left, it is usually in the form of only a few urine scent marks (see results from Experiments 2C and 3). On the other hand, we have noted scent marking in the form of dragging of the anogenital region on an object and on the floor of the arena, leaving a zigzag preputial gland scent mark (see results from Experiments 1, 2A, and 2B). Nevertheless, this form of scent marking is still generally rare albeit it seems to be relatively frequent in older males housed in breeding groups that probably have had some sexual experience.

Contrary to laboratory rats, we have not seen any directed urine marking or any specific postures related to marking with urine (as described by Brown, 1975). Moreover, the number of urine scent marks left in an arena is much smaller for black rats than what has been reported for laboratory rats and meadow voles—most animals did not leave any urine scent marks; the number of pools of urine left in the arena ranged from 0 to 7 in a 40-minute experiment. In sharp contrast with our results, Brown (1975) describes repeated marking of the laboratory rats with urine resulting in 1–7 cm^2 of the experimental cage covered in urine, and Matochik and Barfield (1991) report approximately 6–12 marks within a 10-minute experiment with laboratory rats. Ferkin (2007) reports that meadow voles (a much smaller species compared to black rats) leave around 16 drops of urine per 10-minute experiments and up to 30 drops when odours of different individuals are present. Hurst (1990) reports that even subadult and juvenile male wild mice (*Mus domesticus*) deposit urine marks frequently, even though she does not provide any details.

Based on our results, we can firmly conclude that black rats deposit a much lower number of urine marks in the experiments compared to the above-mentioned species. This begs the question whether urination in the black rats is a stress response rather than scent marking behaviour. The so-called emotionality, assessed through the amount of defaecation and urination in a novel environment, is used to characterize individual propensity to stress (Hall, 1934). Nonetheless, if urination was indeed only a stress response, we should expect habituation across repeated trials with decreased urination in later trials. However, we saw an opposite trend (Fig. 1) suggesting the urine spots were indeed scent marks. Interestingly, the striking difference between the frequency of urine marking of black rats and laboratory rats is still apparent even in those trials in which black rats marked the most frequently.

In accordance with existing literature, we have detected anogenital dragging associated with preputial gland marking although it must be stressed once again that this behaviour was observed only in a minority of the tested black rats. This behaviour was observed only in 30% of black rats, even if considering only the cohort most prone to this behaviour (see below). Previous research (Archunan et al., 2004; Kamalakkannan et al., 2006; Bala & Babbar, 2024) indicates that preputial marking might be an important intraspecific signal as it is linked with the release of both urine and species-specific proteins that might serve as pheromones in intra-specific communication. Nevertheless, even though this behaviour is present in black rats and we have detected it in both sexes, it is not very common (it was present only in 22 of the total of 400 experimental trials). Our results show that this behaviour is more common in older (potentially sexually active) males, which is in accordance with previous studies on laboratory rats (Brown, 1977, 1991). Interestingly, we have detected preputial marking in some of the females, and two of the females (one in Experiment 1 and one in Experiment 2B) exhibited extreme marking of the ground (over 100 and 50 seconds spent marking, 67 and 31 marking bouts, respectively). As some of the previous studies report a rapid increase in marking when the female is in oestrus (Birke, 1978; Birke & Sadler, 1984; Huck et al., 1985; Drickamer, 1995; Gromov, 2015; but see, e.g. Gómez-Baena et al., 2023), we suggest that it might be possible to detect oaestrus from marking behaviour of female black rats. Nevertheless, in order to test this assumption, a validation study controlling for the phase of the female cycle would be needed.

Finally, we did not detect any countermarking in Experiment 3. It is important to point out that in this experiment, the animals were not allowed direct contact with the odour sample. However, several studies show that the reaction of the animal to a scent mark may differ substantially when it has direct contact with the scent mark or urine (Humphries et al., 1999; Nevison et al., 2003; Gómez-Baena et al., 2023). The direct contact and thus the possibility to sample both volatile and nonvolatile compounds of the scent may significantly increase time spent investigating the scent mark and the probability of countermarking (Gómez-Baena et al., 2023). We suggest that this, along with the lower propensity to mark with urine in black rats, may be the reason why we did not detect any countermarking in our experiment.

An indirect (literature based) comparison of scent marking in different species is hindered by rather varied behavioural assays testing for scent marking. The experiments differ in several critical aspects—novelty of the environment, presence or absence of scent marks, presence or absence of objects that could be marked, presence or absence of another animal and other factors that might strongly influence the results of the experiment. Scent marking is tested both in home cages and in novel environments. In some cases, the animals are subjected to conspecific odours, but the specific presentation of these odours varies considerably (from soiled bedding, urine-marked objects or paper to spraying an animal with other individual's urine) and so does the origin of the odours (familiar or unfamiliar individuals, same sex or opposite sex, age, etc.). Marking should occur in all of these situations, but its form and frequency may substantially differ depending on the design of the experiment. Several studies show that scent marking may be focused on specific locations within the environment, for example, animals tend to deposit larger amounts of marks near the entrance to a burrow or the entrance to the experimental cage (Eisenberg & Kleiman, 1972; Birke & Sadler, 1984; Peden & Timberlake, 1990; Gromov, 2015) or around the periphery of the cage or territory (Richards & Stevens, 1974; Hurst, 1987; Gromov, 2015). Scent marking occurs in a novel environment when the animal readily marks it as its "new territory" (e.g. Desjardins, 1973; Richards & Stevens, 1974). Nevertheless, it may be even more pronounced when there are odour marks present in the arena (e.g. Eisenberg & Kleiman, 1972; Birke & Sadler, 1984; Ferkin, 2007; Gómez-Baena et al., 2023) although object marking specifically does not seem to be affected by presence of odours (Peden & Timberlake, 1990).

Despite this, we strongly suggest that scent marking and its forms is species-specific and as such may be strongly influenced by ecology of the respective population. Although black rats (*Rattus rattus*) and laboratory rats are closely related, their ecology is quite different. One striking difference that may play a crucial role in scent marking (both in the frequency of marks and the way of their deposition) is the fact that the black rat is a semi-arboreal species (Aplin et al., 2003; Shiels et al., 2014). We suggest that this aspect of its ecology may affect its scent marking—both the low levels of anogenital scent marking and the almost absent urine marking may reflect it. Unfortunately, we are not aware of any studies on scent marking in species with significant arboreal activity. All of the animals previously tested make use of the ground or burry into the ground. In both cases, they can deposit scent marks on the ground level, which may be used by other animals. In contrast, black rats spend only a limited amount of their time on the ground, and thus their scent marking might reflect their way of life. The probability of encountering a scent mark is much lower in the 3D environment; in addition, marking with urine on a vertical substrate may not be very effective. The semi-arboreal way of life may, therefore, be the reason why we see almost solely the marking with preputial gland via anogenital dragging in black rats.

An alternative explanation for the low frequency of scent marking might lie in another difference in black rat and laboratory brown rat ecology—their territoriality. While both species are social and capable of living in large multimale-multifemale groups, black rats are generally considered less territorial and aggressive (Barnett, 1975; Takahashi & Blanchard, 1982). While home ranges of both species are comparable in size, they use them differently: wild brown rats have a clear core home range, black rats much less so (Byers et al., 2019). One key function of scent marking—a territorial delineation—might thus be suppressed in black rats, at least for a certain amount of time after introduction to a novel environment.

To conclude, scent marking may vary substantially between different species and experimental conditions. In this chapter we present observations from several experiments with wild-derived black rats (*Rattus rattus*) and describe the overall trends in their scent marking. Black rats use scent marking with urine in a quite different way than what was previously reported for laboratory rats. We have not detected any specific behaviour or postures (e.g. leg-lift position) which would accompany urine marking, and the use of urine for marking is very limited in this species. In several experiments, we have detected scent marking by anogenital dragging, but this behaviour is also not very common in black rats. We suggest that the low propensity to use scent marks in this species may be caused by the ecology of the species as it might be much harder to use scent marking in the 3D environment efficiently, or in their lower propensity for territoriality.

References

Aplin, K. P., Chesser, T., & Have, J. T. (2003). Evolutionary biology of the genus *Rattus*: Profile of an archetypal rodent pest. *Aciar Monograph Series, 96*, 487–498.

Arakawa, H., Blanchard, D. C., Arakawa, K., Dunlap, C., & Blanchard, R. J. (2008). Scent marking behavior as an odorant communication in mice. *Neuroscience & Biobehavioral Reviews, 32*(7), 1236–1248. https://doi.org/10.1016/j.neubiorev.2008.05.012

Archunan, G., Kamalakkannan, S., Achiraman, S., & Rajkumar, R. (2004). Identification of glandular (preputial and clitoral) proteins in house rat (*Rattus rattus*) involved in pheromonal communication. *Indian Journal of Experimental Biology, 42*, 1032–1035.

Bakshi, S. (2010). Scent gland and its behavioral function in small mammals – a review. *Agricultural Reviews, 31*(3), 157–171.

Bala, B., & Babbar, B. K. (2024). Behavioural responses and characterization of preputial gland extract of black rat, Rattus rattus. *Biologia, 79*(7), 2079–2093. https://doi.org/10.1007/s11756-024-01685-7

Barnett, S. A. (1975). *The rat: A study in behavior*. Transaction Publishers.

Becker, E. A., Castelli, F. R., Yohn, C. N., Spencer, L., & Marler, C. A. (2018). Species differences in urine scent-marking and counter-marking in *Peromyscus*. *Behavioural Processes, 146*, 1–9. https://doi.org/10.1016/j.beproc.2017.10.011

Beynon, R. J., & Hurst, J. L. (2004). Urinary proteins and the modulation of chemical scents in mice and rats. *Peptides, 25*(9), 1553–1563. https://doi.org/10.1016/j.peptides.2003.12.025

Birke, L. I. (1978). Scent-marking and the oestrous cycle of the female rat. *Animal Behaviour, 26*, 1165–1166. https://doi.org/10.1016/0003-3472(78)90106-9

Birke, L. I. A., & Sadler, D. (1984). Scent-marking behaviour in response to conspecific odours by the rat, Rattus norvegicus. *Animal Behaviour, 32*(2), 493–500. https://doi.org/10.1016/S0003-3472(84)80286-9

Brown, R. E. (1975). Object-directed urine-marking by male rats (*Rattus norvegicus*). *Behavioral Biology, 15*(2), 251–254. https://doi.org/10.1016/S0091-6773(75)91683-1

Brown, R. E. (1977). Odor preference and urine-marking scales in male and female rats: Effects of gonadectomy and sexual experience on responses to conspecific odors. *Journal of Comparative and Physiological Psychology, 91*(5), 1190–1206. https://doi.org/10.1037/h0077394

Brown, R. E. (1985). Effects of social isolation in adulthood on odor preferences and urine-marking in male rats. *Behavioral and Neural Biology, 44*(1), 139–143. https://doi.org/10.1016/S0163-1047(85)91301-9

Brown, R. E. (1991). Effects of rearing condition, gender, and sexual experience on odor preferences and urine marking in long-Evans rats. *Animal Learning & Behavior, 19*(1), 18–28. https://doi.org/10.3758/BF03197856

Brown, R. E. (1992). Responses of dominant and subordinate male rats to the odors of male and female conspecifics. *Aggressive Behavior, 18*(2), 129–138. https://doi.org/10.1002/1098-2337(1992)18:2<129::AID-AB2480180206>3.0.CO;2-U

Byers, K. A., Lee, M. J., Patrick, D. M., & Himsworth, C. G. (2019). Rats about town: A systematic review of rat movement in urban ecosystems. *Frontiers in Ecology and Evolution, 7*, 13. https://doi.org/10.3389/fevo.2019.00013

Desjardins, C., Maruniak, J. A., & Bronson, F. H. (1973). Social rank in house mice: Differentiation revealed by ultraviolet visualization of urinary marking patterns. *Science, 182*(4115), 939–941. https://doi.org/10.1126/science.182.4115.939

Docherty, S. J. (2009). *Scent mark communication and competitive behaviour of the harvest mouse (Micromys minutus)*. Doctoral dissertation, University of Liverpool.

Drickamer, L. C. (1995). Rates of urine excretion by house mouse (*Mus domesticus*): Differences by age, sex, social status, and reproductive condition. *Journal of Chemical Ecology, 21*, 1481–1493. https://doi.org/10.1007/BF02035147

Eisenberg, J. F., & Kleiman, D. G. (1972). Olfactory communication in mammals. *Annual Review of Ecology and Systematics, 3*, 1–32. https://www.jstor.org/stable/2096840

Ferkin, M. H. (2007). Effects of previous interactions and sex on over-marking in meadow voles. *Behaviour, 144*(10), 1297–1313. https://doi.org/10.1163/156853907781890913

Ferkin, M. H. (2010). Age affects over-marking of opposite-sex scent marks in meadow voles, Microtus pennsylvanicus. *Ethology, 116*(1), 24–31. https://doi.org/10.1111/j.1439-0310.2009.01711.x

Ferkin, M. H. (2015). The response of rodents to scent marks: Four broad hypotheses. *Hormones and Behavior, 68*, 43–52. https://doi.org/10.1016/j.yhbeh.2014.10.002

Ferkin, M. H. (2018). Odor communication and mate choice in rodents. *Biology, 7*(1), 13. https://doi.org/10.3390/biology7010013

Ferkin, M. H. (2019). Scent marks of rodents can provide information to conspecifics. *Animal Cognition, 22*(3), 445–452. https://doi.org/10.1007/s10071-019-01250-9

Ferkin, M. H., & Johnston, R. E. (1995). Meadow voles, *Microtus pennsylvanicus*, use multiple sources of scent for sex recognition. *Animal Behaviour, 49*(1), 37–44. https://doi.org/10.1016/0003-3472(95)80151-0

Ferris, C. F., Axelson, J. F., Shinto, L. H., & Albers, H. E. (1987). Scent marking and the maintenance of dominant/subordinate status in male golden hamsters. *Physiology & Behavior, 40*(5), 661–664. https://doi.org/10.1016/0031-9384(87)90114-4

Freeman, A. R., & Ophir, A. G. (2018). Scent-marking behavior of the southern giant pouched rat (*Cricetomys ansorgei*). *Journal of Mammalogy, 99*(6), 1430–1435. https://doi.org/10.1093/jmammal/gyy137

Freeman, A. R., & Ophir, A. G. (2021). Sex differences in social odor discrimination by southern giant pouched rats (*Cricetomys ansorgei*). *Ethology, 127*(12), 1019–1032. https://doi.org/10.1111/eth.13223

Freeman, A. R., Ophir, A. G., & Sheehan, M. J. (2020). The giant pouched rat (*Cricetomys ansorgei*) olfactory receptor repertoire. *PLoS One, 15*(4), e0221981. https://doi.org/10.1371/journal.pone.0221981

Friard, O., & Gamba, M. (2016). BORIS: A free, versatile open-source event-logging software for video/audio coding and live observations. *Methods in Ecology and Evolution, 7*(11), 1325–1330. https://doi.org/10.1111/2041-210X.12584

Gómez-Baena, G., Armstrong, S. D., Phelan, M. M., Hurst, J. L., & Beynon, R. J. (2014). The major urinary protein system in the rat. *Biochemical Society Transactions, 42*(4), 886–892. https://doi.org/10.1042/BST20140083

Gómez-Baena, G., Armstrong, S. D., Halstead, J. O., Prescott, M., Roberts, S. A., McLean, L., Mudge, J. M., Hurst, J. L., & Beynon, R. J. (2019). Molecular complexity of the major urinary protein system of the Norway rat, Rattus norvegicus. *Scientific Reports, 9*(1), 10757. https://doi.org/10.1038/s41598-019-46950-x

Gómez-Baena, G., Pounder, K. C., Halstead, J. O., Roberts, S. A., Davidson, A. J., Prescott, M., Beynon, R. J., & Hurst, J. L. (2023). Unraveling female communication through scent marks in the Norway rat. *Proceedings of the National Academy of Sciences of the United States of America, 120*(25), e2300794120. https://doi.org/10.1073/pnas.2300794120

Gosling, L. M. (1982). A reassessment of the function of scent marking in territories. *Zeitschrift für Tierpsychologie, 60*(2), 89–118. https://doi.org/10.1111/j.1439-0310.1982.tb00492.x

Gosling, L. M., & Roberts, S. C. (2001a). Scent-marking by male mammals: Cheat-proof signals to competitors and mates. In P. J. B. Slater, J. S. Rosenblatt, C. T. Snowdon, & T. J. Roper (Eds.), *Advances in the study of behavior 30* (pp. 169–217). Elsevier. https://doi.org/10.1016/S0065-3454(01)80007-3

Gosling, L. M., & Roberts, S. C. (2001b). Testing ideas about the function of scent marks in territories from spatial patterns. *Animal Behaviour, 62*(3), F7–F10. https://doi.org/10.1006/anbe.2001.1802

Gromov, V. S. (2015). Scent marking in gerbils and its possible functions. *Russian Journal of Theriology, 14*(1), 113–126. https://doi.org/10.15298/rusjtheriol.14.1.06

Hall, C. S. (1934). Drive and emotionality: Factors associated with adjustment in the rat. *Journal of Comparative Psychology, 17*(1), 89–108. https://doi.org/10.1037/h0073676

Humphries, R. E., Robertson, D. H. L., Beynon, R. J., & Hurst, J. L. (1999). Unravelling the chemical basis of competitive scent marking in house mice. *Animal Behaviour, 58*(6), 1177–1190. https://doi.org/10.1006/anbe.1999.1252

Hurst, J. L. (1987). The functions of urine marking in a free-living population of house mice, *Mus domesticus* Rutty. *Animal Behaviour, 35*(5), 1433–1442. https://doi.org/10.1016/S0003-3472(87)80016-7

Hurst, J. L. (1990). Urine marking in populations of wild house mice *Mus domesticus* Rutty. I. Communication between males. *Animal Behaviour, 40*(2), 209–222. https://doi.org/10.1016/S0003-3472(05)80916-9

Hurst, J. L. (2009). Female recognition and assessment of males through scent. *Behavioural Brain Research, 200*(2), 295–303. https://doi.org/10.1016/j.bbr.2008.12.020

Hurst, J. L., & Beynon, R. J. (2004). Scent wars: The chemobiology of competitive signalling in mice. *BioEssays, 26*(12), 1288–1298. https://doi.org/10.1002/bies.20147

Hurst, J. L., Beynon, R. J., Humphries, R. E., Malone, N., Nevison, C. M., Payne, C. E., Robertson, D. H. L., & Veggerby, C. (2001). Information in scent signals of competitive social status: The interface between behaviour and chemistry. In A. Marchlewska-Koj, J. J. Lepri, & D. Müller-Schwarze (Eds.), *Chemical signals in vertebrates 9* (pp. 43–52). Springer. https://doi.org/10.1007/978-1-4615-0671-3_6

Hurst, J. L., Thom, M. D., Nevison, C. M., Humphries, R. E., & Beynon, R. J. (2005). MHC odours are not required or sufficient for recognition of individual scent owners. *Proceedings of the Royal Society B: Biological Sciences, 272*(1564), 715–724. https://doi.org/10.1098/rspb.2004.3004

Johnston, R. E. (2003). Chemical communication in rodents: From pheromones to individual recognition. *Journal of Mammalogy, 84*(4), 1141–1162. https://doi.org/10.1644/BLe-010

Johnston, R. E. (2008). Individual odors and social communication. In H. J. Brockmann, T. J. Roper, M. Naguib, K. E. Wynne-Edwards, C. Barnard, & J. C. Mitani (Eds.), *Advances in the study of behavior* (Vol. 38, pp. 439–505). Elsevier. https://doi.org/10.1016/S0065-3454(08)00009-0

Johnston, R. E., & Lee, N. A. (1976). Persistence of the odor deposited by two functionally distinct scent marking behaviors of golden hamsters. *Behavioral Biology, 16*(2), 199–210. https://doi.org/10.1016/S0091-6773(76)91310-9

Johnston, R. E., Sorokin, E. S., & Ferkin, M. H. (1997). Scent counter-marking by male meadow voles: Females prefer the top-scent male. *Ethology, 103*(6), 443–453. https://doi.org/10.1111/j.1439-0310.1997.tb00159.x

Kamalakkannan, S., Achiraman, S., Rajkumar, R., Ramesh Kumar, K., & Archunan, G. (2006). Identification of sex-associated protein in the preputial gland of house rat (a new insight in rodent pest management). *Acta Physiologica Hungarica, 93*(2–3), 145–152. https://doi.org/10.1556/APhysiol.93.2006.2-3.5

Macdonald, D. W., Herrera, E. A., Taber, A. B., Moreira, J. R., Wolff, J. O., & Sherman, P. W. (2007). Social organization and resource use in capybaras and maras. In J. O. Wolff & P. W. Sherman (Eds.), *Rodent societies: An ecological and evolutionary perspective* (pp. 393–402). University of Chicago Press. https://doi.org/10.7208/9780226905389-035

Matochik, J. A., & Barfield, R. J. (1991). Hormonal control of precopulatory sebaceous scent marking and ultrasonic mating vocalizations in male rats. *Hormones and Behavior, 25*(4), 445–460. https://doi.org/10.1016/0018-506X(91)90013-8

Nevison, C. M., Armstrong, S., Beynon, R. J., Humphries, R. E., & Hurst, J. L. (2003). The ownership signature in mouse scent marks is involatile. *Proceedings of the Royal Society of London. Series B: Biological Sciences, 270*(1527), 1957–1963. https://doi.org/10.1098/rspb.2003.2452

Peden, B. F., & Timberlake, W. (1990). Environmental influences on flank marking and urine marking by female and male rats (*Rattus norvegicus*). *Journal of Comparative Psychology, 104*(2), 122. https://doi.org/10.1037/0735-7036.104.2.122

Richards, D. B., & Stevens, D. A. (1974). Evidence for marking with urine by rats. *Behavioral Biology, 12*(4), 517–523. https://doi.org/10.1016/S0091-6773(74)92371-2

Roberts, S. C. (2007). Scent marking. In J. O. Wolff & P. W. Sherman (Eds.), *Rodent societies: An ecological and evolutionary perspective* (pp. 255–266). University of Chicago Press.

Roberts, S. C., & Gosling, L. M. (2001). The economic consequences of advertising scent mark location on territories. In A. Marchlewska-Koj, J. J. Lepri, & D. Müller-Schwarze (Eds.), *Chemical signals in vertebrates 9* (pp. 11–17). Springer. https://doi.org/10.1007/978-1-4615-0671-3_2

Roberts, S. C., Gosling, L. M., Thornton, E. A., & McClung, J. (2001). Scent-marking by male mice under the risk of predation. *Behavioral Ecology, 12*(6), 698–705.

Roberts, S. C., & Gosling, L. M. (2004). Manipulation of olfactory signaling and mate choice for conservation breeding: A case study of harvest mice. *Conservation Biology, 18*(2), 548–556. https://doi.org/10.1111/j.1523-1739.2004.00514.x

Shiels, A. B., Pitt, W. C., Sugihara, R. T., & Witmer, G. W. (2014). Biology and impacts of Pacific Island invasive species. 11. *Rattus rattus*, the black rat (Rodentia: Muridae). *Pacific Science, 68*(2), 145–184. https://doi.org/10.2984/68.2.1

Steiner, A. L. (1974). Body-rubbing, marking, and other scent-related behavior in some ground squirrels (Sciuridae), a descriptive study. *Canadian Journal of Zoology, 52*(7), 889–906. https://doi.org/10.1139/z74-120

Štolhoferová, I., Rudolfová, V., Skalíková, H., Vobrubová, B., & Frynta, D. (2024). How to explore a new environment: Exploratory tactics of the black rat (*Rattus rattus*). *Current Zoology, 70*(3), 371–382. https://doi.org/10.1093/cz/zoae015

Swanson, H. H. (1980). Social and hormonal influences on scent marking in the mongolian gerbil. *Physiology & Behavior, 24*(5), 839–842. https://doi.org/10.1016/0031-9384(80)90137-7

Takahashi, L. K., & Blanchard, R. J. (1982). Attack and defense in laboratory and wild Norway and black rats. *Behavioural Processes, 7*(1), 49–62. https://doi.org/10.1016/0376-6357(82)90052-3

Thiessen, D. D., Blum, S. L., & Lindzey, G. (1970). A scent marking response associated with the ventral sebaceous gland of the Mongolian gerbil (*Meriones unguiculatus*). *Animal Behaviour, 18*, 26–30. https://doi.org/10.1016/0003-3472(70)90065-5

Thiessen, D. D., Lindzey, G., Blum, S. L., & Wallace, P. (1971). Social interactions and scent marking in the Mongolian gerbil (*Meriones unguiculatus*). *Animal Behaviour, 19*(3), 505–513. https://doi.org/10.1016/S0003-3472(71)80105-7

Thom, M. D., & Hurst, J. L. (2004). Individual recognition by scent. *Annales Zoologici Fennici, 41*(6), 765–787. http://www.jstor.org/stable/23736142

Thomas, S. A., & Wolff, J. O. (2002). Scent marking in voles: A reassessment of over marking, counter marking, and self-advertisement. *Ethology, 108*(1), 51–62. https://doi.org/10.1046/j.1439-0310.2002.00753.x

Vettorazzi, A., Wait, R., Nagy, J., Monreal, J. I., & Mantle, P. (2013). Changes in male rat urinary protein profile during puberty: A pilot study. *BMC Research Notes, 6*(1), 232. https://doi.org/10.1186/1756-0500-6-232

William Huck, U., Lisk, R. D., & Gore, A. C. (1985). Scent marking and mate choice in the golden hamster. *Physiology & Behavior, 35*(3), 389–393. https://doi.org/10.1016/0031-9384(85)90314-2

Yoshimura, H. (1981). Behavioral characteristics of scent marking behavior in the Mongolian gerbil (*Meriones unguiculatus*). *Experimental Animals, 30*(2), 107–112. https://doi.org/10.1538/expanim1978.30.2_107

Žampachová, B., Kaftanová, B., Šimánková, H., Landová, E., & Frynta, D. (2017). Consistent individual differences in standard exploration tasks in the black rat (*Rattus rattus*). *Journal of Comparative Psychology, 131*(2), 150–162. https://doi.org/10.1037/com0000070

Chemical Communication Using the Urinary Compounds in the Domestic Cat

Masao Miyazaki[1](✉)

[1] Laboratory of Biomolecular Science, Faculty of Agriculture, Iwate University, Iwate, Japan
mmasao@iwate-u.ac.jp

Abstract. Many mammals use scent marking to establish territory ownership among conspecifics. Domestic cats (*Felis silvestris catus*) mark their territory by spraying urine and depositing feces along the borders of their home range. When other cats encounter these excretions, they approach and sniff them, suggesting that feline excretions convey information about the scent owner's identity and territory. However, volatile chemical profiles exhibit complex, individual-specific variations and are subject to instability and daily fluctuations, which may limit their reliability as long-term scent signals. In this chapter, I review the chemical basis of intraspecies communication among cats. I begin by discussing the biological significance of proteinuria in cats, followed by an exploration of chemical communication through urine, feces, and anal sac secretions. Additionally, I examine the implications of chronic kidney disease (CKD) on urinary scent marking, demonstrating how renal dysfunction disrupts the metabolism of cat-specific compounds, leading to a reduction in the characteristic feline urinary odor. Finally, I discuss the responsiveness of cats to urinary VOCs and their role in social interactions. These findings contribute to a broader understanding of chemical communication in Felidae, shedding light on the ecological and evolutionary significance of scent-marking behavior in cats.

Keywords: Domestic cats · Carboxylesterase · Chemical communication · Felinine · Olfaction

1 The Finding and Function of Cauxin, a Major Urinary Protein of the Domestic Cat

Most mammals normally excrete minimal amounts of proteins in their urine to reduce protein loss. However, this principle does not apply to domestic cats (*Felis silvestris catus*) (Miyazaki et al., 2003), as is the case in mice, which exhibit physiological proteinuria due to high level excretions of 19 kDa lipocalin family proteins termed major urinary proteins (MUPs) (Hurst & Beynon, 2004). SDS-PAGE analysis of cat urine revealed that a 70 kDa protein is a major protein component in cats (Fig. 1a). Its excretion level is higher in intact males than in castrated males and females. This protein was found in high concentrations exclusively in the urine of certain Felidae species (e.g.,

L. M. Schulte (ed.), *Chemical Signals in Vertebrates 16*,
https://doi.org/10.1007/978-3-032-30752-1_8

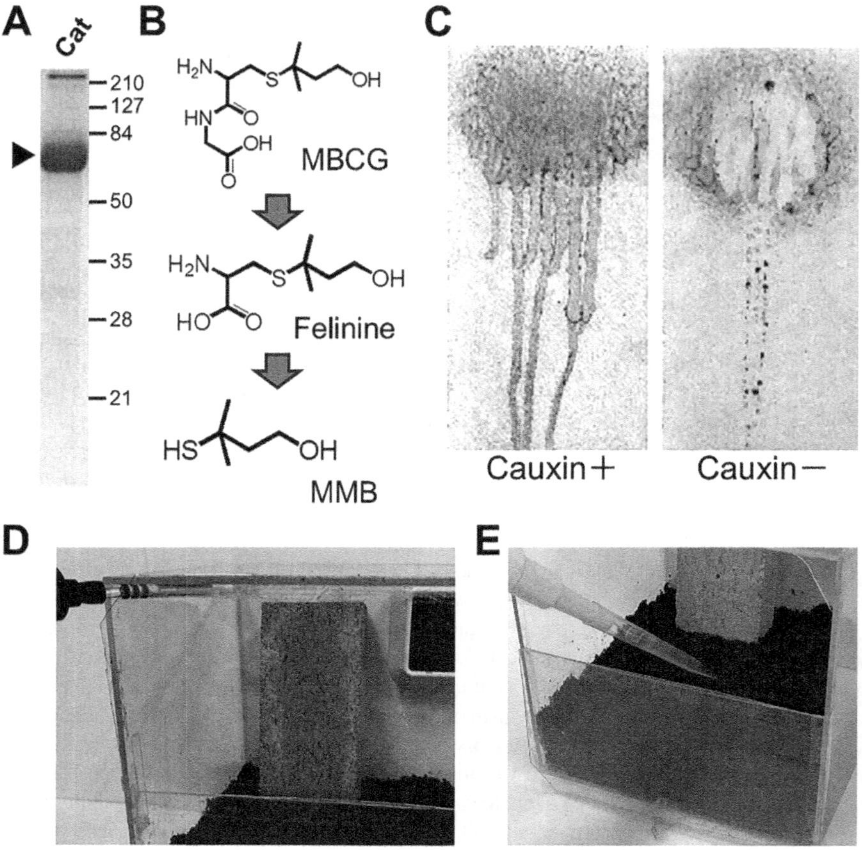

Fig. 1. Characterization and functions of a major urinary protein in the domestic cat. (**a**) A urine sample (10 μL) from a male cat was analyzed by SDS-PAGE under nonreducing conditions, and the gel was stained with Coomassie Brilliant Blue solution. An arrow indicates a band of cauxin. (**b**) Cauxin hydrolyzes 3-methylbutanol cysteinyl-glycine (MBCG) and produces felinine which is further degraded into volatile 3-mercapto-3-methyl-1-butanol (MMB). (**c**) Surface tension comparison between urine samples with and without urinary proteins. Vertically standing glass plates were sprayed with pooled excreted urine or its deproteinized counterpart, followed by staining with ninhydrin solution. (**d** and **e**) To mimic natural environments where urine is either sprayed or excreted, two types of miniature gardens were prepared: in one, 5 ml of pooled feline urine was artificially sprayed onto a block (**d**; spray garden), and in the other, it was poured directly into the soil (**e**; excreted garden).

species within the genus *Lynx*, such as bobcats and Eurasian lynxes) but was absent in other mammals, including humans and dogs. Western blotting and immunohistochemistry using antibodies against the purified protein revealed its tissue-specific expression in the proximal straight tubular epithelial cells of the kidneys and its abundant secretion into the urine of cats. Molecular cloning of its cDNA identified the protein as a novel member of the mammalian carboxylesterase family in the early twenty-first century.

Based on these findings, it was named cauxin (carboxylesterase-like urinary excreted protein).

Given that cauxin excretion is species-, sex-, and age-dependent (Miyazaki et al., 2006a) and that carboxylesterases hydrolyze a broad range of substrates with ester, thioester, and amide bonds, its physiological function was investigated, particularly its role in metabolizing low-molecular-weight urinary compounds, whose excretion correlates with cauxin levels. Screening of candidate substrates and products revealed a strong positive correlation between cauxin and felinine (2-amino-7-hydroxy-5,5-dimethyl-4-thiaheptanoic acid; Fig. 1b, upper part) excretion in mature cats (Miyazaki et al., 2006b). Felinine, a sulfur-containing amino acid unique to cats, was first identified in the 1950s (Hendriks et al., 1995). In vitro enzyme assays demonstrated that cauxin hydrolyzes the peptide bond of a dipeptide, 3-methylbutanol-cysteinylglycine (Fig. 1b; MBCG), producing felinine and glycine in cat urine. Furthermore, despite the presence of 3-methylbutanol-cysteinylglycine, felinine remains below the detection limit in immature cats younger than 3 months. Based on these findings, we concluded that cauxin plays an essential role in the felinine production in cats (Miyazaki et al., 2006b).

Felinine is further metabolized into cat-specific, sulfur-containing volatile organic compounds (VOCs) in the urine. The felinine-derived volatile compounds detected in cat urine include 3-mercapto-3-methyl-1-butanol (Fig. 1b; MMB), 3-mercapto-3-methylbutyl formate, 3-methyl-3-methylthio-1-butanol, and 3-methyl-3-(2-methyldisulfanyl)-1-butanol. Especially, mercaptans, 3-mercapto-3-methyl-1-butanol and 3-mercapto-3-methylbutyl formate, exhibit a characteristic catty odor perceptible to human olfaction. The domestic cat is a territorial animal that marks its environment through scent marking (Bradshaw & Cameron-Beaumont, 2000). Urinary VOCs serve as chemical signals for both intraspecies and interspecies communication in cats (Case, 2003). Cats use urine marking as a primary means of chemical communication, conveying identity, territory boundaries, and reproductive status. Compounds detected exclusively in cats and showing sex-dependent variation are strong candidates for chemical signals involved in conspecific and sex recognition. Since sulfur-containing VOCs contribute to species-, sex-, and age-specific urinary odor, they are likely to function as chemical signals for conspecific recognition or reproductive purposes in cats. These findings led me to further investigate the role of chemical communication in cats and other felids.

Genomic analyses revealed that the cauxin gene has undergone accelerated evolution in small felids compared to large *Panthera* species (Li et al., 2011). This rapid evolution is characterized by strong positive selection, with the dN/dS ratio exceeding 1 in small felids. Notably, several positively selected amino acid sites are located in key protein regions, including disulfide bond formation sites and helix structures, which are crucial for protein function. In contrast to small felids, big cats (*Panthera* spp.) exhibit significantly lower urinary cauxin and felinine levels (Hendriks et al., 1995; McLean et al., 2007). Cauxin is predominantly expressed in cat kidneys but has also been detected in the testis in various mammals (Ecroyd et al., 2006; Moura et al., 2010; Miyazaki et al., 2011; Ru et al., 2015), suggesting an additional role in male reproductive fitness and behavior. These findings indicate that the evolution of CES7 in small felids may have contributed to their unique urinary communication system, influencing territorial marking, individual recognition, and mating behaviors. This evolutionary adaptation could have played a key role in their ecological and reproductive strategies.

2 Urinary Cauxin Increases Adhesion to Vertical Objects

Urine spraying is a common scent-marking behavior in which cats spray urine against vertical surfaces while raising their tails. Males spray more frequently than females, with increased activity during the breeding season, which occurs from late winter to early spring (Feldman, 1994). Compared to urine excreted on the ground, sprayed urine has a stronger odor (Linda, 2003). Studies have shown that cats spend more time sniffing sprayed urine from unfamiliar males than urine eliminated by squatting (Natoli, 1985). However, the chemical composition of sprayed urine remained largely unknown. Our recent studies revealed that cauxin plays a key role in producing the strong odor of sprayed urine in cats.

Chemical comparisons of VOCs emitted from sprayed urine, normal urine, and bladder urine revealed a high degree of similarity in their profiles within the same individuals (Uenoyama et al., 2024). Behavioral analyses further demonstrated that cats perceived sprayed urine and residual bladder urine after spraying as having a similar odor, whereas the urine of another cat is perceived as distinctly different. These findings indicate that sprayed urine originates directly from bladder urine, without supplementation from other secretory glands, such as anal glands.

Given the high similarity in volatile chemical profiles between sprayed and non-sprayed urine, we sought to explore alternative explanations for the strong odor of sprayed urine. During our analyses, we observed that cat urine adhered easily to the inner surface of plastic syringes when transferred into glass vials for urinary volatile analyses. This observation led us to investigate the underlying mechanisms. In general, a liquid's wettability on a solid surface increases as surface tension decreases. Based on this principle, we hypothesized that the high cauxin concentration in cat urine may reduce the surface tension, thereby enhancing the emission of urinary VOCs by increasing the spread of urine over a large vertical surface area.

As anticipated, surface tension of Milli-Q water (average 70.3 mN/m) decreased as cauxin concentration increased (1 mg/ml cauxin concentration: average 60.9 mN/m). At the same protein concentration, cauxin solution exhibited higher wettability than the control solution containing albumin (1 mg/ml albumin concentration: average 65.0 mN/m), the major protein in mammalian blood. When comparing the surface tension of cat urine with and without cauxin, urine containing cauxin showed significantly lower surface tension than deproteinized urine. Interestingly, cauxin-containing urine also demonstrated greater adhesion to vertically positioned glass plates compared to deproteinized urine (Fig. 1c).

Additionally, odorants responsible for the distinctive catty smell were detectable in an artificial miniature garden designed to mimic natural environments, where a block sprayed with tomcat urine was placed (Fig. 1d). In contrast, these odorants were below the detection limit in another garden where the same urine was poured directly into sandy soil and then covered (Fig. 1e). The difference in environmental odors between the two gardens, despite using the same urine sample, can be explained by most of the urinary volatile chemicals being trapped in the porous structure of the sandy soil. This phenomenon did not occur in the urine adhering to the surface of the block. Furthermore, liquid droplets of sprayed urine dried easily on the surface of the block, leading to greater emissions of volatile chemicals from the scent mark compared to normal urine.

In conclusion, feline sprayed urine originates solely from the bladder, without contributions from other secretions. However, despite this exclusive source, sprayed urine emits a strong, pungent odor due to its enhanced adhesion to vertical surfaces.

The urinary protein cauxin plays a crucial role in scent marking by not only producing cat-specific odorants but also enhancing the emission of urinary volatile chemicals through its ability to increase the wettability of sprayed urine.

3 Aging of Urinary Volatile Compounds and Their Olfactory Discrimination by Cats

The chemical composition of cat urine consists of a complex mixture of volatile and nonvolatile metabolites. While previous studies have focused on identifying specific VOCs, such as 3-mercapto-3-methyl-1-butanol as described above, little is known about the broader volatile chemical profiles of cat urine, their temporal changes, and how individual differences are maintained. Comprehensive gas chromatography-mass spectrometry (GC × GC/MS) and behavioral assays revealed olfactory abilities to discriminate urine odors in cats (Suzuki et al., 2019).

The volatile profiles of cat urine were analyzed using GC × GC/MS with a VOC preconcentrator, allowing the detection of hundreds of VOCs in both fresh and aged urine samples (Fig. 2a). The analysis revealed that urine VOCs change dynamically over time, particularly within the first few hours (Fig. 2b). In fresh urine, over 900 VOC peaks were detected, but this number decreased significantly to around 800 VOC peaks within 24 h. Interestingly, some compounds temporarily increased between 3 and 9 h after urination. Fresh urine at 0 h was distinctly separated from the others, indicating significant chemical changes during the early aging phase. These findings indicate that cat urine undergoes distinct, time-dependent changes in its VOC profiles. Fresh urine contains a greater number of VOCs, many of which gradually decrease, resulting in a more uniform VOC profile after 24 h.

Fresh urine from different individual cats exhibited distinct volatile profiles. However, after aging for 24 h, individual differences in volatile composition diminished. This dynamic nature of cat urine VOCs may play a role in chemical communication and scent-marking behavior among felines.

To investigate cats' olfactory discrimination ability toward urinary odor, olfactory habituation-dishabituation tests were conducted. Cats successfully distinguished between fresh and aged urine samples from the same individual and could differentiate fresh urine from different individuals. However, after 24 h of aging, some cats failed to distinguish between urine from different individuals, suggesting that the VOCs responsible for individual recognition may be short-lived.

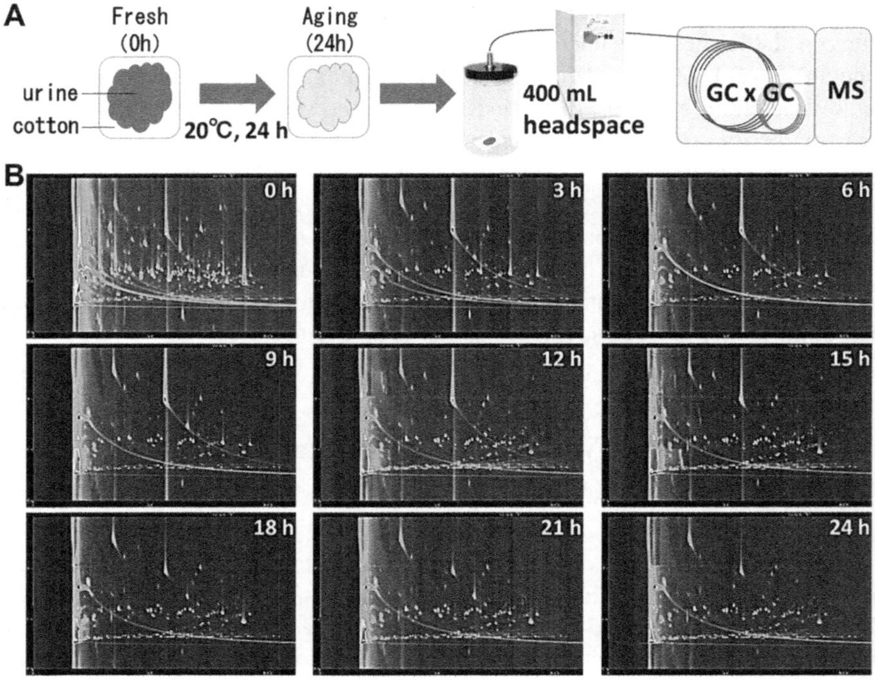

Fig. 2. Temporal changes and individual variation in volatile compounds from cat urine. (**a**) Schematic illustration of the analytical method for cat urine odor. A piece of absorbent cotton soaked with urine was placed in a 1 L glass bottle, and its headspace was analyzed using a two-dimensional gas chromatography-mass spectrometer equipped with a large-capacity automatic gas sampler and concentrator. (**b**) Contour plot of total ion chromatograms of volatile compounds detected from urine of a male cat air-dried for 0–24 h. Each spot represents an individual compound.

These findings provide evidence that cats rely on urine scent for both temporal and individual recognition. The dynamic changes in urine volatiles may help cats determine when another cat has marked a territory, facilitating conflict avoidance. Behavioral tests confirm that cats can distinguish between fresh and aged urine, as well as between individual scent marks, but their ability to recognize individual identity diminishes over time.

The study highlights the complexity of feline chemical communication and suggests that some VOCs may function as individual recognition signals. However, unlike certain mammals, cats do not appear to use stable protein-based markers for long-term scent recognition, such as major urinary proteins (MUPs) in rodents (Hurst et al., 2001). Instead, their recognition system relies on rapidly changing volatile cues. In conclusion, cats can distinguish between fresh and aged urine and recognize individual VOC differences, but only within a limited timeframe.

4 Urinary Odor Regulates Feline Scent-Marking Behavior

Next, I describe how cats respond to urinary VOCs from conspecifics. First, we separated urinary components into an organic solvent phase, which contains most VOCs, and an aqueous phase, which contains water-soluble compounds (Miyazaki et al., 2017a). The organic solvent phase included volatile sulfur-containing chemicals, such as 3-mercapto-3-methyl-1-butanol, while the aqueous phase contained hydrophilic compounds, including urea, a precursor to ammonia.

When exposed to these urinary extracts, cats exhibited differential behavioral responses to the two phases. The organic solvent phase elicited a significantly stronger response, characterized by prolonged sniffing and a higher frequency of the flehmen response—a behavior in which cats lift their heads and curl their upper lips to enhance pheromone detection via the vomeronasal organ. In contrast, the aqueous phase did not elicit a significant flehmen response, suggesting that VOCs are the key bioactive components in cat scent communication.

In outdoor field experiments, when applied to glass plates, the organic solvent phase effectively attracted cats, leading them to spend more time sniffing the treated areas. In contrast, decayed urine, which had lost its lipid-based volatile compounds and emitted a strong ammonia odor, was less attractive to cats. This finding underscores the crucial role of urinary VOCs in scent communication and attraction in cats.

Interestingly, cats did not urinate or defecate in areas treated with urinary lipids, possibly because they perceived these areas as already scent-marked by another cat. This study highlights the significant role of urinary lipids in feline chemical communication. The organic solvent extract, which contains lipophilic volatile compounds, strongly influences cat behavior by inducing attraction and investigation, triggering the flehmen response, and potentially inhibiting urine and fecal marking.

These findings suggest that urinary VOCs can be utilized to manage free-roaming cat populations and support wild felid conservation. The urinary extract effectively attracts cats to optimal locations for video camera trapping and can also be used to deter urination and defecation in gardens and public spaces. These applications provide valuable tools for mitigating issues caused by free-roaming cats and may also contribute to the conservation of small, endangered felids.

5 Urinary Odor in Cats with Kidney Disease

Chronic kidney disease (CKD) is a common condition in domestic cats, particularly in older individuals, affecting approximately 27% of cats over 10 years old (Chakrabarti et al., 2012; Elliott & Barber, 1998). As CKD progresses, cats exhibit symptoms such as weight loss, poor appetite, increased water intake, and polyuria. One notable but underexplored phenomenon is the weakening of urinary odor in cats with renal disease.

Normal cat urine contains sulfur-containing volatile organic compounds (VOCs) as mentioned above. These compounds originate from felinine, an amino acid metabolized

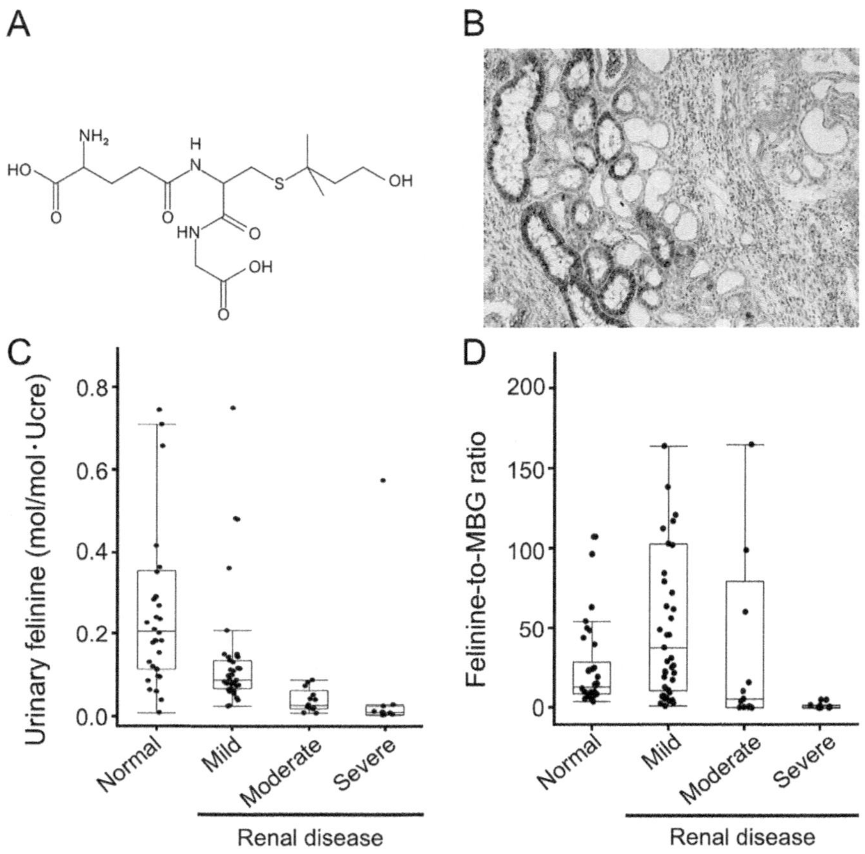

Fig. 3. Urinary excretion of felinine and its precursors in cats with renal disease. (**a**) Chemical structure of 3-methylbutanol-glutathione, an upstream precursor of felinine in cats. (**b**) Immunohistochemical images of cauxin in paraffin-embedded renal sections from a cat with renal disease. (**c**) Urinary concentrations of felinine were measured in 34 healthy cats and 66 cats with renal disease at different stages: mild (42 subjects), moderate (13 subjects), and severe (11 subjects). The concentrations were normalized to urinary creatinine (UCre). Individual values are plotted as points. (**d**) The ratio of felinine to 3-methylbutanol-glutathione (MBG) in cats with renal disease.

from 3-methylbutanol-glutathione (MBG; Fig. 3a) via 3-methylbutanol-cysteinylglycine (MBCG) (Miyazaki et al., 2006c). Felinine production is regulated by testosterone, resulting in higher levels in intact male cats.

The renal expression of cauxin, a key enzyme, decreases in cats with CKD due to the reduction of its production in renal tubular epithelial cells, which are replaced by inflammatory cells such as macrophages and fibroblasts (Fig. 3b). Since renal enzymes, including cauxin and γ-glutamyltransferase (γ-GTP), are involved in felinine metabolism in cats, kidney dysfunction is likely to alter felinine metabolism, leading to reduced felinine levels and a decrease in its associated volatile odorants. To test this hypothesis, we examined the urinary levels of felinine, its precursors, and its derivatives in cats with CKD.

Urinary felinine levels were significantly lower in cats with moderate to severe renal disease than in normal cats (Fig. 3c). MBG levels were significantly higher in cats with severe renal disease, resulting in a significant reduction in the felinine-to-MBG ratio (Fig. 3d). This finding indicates an impaired conversion of precursors to felinine.

Emission levels of felinine-derived sulfurous VOCs were also significantly lower in cats with CKD (Suka et al., 2025). Among these, 3-methyl-3-methylthio-1-butanol showed a strong correlation with urinary felinine levels, confirming its role as a felinine-derived odorant. 3-Mercapto-3-methyl-1-butanol, a major contributor to the characteristic "catty" odor, was detectable in only a few CKD-affected cats, particularly those with mild disease. In contrast, severely affected cats exhibited almost no detectable felinine-derived VOCs, indicating a drastic reduction in urine odor.

This study confirms that urinary felinine levels decrease as renal disease progresses, leading to a marked reduction in felinine-derived VOCs, which are responsible for the characteristic feline urinary odor. The decrease in felinine excretion is likely due to impaired enzymatic conversion of MBG to felinine, as cauxin and γ-GTP are localized in the proximal tubules, which are commonly damaged in CKD. The accumulation of MBG suggests that renal dysfunction disrupts the final steps of felinine metabolism.

The absence of key sulfur-containing VOCs suggests that the weakened urinary odor in CKD-affected cats is not due to urine dilution but rather a metabolic disruption. Even when normal urine was experimentally diluted, VOCs remained detectable, further supporting this conclusion.

The reduction of felinine-derived urinary odorants in CKD-affected cats suggests that these cats may be less effective at scent marking their territory. For feral cats, this could pose a survival disadvantage, as scent marking plays a crucial role in social communication and territorial defense. CKD-induced changes in urinary odor may further impact their ability to establish and maintain territories, potentially affecting their survival in the wild.

6 Chemical Compositions and Biological Significances of Feces and Anal Sac Secretions in Cats

Domestic cats use feces as territorial markers, particularly in the periphery of their home ranges, where unburied feces may signal territory occupancy (Feldman, 1994; Crowell-Davis et al., 2004). Compared to the extensive studies of urinary components, the role of feces and anal sac secretions in chemical signaling has only recently gained attention (Nakabayashi et al., 2012). Here, I summarize the knowledge on the chemical basis of species, sex, and individual recognition using feline feces and anal sac secretions.

GC/MS analysis of fresh cat feces revealed that 3-mercapto-3-methyl-1-butanol is emitted at significantly higher levels in male cats than in females (Miyazaki et al., 2018). Temporal analysis showed that 3-mercapto-3-methyl-1-butanol emission peaks

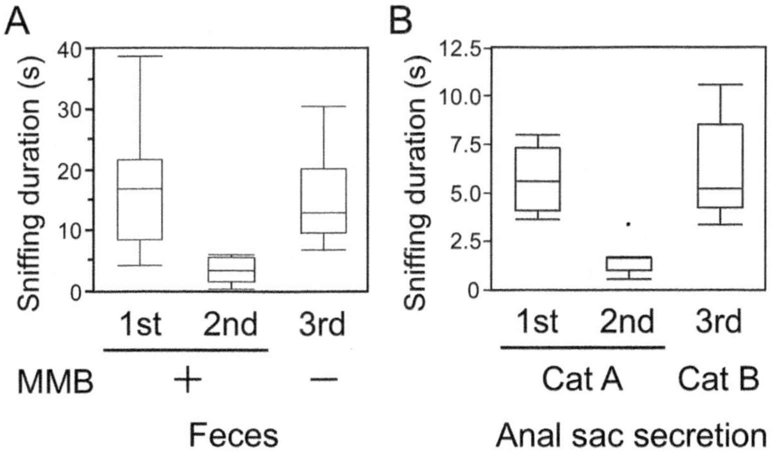

Fig. 4. Olfactory discrimination of fecal and anal sac-derived odors in male cats. (**a**) Olfactory habituation-dishabituation tests were conducted to assess the ability of male cats to discriminate between feces with or without emission of the cat-specific compound 3-mercapto-3-methyl-1--butanol (MMB). Feces treated with a copper solution to inhibit MMB emission and water-treated feces were used in the tests. (**b**) Habituation-dishabituation tests were conducted in five male cats using anal sac secretions collected from two other male cats (Cat A and Cat B). Original data for A and B are from our previous studies (Miyazaki et al., 2018) and (Miyazaki et al., 2017b), respectively.

within a few hours after defecation and gradually diminishes over time, suggesting its role as a time-sensitive scent signal. Behavioral assays confirmed that cats can distinguish between feces with and without 3-mercapto-3-methyl-1-butanol, indicating that it functions as a male-specific scent marker (Fig. 4a). In addition to 3-mercapto-3-methyl-1-butanol, feces contain a mixture of short-chain fatty acids, including acetic, propanoic, and butanoic acids, which vary among individuals. These fatty acid profiles likely serve as unique olfactory signatures, enabling individual recognition among conspecifics. Unlike 3-mercapto-3-methyl-1-butanol, which degrades over time, short-chain fatty acid compositions remain stable, making them reliable cues for long-term scent marking.

Anal sac secretions are another key source of scent-based information in cats. Cats prefer to sniff the perianal area of other cats when interacting in a friendly manner (Crowell-Davis et al., 2004). Anal sac secretions likely contribute to the odor of the perianal area. Unlike feces, which primarily function as species and sex markers, anal sac secretions serve as individual identifiers. GC/MS analysis revealed that short-chain fatty acids are the predominant volatile compounds in feline anal sac secretions, including acetic, propanoic, and butanoic acids, which are also found in the anal sac secretions of other carnivores (Miyazaki et al., 2017b). Notably, no sex differences were observed in the chemical profiles of anal sac secretions. However, individual variation was pronounced, with each cat exhibiting a unique combination of fatty acids. Longitudinal studies confirmed that these individual scent signatures remain stable over time, reinforcing their role in individual recognition. Behavioral experiments using habituation-dishabituation tests demonstrated that cats could distinguish between the anal sac secretions of different individuals (Fig. 4b). Male cats exhibited a clear dishabituation response when presented with a novel anal sac secretion sample, indicating their ability to recognize individual-specific chemical cues.

While both feces and anal sac secretions contribute to feline chemical communication, their roles differ in function and context. Feces primarily encode species and sex information, particularly through the emission of 3-mercapto-3-methyl-1-butanol, which serves as a male-specific pheromone. In contrast, anal sac secretions primarily encode individual identity through variations in fatty acid composition. One possible explanation for this functional divergence lies in the context of scent detection. When encountering scent marks left on the ground (feces), cats lack visual confirmation of the scent owner's identity and rely on chemical cues for species and sex recognition. Anal sac secretions might contribute to the odor of the perianal area and convey individual information in addition to visual information, regarding, for example, the face or body shape of the other cat. Such double-checking of individual identity using both visual and olfactory information may enable cats to distinguish among individuals, even if their appearances are similar due to close relatedness within a family.

Another key distinction is the temporal stability of these scent marks. 3-Mercapto-3-methyl-1-butanol in feces follows a dynamic emission profile, serving as a time-sensitive signal that helps territorial males assess the presence and activity of other males in their territory. In contrast, fatty acids in both feces and anal sac secretions remain stable, allowing for long-term individual identification. While feces provide species-, sex-, and time-sensitive territorial information, anal sac secretions function as stable individual identifiers. Understanding these scent communication mechanisms not only enhances our knowledge of feline behavior but also offers practical applications for pet owners and animal behaviorists.

7 Conclusion

Scent marking is a fundamental aspect of feline communication, playing a crucial role in territoriality, individual recognition, and reproductive signaling. This review examined the chemical composition and biological significance of feline urine, feces, and anal sac secretions, providing insights into how cats use these excretions for intraspecies communication. Urine, feces, and anal sac secretions serve complementary roles in the chemical communication system of cats.

Urinary scent marking is mediated by cauxin, a major urinary protein that facilitates the production of felinine-derived sulfur-containing VOCs. These VOCs contribute to the characteristic "catty" odor, which varies by species, sex, and age. The evolutionary adaptation of cauxin in small felids suggests that urinary scent marking is integral to territorial and social behaviors. Additionally, cauxin enhances urine adhesion to vertical surfaces, increasing odor longevity and marking effectiveness.

Feces and anal sac secretions also contribute to scent communication but serve distinct functions. Fecal VOCs, particularly 3-mercapto-3-methyl-1-butanol, act as species- and male-specific scent markers, while short-chain fatty acids (SCFAs) provide stable individual recognition cues. Anal sac secretions, also rich in SCFAs, serve as long-term individual identifiers, reinforcing social recognition. These findings indicate that cats rely on a combination of volatile and stable chemical cues to convey species, sex, and identity information.

Furthermore, renal disease significantly impacts urinary scent marking by disrupting felinine metabolism and reducing VOC emissions. The resulting weakened urinary odor may impair territorial and social interactions, particularly in feral populations, where scent marking is essential for territory establishment and mate selection. Behavioral studies demonstrate that cats respond strongly to urinary VOCs, using them for territorial assessment and individual recognition. The ability to distinguish fresh from aged urine and recognize individual scent differences highlights the dynamic nature of feline olfactory communication.

Understanding these scent communication mechanisms not only advances knowledge of feline behavior and evolution but also has practical applications in managing free-roaming cat populations, conserving wild felids, and improving domestic cat welfare. Future research should further explore molecular mechanisms of the species-specific felinine biosynthesis and their implications for social dynamics and ecological adaptation in Felidae.

Acknowledgments. I thank Reiko Uenoyama for critically reading the manuscript and giving constructive feedback.

References

Bradshaw, J., & Cameron-Beaumont, C. (2000). The signalling repertoire of the domestic cat and its undomesticated relatives. In D. C. Turner & P. Bateson (Eds.), *The domestic cat: The biology of its behaviour* (pp. 67–93). Cambridge University Press.

Chakrabarti, S., Syme, H. M., & Elliott, J. (2012). Clinicopathological variables predicting progression of azotemia in cats with chronic kidney disease. *Journal of Veterinary Internal Medicine, 26*, 275–281. https://doi.org/10.1111/j.1939-1676.2011.00874.x

Crowell-Davis, S. L., Curtis, T. M., & Knowles, R. J. (2004). Social organization in the cat: A modern understanding. *Journal of Feline Medicine and Surgery, 6*, 19–28. https://doi.org/10.1016/j.jfms.2003.09.013

Ecroyd, H., Belghazi, M., Dacheux, J. L., Miyazaki, M., Yamashita, T., & Gatti, J. L. (2006). An epididymal form of cauxin, a carboxylesterase-like enzyme, is present and active in mammalian male reproductive fluids. *Biology of Reproduction, 74*, 439–447. https://doi.org/10.1095/biolreprod.105.046847

Elliott, J., & Barber, P. J. (1998). Feline chronic renal failure: Clinical findings in 80 cases diagnosed between 1992 and 1995. *Journal of Small Animal Practice, 39*, 78–85. https://doi.org/10.1111/j.1748-5827.1998.tb03598.x

Feldman, H. N. (1994). Methods of scent marking in the domestic cat. *Canadian Journal of Zoology, 72*, 1093–1099. https://doi.org/10.1139/z94-147

Hendriks, W. H., Moughan, P. J., Tarttelin, M. F., & Woolhouse, A. D. (1995). Felinine: A urinary amino acid of Felidae. *Comparative Biochemistry and Physiology – B Biochemistry and Molecular Biology, 112*, 581–588. https://doi.org/10.1016/0305-0491(95)00130-1

Hurst, J. L., & Beynon, R. J. (2004). Scent wars: The chemobiology of competitive signalling in mice. *BioEssays, 26*, 1288–1298. https://doi.org/10.1002/bies.20147

Hurst, J. L., Payne, C. E., Nevison, C. M., Marie, A. D., Humphries, R. E., Robertson, D. H. L., Cavaggioni, A., & Beynon, R. J. (2001). Individual recognition in mice mediated by major urinary proteins. *Nature, 414*, 631–634. https://doi.org/10.1038/414631a

Li, G., Janecka, J. E., & Murphy, W. J. (2011). Accelerated evolution of CES7, a gene encoding a novel major urinary protein in the cat family. *Molecular Biology and Evolution, 28*, 911–920. https://doi.org/10.1093/molbev/msq281

Linda, C. (2003). *The cat: Its behavior, nutrition and health*. Iowa State Press.

McLean, L., Hurst, J. L., Gaskell, C. J., Lewis, J. C. M., & Beynon, R. J. (2007). Characterization of cauxin in the urine of domestic and big cats. *Journal of Chemical Ecology, 33*, 1997–2009. https://doi.org/10.1007/s10886-007-9354-6

Miyazaki, M., Kamiie, K., Soeta, S., Taira, H., & Yamashita, T. (2003). Molecular cloning and characterization of a novel carboxylesterase-like protein that is physiologically present at high concentrations in the urine of domestic cats (*Felis catus*). *Biochemical Journal, 370*, 101–110. https://doi.org/10.1042/BJ20021446

Miyazaki, M., Yamashita, T., Hosokawa, M., Taira, H., & Suzuki, A. (2006a). Species-, sex-, and age-dependent urinary excretion of cauxin, a mammalian carboxylesterase. *Comparative Biochemistry and Physiology Part B: Biochemistry and Molecular Biology, 145*, 270–277. https://doi.org/10.1016/j.cbpb.2006.05.015

Miyazaki, M., Yamashita, T., Suzuki, Y., Saito, Y., Soeta, S., Taira, H., & Suzuki, A. (2006b). A major urinary protein of the domestic cat regulates the production of felinine, a putative pheromone precursor. *Chemistry & Biology, 13*, 1071–1079. https://doi.org/10.1016/j.chembiol.2006.08.013

Miyazaki, M., Yamashita, T., Taira, H., & Suzuki, A. (2006c). The biological function of cauxin, a major urinary protein of the domestic cat (*Felis catus*). In J. Hurst, R. Beynon, S. Roberts, & T. Wyatt (Eds.), *Chemical signals in vertebrates 11* (pp. 67–93). Springer.

Miyazaki, M., Miyazaki, T., Toyonaga, M., Tsutsui, T., Taira, H., Yamashita, T., & Suzuki, A. (2011). Characterisation of the carboxylesterase enzyme cauxin in the seminal fluid of the cat. *The Veterinary Journal, 190*, 378–382. https://doi.org/10.1016/j.tvjl.2010.11.014

Miyazaki, M., Nishimura, T., Hojo, W., Miyazaki, T., & Laine, R. A. (2017a). Potential use of domestic cat (*Felis catus*) urinary extracts for manipulating the behavior of free-roaming cats and wild small felids. *Applied Animal Behaviour Science, 196*, 52–60. https://doi.org/10.1016/j.applanim.2017.07.003

Miyazaki, T., Nishimura, T., Yamashita, T., & Miyazaki, M. (2017b). Olfactory discrimination of anal sac secretions in the domestic cat and the chemical profiles of the volatile compounds. *Journal of Ethology, 36*, 99–105. https://doi.org/10.1007/s10164-017-0532-x

Miyazaki, M., Miyazaki, T., Nishimura, T., Hojo, W., & Yamashita, T. (2018). The chemical basis of species, sex, and individual recognition using feces in the domestic cat. *Journal of Chemical Ecology, 44*, 364–373. https://doi.org/10.1007/s10886-018-0951-3

Moura, A. A., Souza, C. E., Stanley, B. A., Chapman, D. A., & Killian, G. J. (2010). Proteomics of cauda epididymal fluid from mature Holstein bulls. *Journal of Proteomics, 73*, 2006–2020. https://doi.org/10.1016/j.jprot.2010.06.005

Nakabayashi, M., Yamaoka, R., & Nakashima, Y. (2012). Do faecal odours enable domestic cats (*Felis catus*) to distinguish familiarity of the donors? *Journal of Ethology, 30*, 325–329. https://doi.org/10.1007/s10164-011-0321-x

Natoli, E. (1985). Behavioural responses of urban feral cats to different types of urine marks. *Behaviour, 94*, 234–243. https://doi.org/10.1163/156853985X00208

Ru, Y. F., Xue, H. M., Ni, Z. M., Xia, D., Zhou, Y. C., & Zhang, Y. L. (2015). An epididymis-specific carboxyl esterase CES5A is required for sperm capacitation and male fertility in the rat. *Asian Journal of Andrology, 17*, 292–297. https://doi.org/10.4103/1008-682X.143314

Suka, A., Uenoyama, R., Ichizawa, S., Katayama, M., Miyazaki, M., & Miyazaki, T. (2025). Reduction of urinary felinine in domestic cats with renal diseases leads to decreased catty odor. *Veterinary Medicine and Science, 87*, 61–67. https://doi.org/10.1292/jvms.24-0370

Suzuki, C., Miyazaki, T., Yamashita, T., & Miyazaki, M. (2019). GC×GC-MS-based volatile profiling of male domestic cat urine and the olfactory abilities of cats to discriminate temporal changes and individual differences in urine. *Journal of Chemical Ecology, 45*, 579–587. https://doi.org/10.1007/s10886-019-01083-3

Uenoyama, R., Zhu, W., Miura, M., Miyazaki, T., & Miyazaki, M. (2024). Sprayed urine emits a pungent odor due to its increased adhesion to vertical objects via urinary proteins rather than to changes in its volatile chemical profile in domestic cats. *Journal of Chemical Ecology, 50*, 978–993. https://doi.org/10.1007/s10886-024-01490-1

Exploring the Molecular Basis of Chemical Communication in the Iberian Lynx (*Lynx pardinus*)

Beatriz Ortiz-Guisado[1], Elisa María Espinosa-López[1], María Ponte-González[2], Mar Gómez-Marín[1], Evangelina Rodero-Serrano[3], Ana María Molina-López[4], José Guerrero-Casado[2], and Guadalupe Gómez-Baena[1](✉)

[1] Departamento de Bioquímica y Biología Molecular, Universidad de Córdoba, Córdoba, Spain
v52gobag@uco.es
[2] Departamento de Zoología, Universidad de Córdoba, Córdoba, Spain
[3] Departamento de Producción Animal, Universidad de Córdoba, Córdoba, Spain
[4] Departamento de Anatomía y Anatomía Patológica Comparadas, y Toxicología, Universidad de Córdoba, Córdoba, Spain

Abstract. The Iberian lynx (*Lynx pardinus*) is a keystone species of the Mediterranean ecosystem. Having once been on the brink of extinction in 2002, with fewer than 100 individuals, the population has now recovered to over 2400 individuals in 2024. Despite this progress, key aspects of its physiology, particularly chemical communication, remain poorly understood. Our project aims to uncover the molecular foundations of intraspecies communication in the Iberian lynx with the goal of exploiting natural chemical signals to enhance conservation efforts. This review explores the role of chemical communication in the Iberian lynx, focusing on territorial marking, sexual signaling during the mating season, social communication, and potential navigation cues. We discuss the molecular composition of scent marks, including volatile compounds and urinary proteins like cauxin, which may play a role in territorial marking and sexual communication. Furthermore, we examine the implications of these chemical signals for the conservation and management of the species, including their potential applications in reducing human-wildlife conflicts, enhancing captive breeding programs, and facilitating successful reintroduction efforts. A deeper understanding of chemical communication in the Iberian lynx is essential for optimizing conservation strategies and ensuring the long-term survival of this endangered species.

Keywords: Ibèrian lynx · Scent marking · Semiochemical communication

1 Introduction

The Iberian lynx (*Lynx pardinus*) is a keystone species endemic to the Mediterranean ecosystem of the Iberian Peninsula. As a top predator, its presence can limit mesopredator's abundance through top-down effects (Jiménez et al., 2019; Burgos et al., 2023). This limitation of mesopredators can, in turn, increase the abundance of some prey

L. M. Schulte (ed.), *Chemical Signals in Vertebrates 16*,
https://doi.org/10.1007/978-3-032-30752-1_9

species such as the European rabbit (*Oryctolagus cuniculus*) and red-legged partridges (*Alectoris rufa*) (Jiménez et al., 2019), therefore inducing some trophic cascade effects across food webs (Burgos et al., 2024) and thereby influencing the structural complexity and functionality of Mediterranean ecosystems. However, the ecological role of Iberian lynx was almost negligible due to its low abundance and limited distribution at the end of the twentieth century. By 2002, only two isolated populations remained, one in Doñana and another in Sierra Morena, with an estimated overall size of fewer than 100 individuals. Consequently, the Iberian lynx was then listed as Critically Endangered (CR) on the IUCN red list (Rodríguez, 2024), making it the most threatened feline species in the world. Fortunately, since the beginning of the twenty-first century, various conservation programs have been implemented to recover their populations, and the species is now present in many areas of its historical distribution range. The 2023 survey estimated a population size exceeding 2000 individuals, and the species is now listed as Vulnerable (VU) in the IUCN red list (Rodríguez, 2024).

Numerous scientific studies have been conducted as part of these conservation projects which helped to better understand various aspects of the Iberian lynx physiology. However, chemical communication, a key aspect of the evolution and conservation of the species, is poorly understood. This review aims to summarize the current state of knowledge on chemical communication and discuss its implications for the conservation and management in the Iberian lynx.

2 Scent Marking Behavior in the Iberian Lynx

Chemical communication plays a key role in the conservation of species, as it reflects physiology and behavioral processes critical to survival and reproduction. In mammals, olfactory signals deposited in the environment by various scent marking strategies convey information about sex, age, health, reproductive state, and social status of the scent owners. Receptor individuals modify their physiology and behavior in response to such signals and initiate a learning process, which helps them to recognize a territory, a social group, an individual, and their relatedness with the owner (Wyatt, 2014). Intraspecific communication is crucial to the behavioral ecology of solitary carnivores. However, despite its importance, chemical communication remains largely unexplored in many felid species, including the Iberian lynx.

Scent marking with feces, urine, or glandular secretions is common among felids, which are known to leave scent marks at visually prominent locations (Mellen, 1993; Allen et al., 2016). The marking behavior of the Iberian lynx has been extensively documented in captivity (Rivas et al., 2016), revealing a range of marking behaviors including urine spraying, head or cheek rubbing, scraping, claw raking, and licking. Since elusive and solitary species are difficult to observe in their natural habitats, studying scent marking behaviors in the wild presents significant challenges (Allen et al., 2016), and only a few studies have examined the socio-temporal patterns of scent marking behaviors of Iberian lynxes in their natural habitats (Robinson & Delibes, 1988; Ferreras et al., 1997; Burgos et al., 2018). In the following sections of the chapter, we will explore the role of these marking behaviors in territorial defense, social interactions, sexual communication, and navigation.

2.1 Territorial Marking

The Iberian lynx is considered a solitary species that exhibits a territorial structure (Ferreras et al., 1997; Garrote et al., 2010; Sarmento et al., 2017) with well-defined individual home ranges. It occupies territories in regions characterized mainly by shrubs and grasses, with minimal tree cover (Palomares et al., 2001). It is a specialist predator feeding almost exclusively on wild rabbits, and the abundance of its prey seems to be a major driver for the behavior and survival of the species (Palomares et al., 2001). Female territory size depends on food availability and is smaller in areas with high density of rabbits (Palomares et al., 2001). Additionally, female territory size also seems related to defense of offspring from infanticide (López-Bao et al., 2010). Male territory size is larger than that of females (Ferreras et al., 1997; Rueda et al., 2021), with a tendency to overlap with those of one or several females (Ferreras et al., 1997; Palomares et al., 2001; Palomares et al., 2017). The proximity of males to females increases during the summer (Ferreras et al., 1997), which may be related to an effort by males to protect their offspring from potential infanticide by incoming males (López et al., 2010). Both male and female lynxes intensively and exclusively use core areas, which represents about 35% of the home range (Ferreras et al., 1997).

Lynx home ranges are intensely patrolled and marked with feces suggesting a role in advertising the presence of the territory owner to potential competitors (Robinson & Delibes, 1988; Ferreras et al., 1997). Because scent marking requires energy, the Iberian lynx chooses locations in the landscape that maximize the chances of detection by other individuals and promote the spread of its scent. Iberian lynxes place their feces in latrines, typically located in accessible, visible, and transit spots, likely to increase the probability of being detected by conspecifics (Robinson & Delibes, 1988; Burgos et al., 2018). In fact, boundaries between home ranges of neighboring individuals are usually located along significant features of the landscape (i.e. roads; Fig. 1), and these limits are intensively marked with feces (Robinson & Delibes, 1988; Ferreras et al., 1997). Particularly, marking roads, trails, and crossroads may enhance scent mark encounter if the receivers use these human infrastructures for traveling (Rafiq et al., 2020), whereas sheltered places can prolong the persistence of chemical components and reduce weather-related decay (Burgos et al., 2018).

It has been also reported that, related to territorial marking, fecal scent marking may play roles in food delimitation and protection, since scent marking in the Iberian lynx is positively related to the abundance and distribution of its main prey, the wild rabbit (Palomares et al., 2001; Burgos et al., 2018). When food is scarce, fecal marks are concentrated in areas with high rabbit populations, which the lynxes revisit frequently.

In solitary felids, urine spraying is also a form of scent marking for intraspecific communication (Allen et al., 2016). Scent marking with urine has been recorded in natural populations of the Iberian lynx (Ferreras et al., 1997) probably as a behavior related to the establishment and maintenance of the territory. However, further studies are needed to determine sex, age, and seasonal variation in spatial patterns and molecular composition.

Fig. 1. A latrine of Iberian lynx located on a rock on the side of a road, a good example of a conspicuous visible place, easily detectable by other conspecifics (J. Guerrero-Casado).

2.2 Sexual Communication

The Iberian lynx, as a solitary species with large home ranges (Ferreras et al., 1997), relies on scent marking for sexual communication. In this species, the mating season occurs from December to February or March, with female receptivity (estrus) peaking in January for only 2–3 days (Palomares et al., 2005). During the mating season, intra-species interactions are very intense. The Iberian lynx differs from other felines in that it is seasonal monoestrous due to the existence of a persistent corpus luteum (CL), which can remain active for up to 2 years (Amelkina et al., 2016). It is believed that the functionality of this persistent CL is to ensure the monoestrous cycle by suppressing ovarian activity, ensuring birth and lactation in the most favorable period of the year (Amelkina et al., 2016). Births are primarily concentrated in the last 2 weeks of March and the first days of April (Palomares et al., 2017). Some females give birth annually, while others do so in alternate years, although the reason for this is currently unknown. Births usually coincide with the period of the highest rabbit abundance (Gañan et al., 2010). The cubs start eating meat at 30 days old, with March representing the period of highest maternal energetic demand (López-Bao et al., 2009). Results indicate that multiple paternity is a rare event in the Iberian lynx and that males do not participate in the care of the cubs (Palomares et al., 2017).

The traditional perspective assumes that the main role of chemical signals in sexual communication is to enable sex and species recognition and advertise reproductive status. However, sexual selection drives the evolution of signals that maximize conveyed information and assist as predictors of offspring fitness (Steiger et al., 2011). Additionally, chemical signals serve as indicators of mate quality and play a key role in mate selection, directly impacting mating success (Buchinger & Li, 2023). The mechanisms underlying mate choice in wild felids have been largely overlooked, from the assumption that it is not a significant factor affecting reproductive success in solitary species (Johansson & Jones, 2007). However, territorial exclusion of sexual competitors during the mating season might not be complete, and, in species where intrasexual territories overlap, with one female having access to multiple males, there is a potential for female choice (López-Bao et al., 2008), although this hypothesis remains untested.

Since scent marking could act as a method of attracting and facilitating interactions between potential mates, it is expected to occur most frequently between individuals of the opposite sex and during periods when females are receptive. The range of scent marking behaviors shown by the Iberian lynx during the mating season has been extensively studied in captivity as part of the ex situ conservation program (Rivas et al., 2016). The marking activity of Iberian lynxes intensifies before and during the mating season. It is common to observe individuals marking each other with urinary sprays and glandular secretions through head rubbing. These observations suggest that urine and facial glandular secretions may carry scent signals involved in chemical communication during estrus and that their chemical profile are expected to change during the mating season in the Iberian lynx, as described for urine in other Lynx species (Vogt et al., 2016). In fact, during the mating season, the cheek area appears wet and oily (Rivas et al., 2016), suggesting that semiochemicals secreted by the facial glands may have a function in sexual communication, although the chemical composition of this material has not been studied yet.

2.3 Social Communication

Social interactions are crucial factors for survival, dispersal, and mating opportunities, making them essential for maintaining the viability of a population. A common feature in the socioecology of most Felidae species is their solitary nature, with intraspecific interactions generally restricted to those related to reproduction (Bekoff et al., 1984). However, it is well established that even solitary species can form various types of social relationships and exhibit complex social structures. Understanding the social organization and the factors that influence social behavior in endangered species is crucial for their management and conservation (Ferreras et al., 1997).

As described earlier, the Iberian lynx is a solitary species, with interactions restricted to rearing activities by females and infrequent interactions between adult individuals (Ferreras et al., 1997; López-Bao et al., 2014). The low encounter rates among wide-ranging individuals and the rarity of aggressive conflicts between residents and nonresidents, in various scenarios, including the presence of prey hotspots, suggest the existence of a dynamic interaction with active temporal avoidance (López-Bao et al., 2009; López-Bao et al., 2011; López-Bao et al., 2014). Avoiding direct contacts is probably related to the fitness costs associated with potentially aggressive encounters. In this context, indirect communication is believed to occur through chemical communication. For wide-ranging solitary felids, it is expected that scent signals have a certain longevity, allowing them to be detected by conspecifics who may not come across them for several days or weeks (Vogt et al., 2014).

Additionally, it has been suggested that kinship can explain the higher rate of overlapping observed between home ranges in some individuals (López-Bao et al., 2014) highlighting the need of kinship recognition through scent marks in the Iberian lynx.

2.4 Navigation Cues

Navigation can be defined as the process by which individuals identify where they are located and use that information to move throughout the surrounding environment (Greene, 2022). However, how members of the Felidae family navigate is still an under-explored topic.

Navigation is an important skill for wild felids as individuals reaching adulthood disperse, establish their own territories or home ranges, and move around within these areas. The lynx's exploratory behavior may be adequate to sustain consistent movement of individuals between populations up to 50 km apart, or even more distance, allowing them to integrate with more distant populations. At 3 months, the cubs start traveling with their mother, with whom they stay until they are 10 months old, remaining in the maternal territory until they are 12–24 months old. If they have not left by the age of 2, they usually establish a territory as a sub-territory within the maternal home range (Rueda et al., 2021; Cisneros-Araujo et al., 2024). Reintroduced males have been observed to move more slowly than wild ones, while no differences have been found among females. Wild lynxes have larger territories than reintroduced ones because the latter are in the learning period of the new environment, being more cautious or neo-phobic, which is favorable for a higher success rate in reintroduction (Cisneros-Araujo et al., 2024). Unfortunately, their use of roads and other man-made structures with a higher frequency during their migration (Ferreras et al., 1997) increases the risk of roadkill and brings situations of conflict with human society.

Adult males and females observed marking with urine and feces when traveling (Robinson & Delibes, 1988; Ferreras et al., 1997), supporting the hypothesis that these scent marks may contain cues used for navigation. It has been reported that cheetahs recognize some of their surrounding objects, such as marking locations, and that they use them to orientate while traveling (Eaton, 1970). Since cheetahs move in groups and avoid overlapping, it has been observed that if one group follows a path, another group, even if not on the same day, tends to follow the same route of marking spots previously used by any group of cheetahs (Eaton, 1970).

3 Molecular Basis of Scent Communication in the Iberian Lynx

So far, scent marking in the Iberian lynx has been studied mainly from a behavioral perspective (Robinson & Delibes, 1988; Ferreras et al., 1997; Rivas et al., 2016; Burgos et al., 2018), and, despite the known potential role of feces, urine, and facial glands in the chemical communication of the Iberian lynx, the chemical composition of these scent marks remains poorly understood.

When characterizing the molecular composition of scent marks, most of the studies tend to focus on the volatile components since they can be detected from the distance and, hence, they are regarded as key molecules for chemical communication. However, it has also been described that proteins play a crucial role in communication in particular related to pheromone transport and reproductive behavior (Coureaud et al., 2010;

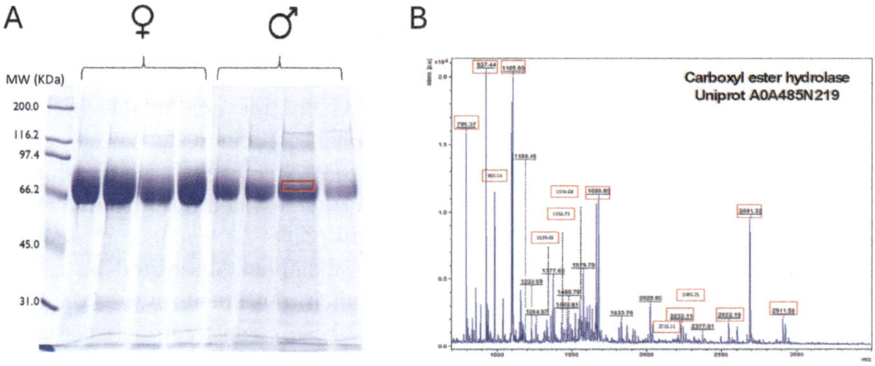

Fig. 2. (**a**) Protein profile of bladder urine from the Iberian lynx. Samples were collected during routine annual medical checkups and 10 μL analyzed by polyacrylamide gel electrophoresis under denaturing and reducing conditions (15% resolving gel). (**b**) The most intense band at 66 kDa (labeled with a red square) was excised and analyzed by peptide mass fingerprinting (PMF) as previously described (Gómez-Baena et al., 2019) revealing that cauxin is one of the major urinary proteins in the Iberian lynx.

Haga et al., 2010; Roberts et al., 2012; Gómez-Baena et al., 2023). While volatile compounds emitted from urine have been well established in several felids (Mattina et al., 1991; Andersen & Vulpius, 1999; Vogt et al., 2016; Suzuki et al., 2019), some limited studies characterized protein characterization in felid scent marks (Miyazaki et al., 2006; McLean et al., 2007; Burger et al., 2008). The Eurasian lynx (*Lynx lynx*) is likely the lynx species most studied (Vogt et al., 2014). In this lynx species, urine contains sex-specific information, reproductive state, and individual identity cues (Vogt et al., 2016). Preliminary data from our group show that one of the main components of the urinary protein fraction in the Iberian lynx is the enzyme cauxin (Fig. 2). This enzyme, first discovered in the domestic cat (Miyazaki et al., 2003) but also present in other felid species (McLean et al., 2007), is responsible for the production of felinine, an amino acid precursor for many volatiles identified in the urine of the domestic cat and other felid species.

Since cauxin concentration is higher in domestic male cats and regulated by testosterone, a potential role for felinine in territorial marking and sexual communication has been suggested (Hendriks et al., 1995). Felinine has also been identified in the urine of the Iberian lynx (data not shown), but its function in territorial marking and sexual communication is still under investigation.

4 Chemical Communication: Implications on Conservation.

Studying chemical signals in wildlife offers valuable insights for the conservation of endangered species by improving several aspects of their ecology and management. In fact, in the recently published "Strategy for the Conservation of the Iberian Lynx in Spain and Portugal" (Estrategia para la conservación del lince ibérico *Lynx pardinus* en

España y Portugal, 2024), a specific action dedicated to the "Study on the Role of Scent Marks" is described. A deeper understanding of the molecules involved in these processes, their function, factors affecting their secretion and function, and being able to replicate their effects in synthetic analogs could have multiple applications in the conservation of the species. By understanding how animals communicate, conservationists can develop strategies to improve conservation plans. Among these applications, the optimization of camera trapping could benefit as urine obtained from captive animals has been used as lure to encourage visits to control points for population monitoring (Garrote et al., 2010). Another important potential application would be reducing human-wildlife conflicts, for example, by guiding their movements away from danger zones such as roads. Additionally, knowledge of these signals can enhance captive breeding programs by promoting natural mating behaviors. Exploring the molecular mechanisms involved in sexual communication in wild felids provides critical insights into the role of these signals in mate choice and their contribution to reproductive success. In release programs, understanding the chemical cues that animals rely on can increase the likelihood of successful reintroduction into the wild by ensuring that released animals are attracted to the right habitats and interact properly with the environment. Lastly, studying chemical signals can greatly improve the welfare of captive animals by providing insights into their social interactions, stress levels, and overall well-being, allowing for more tailored and enriching care. Ensuring animal welfare is an obligation of our society, which must become a priority. In the case of wildlife, and specifically the Iberian lynx kept in captivity, it is one of the main pillars supporting their maintenance and reproductive success, as well as the success of their subsequent release and reintegration into the natural environment. The alteration of animal welfare can negatively influence not only reproduction but also create states of immunosuppression, making them more susceptible to any pathogen. As an example, semiochemicals secreted by the sebaceous glands of the domestic cat's cheeks, in particular fractions 3 and 4 of facial pheromones (F3 and F4) and a cat's maternal appeasing pheromone (CAP), have been employed to improve the well-being of the domestic species and other wild felines in captivity (Bertoni et al., 2023). Similarly, the rabbit maternal neonatal pheromone 2-methyl-2-butenal (2M2B) has also been shown to act as an interomone with domestic cats to enhance the introduction of new home-housed cats by preventing early aggressions (McGlone et al., 2019).

Environmental enrichment is a rather controversial topic in breeding centers for lynxes selected for subsequent release or reintegration into the wild, since the animals cannot have any direct interaction with the caretakers unless they need to be captured for some reason, and no enrichment that differs from the conditions they will encounter in the wild can be used. For breeders, it is also not possible to use enrichment as it could imprint on the offspring. In the ex situ conservation program (Rivas et al., 2016), they have implemented an enrichment program divided into different types of enrichments: physical or structural, occupational, nutritional, social, and sensory systems. The latter is aimed at enhancing the sense of smell through the use of smells in the installation using feces or urine from other Iberian lynxes fundamentally. Therefore, although the techniques used are limited, they are already implemented and routinely used in breeding centers. This type of enrichment, by influencing marking, can interfere with chemical communication. Among the enrichment methods that could have an impact is the use of natural scratching posts, such as surfaces like trees and logs that animals can use to scratch and rub against, marking the area with their claws or whiskers, as well as urine or feces. Another possible method is the use of feces and urine from other animals, spraying some area of the enclosure with the urine of another male or a female in heat, thus generating responses to that stimulus, in addition to direct interaction with another conspecific. In this sense, the study of the molecular composition of lynx urine and how this composition is related to intraspecific communication is essential to carry out a favorable environmental enrichment that is not counterproductive as it is a potential source of stress.

Nowadays, the current challenge of Iberian lynx conservation programs is to interconnect existing populations and increase their range, assimilating it to their historical distribution range. It has been observed that reintroduced Iberian lynxes show nomadic behavior in the early stages or establish multiple transitional residence areas (Cisneros-Araujo et al., 2024). We hypothesized that chemical signals in urine and feces could influence movements in the territory. Identifying navigation cues could therefore be very relevant for the ongoing Iberian lynx conservation project, whose main objective is to connect the existing nuclei, with the final goal of achieving a demographically and ecologically functional Iberian lynx metapopulation.

5 Conclusion

In summary, our understanding of scent marking communication in the Iberian lynx remains limited. Further research is needed to uncover the chemical composition of facial glands, fecal marks, and urine sprays, as well as to explore the role these semiochemicals play in territorial marking, sexual communication, and social interactions in this elusive species.

Acknowledgments. Work funded by the Life Lynxconnect project (LIFE19 NAT/ES/001055). We would like to express our gratitude to the staff at the "La Olivilla" and "El Acebuche" breeding centers, particularly María José Pérez Aspa, Rubén González, Antonio Rivas, and Yasmin Bouyafrouri, for their invaluable assistance in the collection of the samples. We also extend our gratitude to Francisco Javier Salcedo, coordinator of the Life Lynxconnect project, for his ongoing assistance throughout the research process.

References

Allen, M. L., Wittmer, H. U., Setiawan, E., Jaffe, S., & Marshall, A. J. (2016). Scent marking in Sunda clouded leopards (*Neofelis diardi*): Novel observations close a key gap in understanding felid communication behaviours. *Scientific Reports, 6*(1), 35433.

Amelkina, O., Zschockelt, L., Painer, J., Serra, R., Villaespesa, F., Krause, E., Jewgenow, K., & Braun, B. C. (2016). Progesterone, estrogen, and androgen receptors in the corpus luteum of the domestic cat, Iberian lynx (*Lynx pardinus*) and Eurasian lynx (*Lynx lynx*). *Theriogenology, 86*(9), 2107–2118.

Andersen, K. F., & Vulpius, T. (1999). Urinary volatile constituents of the lion, Panthera leo. *Chemical Senses, 24*(2), 179–189.

Bekoff, M., Daniels, T. J., & Gittleman, J. L. (1984). Life history patterns and comparative social ecology of carnivores. *Annual Review of Ecology and Systematics, 15*, 191–232.

Bertoni, V., Regaiolli, B., Cozzi, A., Vaglio, S., & Spiezio, C. (2023). Can an enrichment programme with novel manipulative and scent stimuli change the behaviour of zoo-housed European wildcats? A case study. *Animals, 13*(11), 1762.

Buchinger, T. J., & Li, W. (2023). Chemical communication and its role in sexual selection across Animalia. *Communications Biology, 6*(1), 1178.

Burger, B. V., Viviers, M. Z., Bekker, J. P., le Roux, M., Fish, N., Fourie, W. B., & Weibchen, G. (2008). Chemical characterization of territorial marking fluid of male Bengal tiger, *Panthera tigris*. *Journal of Chemical Ecology, 34*(5), 659–671.

Burgos, T., Virgós, E., Valero, E. S., Arenas-Rojas, R., Rodríguez-Siles, J., & Recio, M. R. (2018). Prey density determines the faecal-marking behaviour of a solitary predator, the Iberian lynx (*Lynx pardinus*). *Ethology, Ecology & Evolution, 31*(3), 219–230.

Burgos, T., Salesa, J., Fedriani, J. M., Escribano-Avila, G., Jimenez, J., Krofel, M., Cancio, I., Hernandez-Hernandez, J., Rodriguez-Siles, J., & Virgos, E. (2023). Top-down and bottom-up effects modulate species co-existence in a context of top predator restoration. *Scientific Reports, 13*(1), 4170.

Burgos, T., Escribano-Ávila, G., Fedriani, J. M., González-Varo, J. P., Illera, J. C., Cancio, I., Hernández-Hernández, J., & Virgós, E. (2024). Apex predators can structure ecosystems through trophic cascades: Linking the frugivorous behaviour and seed dispersal patterns of mesocarnivores. *Functional Ecology, 38*(6), 1407–1419.

Cisneros-Araujo, P., Garrote, G., Corradini, A., Farhadinia, M. S., Robira, B., López, G., Fernández, L., López-Parra, M., García-Tardío, M., Arenas-Rojas, R., del Rey, T., Salcedo, F. J., Sarmento, P., Sánchez, J. F., Palacios, M. J., García-Viñás, J. I., Damiani, M. L., Hachem, F., Gastón, A., & Cagnacci, F. (2024). Born to be wild: Captive-born and wild Iberian lynx (*Lynx pardinus*) reveal space-use similarities when reintroduced for species conservation concerns. *Biological Conservation, 294*, 110646.

Coureaud, G., Charra, R., Datiche, F., Sinding, C., Thomas-Danguin, T., Languille, S., Hars, B., & Schaal, B. (2010). A pheromone to behave, a pheromone to learn: The rabbit mammary pheromone. *Journal of Comparative Physiology A, 196*(10), 779–790.

Eaton, R. L. (1970). Group interactions, spacing and territoriality in cheetahs. *Zeitschrift für Tierpsychologie, 27*(4), 481–491.

Estrategia para la conservación del lince ibérico Lynx pardinus en España y Portugal. (2024).

Ferreras, P., Beltran, J. F., Aldama, J. J., & Delibes, M. (1997). Spatial organization and land tenure system of the endangered Iberian lynx (*Lynx pardinus*). *Journal of Zoology, 243*, 163–189.

Gañan, N., Sestelo, A., Garde, J. J., Martinez, F., Vargas, A., Sanchez, I., Perez-Aspa, M. J., Lopez-Bao, J. V., Palomares, F., Gomendio, M., & Roldan, E. R. (2010). Reproductive traits in captive and free-ranging males of the critically endangered Iberian lynx (*Lynx pardinus*). *Reproduction, 139*(1), 275–285.

Garrote, G., Perez de Ayala, R., Pereira, P., Robles, F., Guzmán, N., García, F. J., Iglesias, M. C., Hervás, J., Fajardo, I., Simón, M., & Barroso, J. L. (2010). Estimation of the Iberian lynx (*Lynx pardinus*) population in the Doñana area, SW Spain, using capture-recapture analysis of camera trapping data. *European Journal of Wildlife Research., 57*, 355–362.

Gómez-Baena, G., Armstrong, S. D., Halstead, J. O., Prescott, M., Roberts, S. A., McLean, L., Mudge, J. M., Hurst, J. L., & Beynon, R. J. (2019). Molecular complexity of the major urinary protein system of the Norway rat, Rattus norvegicus. *Scientific Reports, 9*(1), 10757.

Gómez-Baena, G., Pounder, K. C., Halstead, J. O., Roberts, S. A., Davidson, A. J., Prescott, M., Beynon, R. J., & Hurst, J. L. (2023). Unraveling female communication through scent marks in the Norway rat. *Proccedings of the National Academy of Sciences, 120*(25), e2300794120.
Greene, B. D. (2022). Feline navigation. In J. Vonk & T. K. Shackelford (Eds.), *Encyclopedia of animal cognition and behavior* (pp. 2696–2699). Springer.
Haga, S., Hattori, T., Sato, T., Sato, K., Matsuda, S., Kobayakawa, R., Sakano, H., Yoshihara, Y., Kikusui, T., & Touhara, K. (2010). The male mouse pheromone ESP1 enhances female sexual receptive behaviour through a specific vomeronasal receptor. *Nature, 466*(7302), 118–122.
Hendriks, W. H., Tarttelin, M. F., & Moughan, P. J. (1995). Twenty-four hour felinine excretion patterns in entire and castrated cats. *Physiology & Behavior, 58*(3), 467–469.
Jiménez, J., Nuñez-Arjona, J. C., Mougeot, F., Ferreras, P., González, L. M., García-Domínguez, F., Muñoz-Igualada, J., Palacios, M. J., Pla, S., Rueda, C., Villaespesa, F., Nájera, F., Palomares, F., & López-Bao, J. V. (2019). Restoring apex predators can reduce mesopredator abundances. *Biological Conservation, 238*, 108234.
Johansson, B. G., & Jones, T. M. (2007). The role of chemical communication in mate choice. *Biological Reviews Cambridge Philosophical Society, 82*(2), 265–289.
López, G., López-Parra, M., Fernández, L., Ruiz, G., Gil-Sánchez, J. M., & Simón, M. A. (2010). First evidence of non-parental male infanticide in the endangered Iberian lynx (*Lynx pardinus*). *Wildlife Biology in Practice, 3*(6), 67–74.
López-Bao, J. V., Rodriguez, A., & Alés, E. (2008). Field observation of two males following a female in the Iberian lynx (*Lynx pardinus*) during the mating season. *Mammalian Biology, 73*, 404–406.
Lopez-Bao, J. V., Rodriguez, A., & Palomares, F. (2009). Competitive asymmetries in the use of supplementary food by the endangered Iberian lynx (*Lynx pardinus*). *PLoS One, 4*(10), e7610.
López-Bao, J. V., Palomares, F., Rodríguez, A., & Delibes, M. (2010). Effects of food supplementation on home-range size, reproductive success, productivity and recruitment in a small population of the Iberian lynx. *Animal Conservation, 13*, 35–42.
López-Bao, J. V., Palomares, F., Rodríguez, A., & Ferreras, P. (2011). Intraspecific interference influences the use of prey hotspots. *Oikos, 120*, 1489–1496.
Lopez-Bao, J. V., Rodriguez, A., Delibes, M., Fedriani, J. M., Calzada, J., Ferreras, P., & Palomares, F. (2014). Revisiting food-based models of territoriality in solitary predators. *Journal of Animal Ecology, 83*(4), 934–942.
Mattina, M. J., Pignatello, J. J., & Swihart, R. K. (1991). Identification of volatile components of bobcat (*Lynx Rufus*) urine. *Journal of Chemical Ecology, 17*(2), 451–462.
McGlone, J. J., Garcia, A., Thompson, W. G., & Pirner, G. M. (2019). Maternal-neonatal pheromone/interomone added to cat litter improves litter box use and reduces aggression in pair-housed cats. *Journal of Applied Animal Welfare Science, 22*(2), 127–138.
McLean, L., Hurst, J. L., Gaskell, C. J., Lewis, J. C., & Beynon, R. J. (2007). Characterization of cauxin in the urine of domestic and big cats. *Journal of Chemical Ecology, 33*(10), 1997–2009.
Mellen, J. D. (1993). A comparative analysis of scent-marking, social and reproductive behavior in 20 species of small cats (*Felis*). *American Zoology, 33*, 151–166.
Miyazaki, M., Kamiie, K., Soeta, S., Taira, H., & Yamashita, T. (2003). Molecular cloning and characterization of a novel carboxylesterase-like protein that is physiologically present at high concentrations in the urine of domestic cats (*Felis catus*). *Biochemical Journal, 370*(Pt 1), 101–110.
Miyazaki, M., Yamashita, T., Hosokawa, M., Taira, H., & Suzuki, A. (2006). Species-, sex-, and age-dependent urinary excretion of cauxin, a mammalian carboxylesterase. *Comparative Biochemistry and Physiology. Part B, Biochemistry & Molecular Biology, 145*(3–4), 270–277.

Palomares, F., Delibes, M., Revilla, E., Calzada, J., & Fedriani, J. M. (2001). Spacial ecology of Iberian lynx and abundance of European rabbits in southwestern Spain. *Wildlife Monographs, 148*, 1–136.

Palomares, F., Revilla, E., Calzada, J., Fernández, N., & Delibes, M. (2005). Reproduction and pre-dispersal survival of Iberian lynx in a subpopulation of the Doñana National Park. *Conservation Biology, 122*, 53–59.

Palomares, F., Lucena-Perez, M., Lopez-Bao, J. V., & Godoy, J. A. (2017). Territoriality ensures paternity in a solitary carnivore mammal. *Scientific Reports, 7*(1), 4494.

Rafiq, K., Jordan, N. R., Meloro, C., Wilson, A. M., Hayward, M. W., Wich, S. A., & McNutt, J. W. (2020). Scent-marking strategies of a solitary carnivore: Boundary and road scent-marking in the leopard. *Animal Behavior, 161*, 115–126.

Rivas, A., Boixader, J., Vargas, A., Pérez, M. J., Serra, R., Asensio, V., Sánchez, I., Lobato, V., Vázquez, A., Pardo, A. J., Chaparro, J. M., Penabad, L., Reeves, J., Jiménez, O. M., Ferreira, C., Yerga, J., Gonçalves, N., Azevedo, A., Ferreira, L., et al. (2016). *Manual de Manejo del Lince Ibérico en cautividad* (2nd ed.). Grupo Asesor de Manejo en Cautividad del Comité de Cría del Lince Ibérico. Programa de Conservación Ex-situ del Lince Ibérico.

Roberts, S. A., Davidson, A. J., McLean, L., Beynon, R. J., & Hurst, J. L. (2012). Pheromonal induction of spatial learning in mice. *Science, 338*(6113), 1462–1465.

Robinson, I. H., & Delibes, M. (1988). The distribution of faeces by the Spanish lynx (*Felis pardina*). *Journal of Zoology, 216*(4), 577–582.

Rodríguez, A. (2024). *Lynx pardinus*. The IUCN red list of threatened species.

Rueda, C., Jimenez, J., Palacios, M. J., & Margalida, A. (2021). Exploratory and territorial behavior in a reintroduced population of Iberian lynx. *Scientific Reports, 11*(1), 14148.

Sarmento, P., Carrapato, C., Eira, C., & Silva, P. (2017). Spatial organization and social relations in a reintroduced population of endangered Iberian lynx (*Lynx pardinus*). *Oryx, 53*(2), 344–355.

Steiger, S., Schmitt, T., & Schaefer, H. M. (2011). The origin and dynamic evolution of chemical information transfer. *Proceedings of The Royal Society B, 278*, 970–979.

Suzuki, C., Miyazaki, T., Yamashita, T., & Miyazaki, M. (2019). GC x GC-MS-based volatile profiling of male domestic cat urine and the olfactory abilities of cats to discriminate temporal changes and individual differences in urine. *Journal of Chemical Ecology, 45*(7), 579–587.

Vogt, K., Zimmermann, F., Kolliker, M., & Breitenmoser, U. (2014). Scent-marking behaviour and social dynamics in a wild population of Eurasian lynx *Lynx lynx*. *Behavioural Processes, 106*, 98–106.

Vogt, K., Boos, S., Breitenmoser, U., & Kolliker, M. (2016). Chemical composition of the Eurasian lynx urine conveys information on reproductive state, individual identity, and urine age. *Chemoecology, 26*(6), 205–217.

Wyatt, T. D. (2014). *Pheromones and animal behavior: Chemical signals and signatures*. Cambridge University Press.

Olfactory Stimuli Involved in the Male Effect in Small Ruminants: A Current Perspective

Maxime A. Meunier[1], Hélène Vacher[1], Chantal Porte[1], Philippe Chemineau[1], José-Alberto Delgadillo[2], and Matthieu Keller[1(✉)]

[1] UMR Physiologie de la Reproduction et des Comportements, INRAE, CNRS, IFCE, Université de Tours, Nouzilly, France
matthieu.keller@inrae.fr

[2] Centro de Investigación en Reproducción Caprina (CIRCA), Universidad Autónoma Agraria Antonio Narro, Torreón, Coahuila, Mexico

Abstract. In certain mammals, photoperiodic variations regulate reproductive function, limiting the breeding season to a specific period. In temperate and subtropical latitudes, sheep and goats are considered short-day breeders, with their breeding season typically occurring from late summer to early winter. However, male sociosexual stimuli can override photoperiodic regulation, reactivating female estrus behavior, luteinizing hormone (LH) pulsatility, and ovulation during the anestrous period (i.e., sexual rest)—a phenomenon known as the "male effect" in sheep and goats. Over the last few decades, several investigations have focused on the involvement of various sensory components in the emission, detection, and transduction of the male stimulus in these models. It appeared that olfactory stimuli play a significant role. This review aims to provide an updated understanding of the role of male olfactory stimuli (from rams and bucks) in stimulating female sexual activity (in ewes and does) during the male effect. We will review the evidence regarding the role of these stimuli in this phenomenon, their nature(s), and the involvement of the different olfactory systems (main and accessory). Additionally, we will discuss how to characterize these stimuli and explore future directions related to them.

Keywords: Sheep (*Ovis aries*) · Goats (*Capra hircus*) · Male effect · Olfactory stimuli · Main olfactory system · Accessory olfactory system

1 The "Male Effect": When Sociosexual Interactions Override the Photoperiodic Control of Seasonality

In temperate and subtropical latitudes, sheep and goats are seasonal breeders. Males and females alternate between a phase of reproductive activity, known as the breeding season, and a phase of reproductive rest, referred to as the anestrous period. In these latitudes, sheep (*Ovis aries*) and goats (*Capra hircus*) are considered short-day

L. M. Schulte (ed.), *Chemical Signals in Vertebrates 16*,
https://doi.org/10.1007/978-3-032-30752-1_10

breeders, as their breeding season occurs when the length of daylight decreases from late summer to early winter (Chemineau et al., 1992; Delgadillo, 2011). Photoperiodic variations were initially considered as the primary environmental factors controlling sexual activity in these species (Delgadillo et al., 2020). These variations are processed by the organism and converted into hormonal messages, such as melatonin, which, in turn, regulate the reproductive axis (hypothalamus-pituitary-gonads) (for review: Malpaux et al., 1996; Dardente & Simonneaux, 2022). However, other environmental factors, such as sociosexual stimuli, appear capable of influencing this axis in a manner that overrides the neuronal circuits integrating photoperiod variations (Delgadillo et al., 2020).

In sheep and goats, males can induce ovulation by being introduced into a flock of females, a phenomenon known as the "male effect" (for review: Delgadillo et al., 2009). This effect was first reported in sheep in the 1940s as the "ram effect" (Underwood et al., 1944) and later in goats in the 1960s as the "buck effect" (Shelton, 1960). Research over the past few decades has shown the critical role of male sexual activity in triggering ovulation in anestrous females. Notably, males also exhibit a seasonal decline in sexual activity that coincides with the period of sexual rest in females (bucks: Delgadillo et al., 1991; rams: Lincoln, 1979; Ortavant et al., 1985). However, light treatment mimicking long-day conditions, combined with melatonin implants, can restore sexual activity in treated males (rams: Chemineau et al., 1996; Delgadillo et al., 2001, bucks: 2015). These sexually active males can then induce and maintain ovulatory activity in anestrous females (Abecia et al., 2015, 2019; goats: Delgadillo et al., 2015; Flores et al., 2000).

Numerous studies have explored the general aspects of the male effect (for review, see Delgadillo et al., 2009). However, relatively few have examined the specific nature of the sociosexual stimuli emitted by males. Among the sensory components involved in this phenomenon, olfactory stimuli appear to play an important role.

2 Evidence of the Role of Olfactory Stimuli in the Male Effect

Several experiments highlighted the role of male olfactory stimuli in triggering the robust activity of the reproductive axis in anestrous females. Exposure of anestrous females to ram fleece (i.e., ram odor) (Knight & Lynch, 1980; Knight et al., 1983; Cohen-Tannoudji et al., 1994) or buck hair (i.e., buck odor) (Claus et al., 1990; Walkden-Brown et al., 1993) can induce an increase in luteinizing hormone (LH) pulsatility (short-term response), sometimes leading to ovulation (long-term response) (Fig. 1). Moreover, the destruction of the sense of smell by irrigation of the nasal mucosa with zinc sulfate significantly reduces the ovulatory response of anestrous does to the introduction of bucks (Chemineau et al., 1986). Another study in goats revealed that exposure of females to sedated sexually active bucks (i.e., not displaying sexual behavior) increases LH pulsatility without triggering ovulation (Vielma et al., 2009). Thus, even if successful ovulation is not guaranteed, male olfactory stimuli can trigger a response from the female reproductive axis, at least in terms of LH pulsatility.

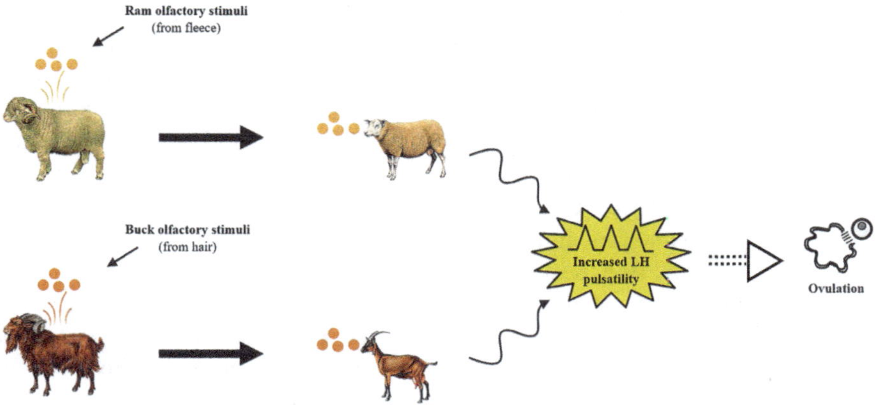

Fig. 1. Schematic representation of the role of olfactory stimuli emitted by rams and bucks in the male effect. *LH* luteinizing hormone. (Drawings © Paul Cann).

In mammals, LH pulsatility is modulated by gonadotropin-releasing hormone (GnRH) discharges into the pituitary portal system (Knobil, 1980, 1988). GnRH discharges can be monitored by electrophysiological recording in the mediobasal hypothalamus's (MBH) multiple-unit activity (MUA) (Mori et al., 1991; Mori & Tanaka, 1995). Interestingly, exposure of Shiba does to buck hair collected from the head, neck, and back regions of intact bucks induces characteristic increases in MUA of the MBH (Hamada et al., 1996; Ichimaru et al., 1999), particularly in the arcuate nucleus (Murata et al., 2011; Sakamoto et al., 2013). This hypothalamic nucleus is known to host kisspeptin-secreting neurons, a neuropeptide considered a potent secretagogue of GnRH (recent review: Sobrino et al., 2022). Thus, the link between olfactory stimuli and reproductive neuropeptides is an essential indicator of the involvement of these stimuli in the male effect.

3 Nature of Olfactory Stimuli Provided by Males

Ram fleece and buck hair contain the volatile or nonvolatile compounds involved in the male effect. This is not the case with male urine. In goats, buck urine can attract estrous does (Fritz et al., 2021); however, this effect has not been studied with anestrous females. However, in both sheep and goats, male urine does not induce any variation in LH secretion or trigger ovulation in anestrous females (sheep: Cohen-Tannoudji et al., 1994; goats: Walkden-Brown et al., 1993). As a result, several studies have investigated the origin and nature of the chemical compounds conveyed by fleece and hair. Unlike rams, whose biologically active compounds are distributed over the entire body (Knight & Lynch, 1980), those of bucks appear to be primarily concentrated in sebaceous glands around the head and neck and far from the rump (Iwata et al., 2000; Wakabayashi et al., 2000). The androgen dependence of stimulus production was only investigated in goats (Walkden-Brown et al., 1994; Iwata et al., 2000). Treatment of Shiba wethers (i.e., castrated bucks) with testosterone induces the development and size of the sebaceous glands, glands associated with hair follicles distributed all over the tegument that synthesize chemical compounds involved in the olfactory stimuli, and the characteristic induction of MUA in the female MBH (Iwata et al., 2000). Only a few studies have investigated the potential olfactory stimuli in ram fleece (Signoret, 1991). The association of synthetic 1,2-hexadecanediol and 1,2-octadecanediol efficiently stimulated LH secretion in anestrous ewes (Cohen-Tannoudji et al., 1994). However, the involvement of these compounds in ram stimuli has not been further investigated. Most work has focused on the stimuli from buck hair.

Several studies have attempted to isolate physiologically active compounds from sexually active buck hair by fractionation approach. Lipids extracted with organic solvents induced an increase in plasmatic LH concentration and pulsatility in females when they were exposed to them (Claus et al., 1990). A recent study by our team found that aqueous and methanol extracts, which contain the most polar, hydrophilic molecules (e.g., aldehydes and proteins), and the chloroform extract, which contains the most apolar, hydrophobic molecules (e.g., lipids), strongly activated cells from the main olfactory epithelium (MOE) and the vomeronasal organ (VNO), including olfactory neurons (Meunier et al., 2025b). Interestingly, aqueous extracts induced more cell responses than methanol and chloroform extracts (Meunier et al., 2025b). However, these findings suggest that a range of compounds can trigger olfactory organ activation. However, since the study did not identify specific compounds, the exact composition of the fractions remains unknown and requires further investigation. One identified compound responsible for the strong "goaty odor" of bucks during the rut is 4-ethyl octanoic acid (Sugiyama et al., 1981). This compound does not increase MUA in the MBH (Iwata et al., 2003) or LH pulsatility in olfactory-exposed females (Claus et al., 1990). However, when stored at room temperature for several months, synthetic 4-ethyl octanoic acid induces MUA increases in the MBH, suggesting that this acid may serve as a precursor of the stimulus provided by the buck (Iwata et al., 2003). Another compound, the 4-ethyloctanal, was found to trigger MUA increases in the arcuate nucleus and was, therefore, considered a strong candidate for the buck stimulus (Murata et al., 2014). However, a mixture of 18 molecules, including 4-ethyloctanal, identified only in sexually active bucks, induced a stronger response than this aldehyde alone, suggesting the involvement of several compounds in the olfactory stimulus (Murata et al., 2014). Moreover, the authors studied only the Shiba goat, a Japanese breed that does not exhibit seasonal reproductive cycles. Therefore, their generalization to the *Capra* species, and specifically their relevance to seasonal breeds like the Alpine goat, remains uncertain.

4 The Main Rather Than the Accessory Olfactory System in the Processing of Male Olfactory Stimulus

In mammals, olfactory stimuli are primarily detected by two sensory structures: the main olfactory epithelium (MOE) and the vomeronasal organ (VNO) (for review, see Bakker et al., 2022). The MOE is a pseudostratified epithelium composed of basal cells, supporting cells, and olfactory sensory neurons (OSNs) detecting odor-active

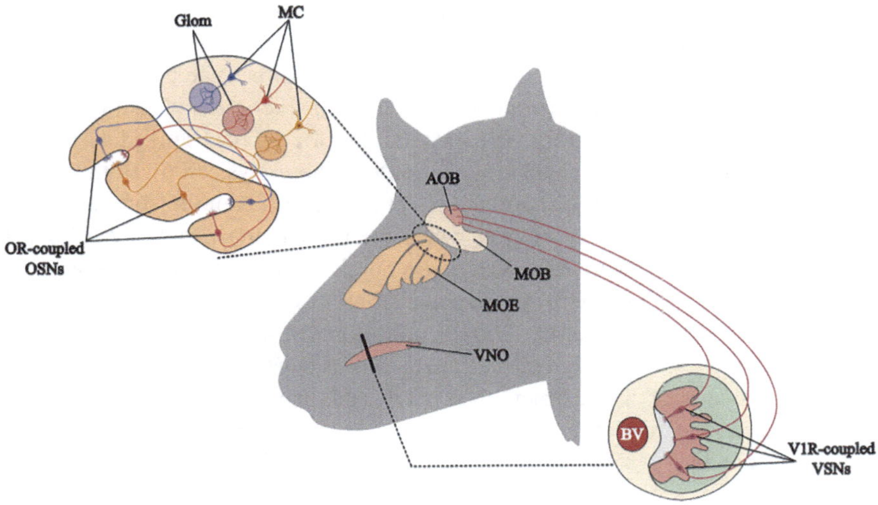

Fig. 2. Schematic representation of the perception of olfactory stimuli by the main and accessory olfactory systems, example of the goat. *AOB* accessory olfactory bulb, *BV* blood vessel, *Glom* glomeruli, *MC* mitral cells, *MOB* main olfactory bulb, *MOE* main olfactory epithelium, *OR* olfactory receptor, *OSN* olfactory sensory neuron, *VNO* vomeronasal organ, *VSN* vomeronasal sensory neuron, *V1R* vomeronasal type-1 receptor.

compounds via G protein-coupled receptors known as olfactory receptors (ORs) (Buck & Axel, 1991; Mombaerts, 2004). The MOE is the sensory organ of the main olfactory system (MOS). The VNO is a tubular structure divided into sensory and non-sensory epithelia (Ichikawa et al., 1999). The VNO is the sensory organ of the accessory olfactory system (AOS) containing vomeronasal sensory neurons (VSNs) that detect olfactory stimuli via vomeronasal receptors (VRs) (Leinders-Zufall et al., 2000). VRs are classified into type 1 (V1Rs) and type 2 receptors (V2Rs). These receptors are regionalized in the VNO of rodents, with apical VSNs expressing V1Rs and basal VSNs expressing V2Rs (Munger et al., 2009; Tirindelli et al., 2009; Liberles, 2014). This regionalization appears to be absent in sheep and goats (Fig. 2), which express V1Rs over the entire sensory epithelium of the VNO, in contrast to V2Rs, which are not or only minimally expressed (Takigami et al., 2000; Wakabayashi, 2002; Salazar et al., 2007; Ohara et al., 2009). While over 200 V1Rs have been identified in mice (Francia et al., 2014), only around 10 (Kondoh et al., 2022) or 20 genes have been identified in sheep and goats (Ohara et al., 2009). The very low number of V2Rs identified corresponds mainly to pseudogenes (Kondoh et al., 2022). Interestingly, the goat MOE also expresses a gene from the V1Rs family (*gV1ra1*) (Wakabayashi, 2002; Wakabayashi et al., 2007).

The involvement of the MOS and AOS has been studied in sheep and goats as part of the male effect. In contrast to what is observed in rodents, the MOS appears to play a more critical role than the AOS in detecting and integrating olfactory stimuli emitted by rams or bucks (for review, see Keller & Lévy, 2012). In ewes, surgical lesioning of the VNO or of its efferent nerves does not inhibit the LH response to ram exposure

(Cohen-Tannoudji et al., 1989). In contrast, the destruction of the MOE with zinc sulfate reduces the number of LH-responding ewes to the ram olfactory stimulus (Gelez & Fabre-Nys, 2004) and the number of does responding directly to bucks (Chemineau et al., 1986). The expression of the neuronal activity marker c-fos is also increased in mitral cells of the main olfactory bulb (MOB) and the cortical amygdala (CoA) in response to ram fleece in anovulatory ewes (Gelez & Fabre-Nys, 2006). Both structures are integral to the MOS. Furthermore, the effect of olfactory stimuli from ram's fleece on LH secretion is completely blocked when the CoA is inactivated by a lidocaine injection (Gelez et al., 2004a). However, this is not the case when the medial amygdala and the ventromedial nucleus of the hypothalamus, two regions associated with the AOS, are inactivated (Gelez et al., 2004b). Additionally, a recent study by our team found that MOE cells, including OSNs, from anestrous does exhibit a stronger response to sexually active buck hair extracts than MOE cells from does sampled during the breeding season (Meunier et al., 2025b). Despite a high percentage of VSNs and non-VSN cells responding to sexually active buck hair extracts (30–75%), no difference was observed between anestrous and cycling females (Meunier et al., 2025b). Taken together, these findings support an essential role of MOS in the male effect but do not rule out the potential involvement of AOS, at least in the Alpine goat.

5 Discussion and Further Research Areas

5.1 Can the Ram/Buck's Olfactory Stimuli Be Considered "Pheromones"?

In this review, we have deliberately not used the term "pheromone" to refer to the olfactory stimuli of rams and bucks. This term is the subject of intense debate in mammalian olfactory communication research, and the male effect is no exception (for review, see Delgadillo et al., 2009; Hawken & Martin, 2012). In this context, olfactory stimuli used by males do not align with the original definition established in insects (Karlson & Lüscher, 1959). The definition proposed by Wyatt is more detailed and appears more appropriate for use in mammals (Wyatt, 2014). Wyatt defined pheromone as "molecules that are evolved signals, in defined ratios in the case of multiple component pheromones, which are emitted by an individual and received by a second individual of the same species, in which they cause a specific reaction, for example, a stereotyped behavior or a developmental process" (Wyatt, 2010, 2014). A key point of these definitions is that pheromones evoke a specific reaction by the recipient individual (i.e., at least an LH response in the case of the male effect). However, studies in ewes show that the LH response to the ram's olfactory stimuli carried by fleece depends on sexual experience (Gelez et al., 2004a). Furthermore, sexually naïve ewes do not fully activate the neuronal pathways observed to respond in sexually experienced females when exposed to this olfactory stimulus (Chanvallon & Fabre-Nys, 2009). These findings suggest that ewes "learn" to recognize ram characteristics. Indeed, ewes preexposed for 2 weeks to lavender-scented rams showed LH responses to lavender alone, while ewes exposed to unscented rams were unable to do so (Fabre-Nys et al., 2015; Gelez et al., 2004b). Males were able to perform courtship but no mating. However, the conditions required for the ewes to learn the necessary male characteristics remain unclear (Chanvallon et al., 2010; Fabre-Nys et al., 2015). Another key property of a so-called pheromone is its species or genus specificity. Interestingly, ram fleece can increase MUA in the MBH and stimulate LH secretion in does (Ichimaru et al., 2008), while buck hair also stimulates LH secretion in ewes (Over et al., 1990). These results highlight similarities between the olfactory stimuli of rams and bucks involved in the male effect. The specificity of the stimulus provided by one or more compounds could be at a higher taxonomic level, such as the subfamily Caprinae (including the genera *Capra* and *Ovis*).

Wyatt also defined another category of semiochemicals that might be of interest in the context of the male effect: signature mixtures (Wyatt, 2014). A signature mixture corresponds to a variable set of molecules that conspecifics learn to recognize an animal as an individual, a class of individuals (e.g., dominance, sex, age), or a physiological or psychological state (e.g., sexually motivated, health status) (Wyatt, 2014). In the context of the male effect, introducing a new male increases the response efficiency of anestrous females (Delgadillo et al., 2009). The hypothesis put forward is that females can detect the signature mixture of an unknown male and that the "pheromone," which is ubiquitously emitted by sexually active males, would then be "authorized" to act (Wyatt, 2014). However, exposure to an intact male was more effective than olfactory stimuli alone in triggering female ovulation, suggesting a role for nonolfactory stimuli, including auditory, visual, and somatosensory (for review see: Delgadillo et al., 2009). However, the involvement of nonolfactory sensory stimuli in the male effect goes beyond the scope of this review.

5.2 New Research Directions and Challenges for Olfactory Stimuli Involved in the Male Effect

One of the most significant discoveries in recent years is the identification of 4-ethyloctanal in sexually active Shiba bucks (Murata et al., 2014). While the authors claim to have discovered the male effect pheromone in goats, we consider this compound to be a promising candidate rather than a definitive finding. In fact, this compound did not elicit a response from the GnRH pulse generator in all individuals tested, although the sample size was limited. Additionally, it remains uncertain whether this compound is present in other breeds or species and if it has the same effect outside the Shiba breed. To better understand the male effect in highly seasonal breeds like the Alpine, further studies are needed to identify the key compounds present in sexually active buck hair. Once these compounds of interest have been identified, it will be essential to confirm their ability to induce a physiological response will be required (Wyatt, 2014). In the context of the male effect, this would be demonstrated by increased LH pulsatility, potentially leading to ovulation in exposed anestrous females. The recent development of in vitro calcium imaging techniques in goats offers a valuable tool for identifying candidate compounds that can be tested in vivo to assess their endocrine and behavioral effects (Meunier et al., 2023, 2025a). In addition, this approach, combined with immunohistochemistry using neuronal activation markers such as c-fos and/or functional magnetic resonance imaging (fMRI), could provide insight into the olfactory sensory pathway(s) involved in stimulus detection and transduction (Meunier et al., 2025b).

In rams, two key compounds—1,2-hexadecanediol and 1,2-octadecanediol—were identified decades ago (Cohen-Tannoudji et al., 1994), but their role has not been thoroughly investigated. A systematic study is needed to characterize these compounds in a larger number of individuals and compare them with sexually inactive rams. Additionally, adapting the calcium imaging approach to sheep would enhance our ability to characterize the olfactory stimuli involved.

The recent advances in small ruminant research tools, such as calcium imaging and fMRI, have reignited interest in understanding the olfactory mechanisms underlying the male effect. These innovative approaches offer new perspectives for screening and identifying relevant compounds.

Acknowledgments. This work was supported by a grant from Région Centre Val de Loire (France) and the Inrae PHASE Department (France).

References

Abecia, J. A., Chemineau, P., Flores, J. A., Keller, M., Duarte, G., Forcada, F., & Delgadillo, J. A. (2015). Continuous exposure to sexually active rams extends estrous activity in ewes in spring. *Theriogenology, 84*(9), 1549–1555. https://doi.org/10.1016/j.theriogenology.2015.08.002

Abecia, J. A., Keller, M., Palacios, C., Chemineau, P., & Delgadillo, J. A. (2019). Light-induced sexually active rams prevent the seasonal inhibition of luteinizing-hormone in ovariectomized estradiol-implanted ewes. *Theriogenology, 136*, 43–46. https://doi.org/10.1016/j.theriogenology.2019.06.032

Bakker, J., Leinders-Zufall, T., & Chamero, P. (2022). The sense of smell: Role of the olfactory system in social behavior. In D. W. Pfaff, N. D. Volkow, & J. L. Rubenstein (Eds.), *Neuroscience in the 21st century* (pp. 1215–1243). Springer International Publishing. https://doi.org/10.1007/978-3-030-88832-9_29

Buck, L., & Axel, R. (1991). A novel multigene family may encode odorant receptors: A molecular basis for odor recognition. *Cell, 65*(1), 175–187. https://doi.org/10.1016/0092-8674(91)90418-X

Chanvallon, A., & Fabre-Nys, C. (2009). In sexually naive anestrous ewes, male odour is unable to induce a complete activation of olfactory systems. *Behavioural Brain Research, 205*(1), 272–279. https://doi.org/10.1016/j.bbr.2009.08.013

Chanvallon, A., Scaramuzzi, R. J., & Fabre-Nys, C. (2010). Early sexual experience and stressful conditions affect the response of young ewes to the male. *Physiology and Behavior, 99*(4), 457–465. https://doi.org/10.1016/j.physbeh.2009.12.014

Chemineau, P., Levy, F., & Thimonier, J. (1986). Effects of anosmia on LH secretion, ovulation and oestrous behaviour induced by males in the anovular creole goat. *Animal Reproduction Science, 10*(2), 125–132. https://doi.org/10.1016/0378-4320(86)90024-2

Chemineau, P., Malpaux, B., Delgadillo, J. A., Guérin, Y., Ravault, J. P., Thimonier, J., & Pelletier, J. (1992). Control of sheep and goat reproduction: Use of light and melatonin. *Animal Reproduction Science, 30*(1–3), 157–184. https://doi.org/10.1016/0378-4320(92)90010-B

Chemineau, P., Malpaux, B., Pelletier, J., Leboeuf, B., Delgadillo, J. A., Deletang, F., Pobel, T., & Brice, G. (1996). Emploi des implants de mélatonine et des traitements photopériodiques pour maîtriser la reproduction saisonnière chez les ovins et les caprins. *INRAE Productions Animales, 9*(1), 45–60. https://doi.org/10.20870/productions-animales.1996.9.1.4034

Claus, R., Over, R., & Dehnhard, M. (1990). Effect of male odour on LH secretion and the induction of ovulation in seasonally anoestrous goats. *Animal Reproduction Science, 22*(1), 27–38. https://doi.org/10.1016/0378-4320(90)90035-E

Cohen-Tannoudji, J., Lavenet, C., Locatelli, A., Tillet, Y., & Signoret, J. P. (1989). Non-involvement of the accessory olfactory system in the LH response of anoestrous ewes to male odour. *Journal of Reproduction and Fertility, 86*(1), 135–144. https://doi.org/10.1530/jrf.0.0860135

Cohen-Tannoudji, J., Einhorn, J., & Signoret, J. P. (1994). Ram sexual pheromone: First approach of chemical identification. *Physiology and Behavior, 56*(5), 955–961. https://doi.org/10.1016/0031-9384(94)90329-8

Dardente, H., & Simonneaux, V. (2022). GnRH and the photoperiodic control of seasonal reproduction: Delegating the task to kisspeptin and RFRP-3. *Journal of Neuroendocrinology, 34*(5), 1–17. https://doi.org/10.1111/jne.13124

Delgadillo, J. A. (2011). Environmental and social cues can be used in combination to develop sustainable breeding techniques for goat reproduction in the subtropics. *Animal, 5*(1), 74–81. https://doi.org/10.1017/S1751731110001400

Delgadillo, J. A., Leboeuf, B., & Chemineau, P. (1991). Decrease in the seasonality of sexual behavior and sperm production in bucks by exposure to short photoperiodic cycles. *Theriogenology, 36*(5), 755–770. https://doi.org/10.1016/0093-691X(91)90341-A

Delgadillo, J. A., Carrillo, E., Morán, J., Duarte, G., Chemineau, P., & Malpaux, B. (2001). Induction of sexual activity of male creole goats in subtropical northern Mexico using long days and melatonin. *Journal of Animal Science, 79*(9), 2245–2252. https://doi.org/10.2527/2001.7992245x

Delgadillo, J. A., Gelez, H., Ungerfeld, R., Hawken, P. A. R., & Martin, G. B. (2009). The 'male effect' in sheep and goats – Revisiting the dogmas. *Behavioural Brain Research, 200*(2), 304–314. https://doi.org/10.1016/j.bbr.2009.02.004

Delgadillo, J. A., Flores, J. A., Hernández, H., Poindron, P., Keller, M., Fitz-Rodríguez, G., Duarte, G., Vielma, J., Fernández, I. G., & Chemineau, P. (2015). Sexually active males prevent the display of seasonal anestrus in female goats. *Hormones and Behavior, 69*, 8–15. https://doi.org/10.1016/j.yhbeh.2014.12.001

Delgadillo, J. A., Hernández, H., Abecia, J. A., Keller, M., & Chemineau, P. (2020). Is it time to reconsider the relative weight of sociosexual relationships compared with photoperiod in the control of reproduction of small ruminant females? *Domestic Animal Endocrinology, 73*, 106468. https://doi.org/10.1016/j.domaniend.2020.106468

Fabre-Nys, C., Kendrick, K. M., & Scaramuzzi, R. J. (2015). The "ram effect": New insights into neural modulation of the gonadotropic axis by male odors and socio-sexual interactions. *Frontiers in Neuroscience, 9*. https://doi.org/10.3389/fnins.2015.00111

Flores, J. A., Véliz, F. G., Pérez-Villanueva, J. A., Martínez De La Escalera, G., Chemineau, P., Poindron, P., Malpaux, B., & Delgadillo, J. A. (2000). Male reproductive condition is the limiting factor of efficiency in the male effect during seasonal anestrus in female goats. *Biology of Reproduction, 62*(5), 1409–1414. https://doi.org/10.1095/biolreprod62.5.1409

Francia, S., Pifferi, S., Menini, A., & Tirindelli, R. (2014). Vomeronasal receptors and signal transduction in the vomeronasal organ of mammals. In C. Mucignat-Caretta (Ed.), *Neurobiology of chemical communication*. CRC Press/Taylor & Francis.

Fritz, W. F., Becker, S. E., & Katz, L. S. (2021). Urine from domesticated male goats (*Capra hircus*) provides attractive olfactory cues to estrous females. *Applied Animal Behaviour Science, 236*, 105252. https://doi.org/10.1016/j.applanim.2021.105252

Gelez, H., & Fabre-Nys, C. (2004). The "male effect" in sheep and goats: A review of the respective roles of the two olfactory systems. *Hormones and Behavior, 46*(3), 257–271. https://doi.org/10.1016/j.yhbeh.2004.05.002

Gelez, H., & Fabre-Nys, C. (2006). Neural pathways involved in the endocrine response of anestrous ewes to the male or its odor. *Neuroscience, 140*(3), 791–800. https://doi.org/10.1016/j.neuroscience.2006.02.066

Gelez, H., Archer, E., Chesneau, D., Campan, R., & Fabre-Nys, C. (2004a). Importance of learning in the response of ewes to male odor. *Chemical Senses, 29*(7), 555–563. https://doi.org/10.1093/chemse/bjh054

Gelez, H., Archer, E., Chesneau, D., Magallon, T., & Fabre-Nys, C. (2004b). Inactivation of the olfactory amygdala prevents the endocrine response to male odour in anoestrus ewes. *The European Journal of Neuroscience, 19*(6), 1581–1590. https://doi.org/10.1111/j.1460-9568.2004.03261.x

Hamada, T., Nakajima, M., Takeuchi, Y., & Mori, Y. (1996). Pheromone-induced stimulation of hypothalamic gonadotropin-releasing hormone pulse generator in ovariectomized, estrogen-primed goats. *Neuroendocrinology, 64*, 313–319. https://doi.org/10.1016/0168-0102(96)89184-X

Hawken, P. A. R., & Martin, G. B. (2012). Sociosexual stimuli and gonadotropin-releasing hormone/luteinizing hormone secretion in sheep and goats. *Domestic Animal Endocrinology, 43*(2), 85–94. https://doi.org/10.1016/j.domaniend.2012.03.005

Ichikawa, M., Shin, T., & Soo Kang, M. (1999). Fine structure of the vomeronasal sensory epithelium of Korean goats (*Capra hircus*). *Journal of Reproduction and Development, 45*(1), 81–89. https://doi.org/10.1262/jrd.45.81

Ichimaru, T., Takeuchi, Y., & Mori, Y. (1999). Stimulation of the GnRH pulse generator activity by continuous exposure to the male pheromones in the female goat. *Journal of Reproduction and Development, 45*(4), 243–248. https://doi.org/10.1262/jrd.45.243

Ichimaru, T., Mogi, K., Ohkura, S., Mori, Y., & Okamura, H. (2008). Exposure to ram wool stimulates gonadotropin-releasing hormone pulse generator activity in the female goat. *Animal Reproduction Science*. https://doi.org/10.1016/J.ANIREPROSCI.2007.05.012

Iwata, E., Wakabayashi, Y., Kakuma, Y., Kikusui, T., Takeuchi, Y., & Mori, Y. (2000). Testosterone-dependent primer pheromone production in the sebaceous gland of male goat. *Biology of Reproduction, 62*(3), 806–810. https://doi.org/10.1095/biolreprod62.3.806

Iwata, E., Kikusui, T., Takeuchi, Y., & Mori, Y. (2003). Substances derived from 4-ethyl octanoic acid account for primer pheromone activity for the "male effect" in goats. *Journal of Veterinary Medical Science, 65*(9), 1019–1021. https://doi.org/10.1292/jvms.65.1019

Karlson, P., & Lüscher, M. (1959). 'Pheromones': A new term for a class of biologically active substances. *Nature, 183*(4653), 55–56. https://doi.org/10.1038/183055a0

Keller, M., & Lévy, F. (2012). The main but not the accessory olfactory system is involved in the processing of socially relevant chemosignals in ungulates. *Frontiers in Neuroanatomy, 6*, 39. https://doi.org/10.3389/fnana.2012.00039

Knight, T. W., & Lynch, P. R. (1980). Source of ram pheromones that stimulate ovulation in the ewe. *Animal Reproduction Science, 3*, 133–136. https://doi.org/10.1016/0378-4320(80)90040-8

Knight, T. W., Tervit, H. R., & Lynch, P. R. (1983). Effects of boar pheromones, ram's wool and presence of bucks on ovarian activity in anovular ewes early in the breeding season. *Animal Reproduction Science, 6*(2), 129–134. https://doi.org/10.1016/0378-4320(83)90017-9

Knobil, E. (1980). The neuroendocrine control of the menstrual cycle. In *Recent progress in hormone research* (Vol. 36). Academic. https://doi.org/10.1016/b978-0-12-571136-4.50008-5

Knobil, E. (1988). The neuroendocrine control of ovulation. *Human Reproduction, 3*(4), 469–472. https://doi.org/10.1093/oxfordjournals.humrep.a136730

Kondoh, D., Kawai, Y. K., Watanabe, K., & Muranishi, Y. (2022). Tissue and cell artiodactyl livestock species have a uniform vomeronasal system with a vomeronasal type 1 receptor (V1R) pathway. *Tissue and Cell, 77*(June), 101863. https://doi.org/10.1016/j.tice.2022.101863

Leinders-Zufall, T., Lane, A. P., Puche, A. C., Ma, W., Novotny, M. V., Shipley, M. T., & Zufall, F. (2000). Ultrasensitive pheromone detection by mammalian vomeronasal neurons. *Nature, 405*(6788), 792–796. https://doi.org/10.1038/35015572

Liberles, S. D. (2014). Mammalian pheromones. *Annual Review of Physiology, 76*, 151–175. https://doi.org/10.1146/annurev-physiol-021113-170334

Lincoln, G. A. (1979). Photoperiodic control of seasonal breeding in the ram: Participation of the cranial sympathetic nervous system. *Journal of Endocrinology, 82*(1), 135–147. https://doi.org/10.1677/joe.0.0820135

Malpaux, B., Viguie, C., Thiery, J., & Chemineau, P. (1996). Photoperiodic control of breeding activity. *Productions Animales, INRA, 9*(1), 9–23.

Meunier, M., Trouillet, A.-C., Keller, M., & Chamero, P. (2023). Detection of male olfactory information by female goats: A calcium imaging approach. In B. Schaal, M. Keller, D. Rekow, & F. Damon (Eds.), *Chemical signals in vertebrates 15* (pp. 165–176). Springer International Publishing. https://doi.org/10.1007/978-3-031-35159-4_9

Meunier, M., Keller, M. & Chamero, P. (2025a). Calcium imaging in olfaction: non-rodent mammalian species. In Olfactory receptors and protocols. *Methods in Molecular Biology, 2915*, 49–60. De Souza F.M.S. and Ma M. (Eds), Humana Press. ISBN 1064-3745. https://doi.org/10.1007/978-1-0716-4466-9_2

Meunier, M., Trives, E., Gerardin, E., Porte, C., Acquistapace, A., Chemineau, P., Chamero, P., Vacher, H. & Keller, M. (2025b). Sensory responses of olfactory and vomeronasal neurons in does are influenced by buck odors and their reproductive status. *Chemical Senses 50*(bjaf012):1–15. https://doi.org/10.1093/chemse/bjaf012

Mombaerts, P. (2004). Genes and ligands for odorant, vomeronasal and taste receptors. *Nature Reviews Neuroscience, 5*(4), 263–278. https://doi.org/10.1038/nrn1365

Mori, Y., & Tanaka, T. (1995). Electrophysiological approach to the hypothalamic GnRH pulse generator. *Journal of Reproduction and Fertility. Supplement, 49*, 231–243.

Mori, Y., Nishihara, M., Tanaka, T., Shimizu, T., Yamaguchi, M., Takeuchi, Y., & Hoshino, K. (1991). Chronic recording of electrophysiological manifestation of the hypothalamic gonadotropin-releasing hormone pulse generator activity in the goat. *Neuroendocrinology, 53*(4), 392–395. https://doi.org/10.1159/000125746

Munger, S. D., Leinders-Zufall, T., & Zufall, F. (2009). Subsystem organization of the mammalian sense of smell. *Annual Review of Physiology, 71*, 115–140. https://doi.org/10.1146/annurev.physiol.70.113006.100608

Murata, K., Wakabayashi, Y., Sakamoto, K., Tanaka, T., Takeuchi, Y., Mori, Y., & Okamura, H. (2011). Effects of brief exposure of male pheromone on multiple-unit activity at close proximity to kisspeptin neurons in the goat arcuate nucleus. *Journal of Reproduction and Development, 57*(2), 197–202. https://doi.org/10.1262/jrd.10-070e

Murata, K., Tamogami, S., Itou, M., Ohkubo, Y., Wakabayashi, Y., Watanabe, H., Okamura, H., Takeuchi, Y., & Mori, Y. (2014). Identification of an olfactory signal molecule that activates the central regulator of reproduction in goats. *Current Biology, 24*(6), 681–686. https://doi.org/10.1016/j.cub.2014.01.073

Ohara, H., Nikaido, M., Date-Ito, A., Mogi, K., Okamura, H., Okada, N., Takeuchi, Y., Mori, Y., & Hagino-Yamagishi, K. (2009). Conserved repertoire of orthologous vomeronasal type 1 receptor genes in ruminant species. *BMC Evolutionary Biology, 9*(1). https://doi.org/10.1186/1471-2148-9-233

Ortavant, R., Pelletier, J., Ravault, J. P., Thimonier, J., & Volland-Nail, P. (1985). Photoperiod: Main proximal and distal factor of the circannual cycle of reproduction in farm mammals. *Oxford Reviews of Reproductive Biology, 7*, 305–345.

Over, R., Cohen-Tannoudji, J., Dehnhard, M., Claus, R., & Signoret, J. P. (1990). Effect of pheromones from male goats on LH-secretion in anoestrous ewes. *Physiology & Behavior, 48*(5), 665–668. https://doi.org/10.1016/0031-9384(90)90208-L

Sakamoto, K., Wakabayashi, Y., Yamamura, T., Tanaka, T., Takeuchi, Y., Mori, Y., & Okamura, H. (2013). A population of kisspeptin/neurokinin B neurons in the arcuate nucleus may be the central target of the male effect phenomenon in goats. *PLoS One, 8*(11), e81017. https://doi.org/10.1371/journal.pone.0081017

Salazar, I., Quinteiro, P. S., Alemañ, N., Cifuentes, J. M., & Troconiz, P. F. (2007). Diversity of the vomeronasal system in mammals: The singularities of the sheep model. *Microscopy Research and Technique, 70*(8), 752–762. https://doi.org/10.1002/jemt.20461

Shelton, M. (1960). Influence of the presence of a male goat on the initiation of estrous cycling and ovulation of angora does. *Journal of Animal Science, 19*(2), 368–375. https://doi.org/10.2527/jas1960.192368x

Signoret, J. P. (1991). Sexual pheromones in the domestic sheep: Importance and limits in the regulation of reproductive physiology. *The Journal of Steroid Biochemistry and Molecular Biology, 39*(4), 639–645. https://doi.org/10.1016/0960-0760(91)90263-5

Sobrino, V., Avendaño, M. S., Perdices-López, C., Jimenez-Puyer, M., & Tena-Sempere, M. (2022). Kisspeptins and the neuroendocrine control of reproduction: Recent progress and new frontiers in kisspeptin research. *Frontiers in Neuroendocrinology, 65*, 100977. https://doi.org/10.1016/j.yfrne.2021.100977

Sugiyama, T., Sasada, H., Masaki, J., & Yamashita, K. (1981). Unusual fatty acids with specific odor from mature male goat. *Agricultural and Biological Chemistry, 45*(11), 2655–2658. https://doi.org/10.1271/bbb1961.45.2655

Takigami, S., Mori, Y., & Ichikawa, M. (2000). Projection pattern of vomeronasal neurons to the accessory olfactory bulb in goats. *Chemical Senses, 25*(4), 387–393. https://doi.org/10.1093/chemse/25.4.387

Tirindelli, R., Dibattista, M., Pifferi, S., & Menini, A. (2009). From pheromones to behavior. *Physiological Reviews, 89*(3), 921–956. https://doi.org/10.1152/physrev.00037.2008

Underwood, E., Shier, F., & Davenport, N. (1944). The breeding season in merino, crossbreed and British reed ewes in the agricultural districts. *Journal of the Department of Agriculture Western Australia, 21*, 135–143.

Vielma, J., Chemineau, P., Poindron, P., Malpaux, B., & Delgadillo, J. A. (2009). Male sexual behavior contributes to the maintenance of high LH pulsatility in anestrous female goats. *Hormones and Behavior, 56*(4), 444–449. https://doi.org/10.1016/j.yhbeh.2009.07.015

Wakabayashi, Y. (2002). A putative pheromone receptor gene is expressed in two distinct olfactory organs in goats. *Chemical Senses, 27*(3), 207–213. https://doi.org/10.1093/chemse/27.3.207

Wakabayashi, Y., Iwata, E., Kikusui, T., Takeuchi, Y., & Mori, Y. (2000). Regional differences of pheromone production in the sebaceous glands of castrated goats treated with testosterone. *Journal of Veterinary Medical Science, 62*(10), 1067–1072. https://doi.org/10.1292/jvms.62.1067

Wakabayashi, Y., Ohkura, S., Okamura, H., Mori, Y., & Ichikawa, M. (2007). Expression of a vomeronasal receptor gene (V1r) and G protein α subunits in goat, Capra hircus, olfactory receptor neurons. *The Journal of Comparative Neurology, 503*(2), 371–380. https://doi.org/10.1002/cne.21394

Walkden-Brown, S. W., Restall, B. J., & Henniawati. (1993). The male effect in the Australian cashmere goat. 2. Role of olfactory cues from the male. *Animal Reproduction Science, 32*(1–2), 55–67. https://doi.org/10.1016/0378-4320(93)90057-X

Walkden-Brown, S. W., Restall, B. J., Norton, B. W., Scaramuzzi, R. J., & Martin, G. B. (1994). Effect of nutrition on seasonal patterns of LH, FSH and testosterone concentration, testicular mass, sebaceous gland volume and odour in Australian cashmere goats. *Reproduction, 102*(2), 351–360. https://doi.org/10.1530/jrf.0.1020351

Wyatt, T. D. (2010). Pheromones and signature mixtures: Defining species-wide signals and variable cues for identity in both invertebrates and vertebrates. *Journal of Comparative Physiology A: Neuroethology, Sensory, Neural, and Behavioral Physiology, 196*(10), 685–700. https://doi.org/10.1007/s00359-010-0564-y

Wyatt, T. D. (2014). *Pheromones and animal behavior: Chemical signals and signatures* (2nd ed.). Cambridge University Press. https://doi.org/10.1017/CBO9781139030748

Part III

Tools, Methods, and Analytical Advances

Contrasting Effects of Tricaine and Liquid Nitrogen on Injury Cue Perception in *Ranitomeya sirensis*

Konrad Lipkowski[1,2] (✉) and Lisa M. Schulte[2]

[1] Museum Wiesbaden, Naturhistorische Sammlung, Wiesbaden, Germany
[2] Wildlife-/Zoo-Animal-Biology and Systematics, Faculty of Biological Sciences, Goethe University Frankfurt, Frankfurt am Main, Hessen, Germany
lipkowski@bio.uni-frankfurt.de

Abstract. Predator and prey interactions are vital for survival and often depend on sensory abilities. Tadpoles use chemical injury cues to assess risk and adjust behavior, life history, and morphology. Such insights are commonly achieved using chemical extracts prepared from donor individuals, presented to receiver tadpoles. One potential pitfall in these studies is the euthanasia method. Studies with gregarious species have shown that especially chemical anesthetics can alter how tadpoles respond to ecologically relevant cues. In taxa with solitary tadpoles, such as poison dart frogs (Dendrobatidae) that develop in tiny, resource-limited phytotelmata, the prevalence of chemically mediated alarm responses and the impact of common laboratory procedures remain unclear. Here, we tested how tadpoles of the dendribatid species *Ranitomeya sirensis* reacted to chemical injury cues of conspecifics when different euthanasia protocols where applied. Predator naive tadpoles were exposed to extracts prepared from donor tadpoles euthanized with blunt force trauma (BFT), liquid nitrogen (LN2), tricaine methanesulfonate (MS222), or water as a control. We quantified the change in activity in each tadpole after addition of the stimuli. Tadpoles increased activity to extracts prepared using blunt force trauma and liquid nitrogen but showed no consistent change to extracts prepared using chemical euthanasia with MS222. These findings indicate that, even though the behavioral reaction to injury cues in *R. sirensis* might vary from gregarious species, anesthetic residues likely interfere with mediating chemical information in a similar manner. When MS222 administration is unavoidable, studies should include anesthetic-only controls to disentangle confounding effects of chemical anesthesia on chemically mediated responses in tadpoles and other aquatic organisms.

Keywords: Dendrobatidae · Chemical communication · Antipredator · Euthanasia · Alarm substance · Tricaine

1 Introduction

Predation is a major selective pressure driving the evolution of sophisticated sensory systems and plastic antipredator strategies (Ferrari et al., 2010; Bradbury & Vehrencamp, 2011). Amphibian larvae are particularly vulnerable (Duellman & Trueb, 1994;

L. M. Schulte (ed.), *Chemical Signals in Vertebrates 16*,
https://doi.org/10.1007/978-3-032-30752-1_11

McDiarmid & Altig, 1999; Wells, 2007) and therefore rely heavily on chemical information to assess ambient predation risk. Predator-borne kairomones can reveal a predator's identity, recent diet, and spatial proximity (Kats & Dill, 1998; Ferrari et al., 2010), while prey-borne substances provide more direct risk information, as they are released when prey is stressed (disturbance substances) or physically injured (injury cues; Mitchell et al., 2017; Crane et al., 2022). Upon acute or prolonged perception, such chemical cues modulate tadpole behavior, life history, and morphology (van Buskirk & Mccollum, 2000; Relyea, 2001; Fraker et al., 2009; Gazzola et al., 2015; Hettyey et al., 2015).

Most experimental work, however, focuses on gregarious ranid, bufonid, and hylid tadpoles from large clutches (Rödin-Mörch et al., 2011; Hossie et al., 2016; Schulte et al., 2023), leaving solitary or small-group taxa in resource-limited microhabitats underrepresented (Schulte et al., 2023). Among these understudied taxa, Neotropical poison dart frogs (Dendrobatoidea) offer valuable insight into the evolution of chemical communication. Tadpoles of many dendrobatoid species develop in phytotelmata, i.e., in small resource-restricted water bodies in plants (Brust, 1993; Caldwell & De Araújo, 1998; Summers, 1999; Poelman & Dicke, 2007; Wells, 2007; Schulte et al., 2010). These habitats generally promote a largely solitary or small-group lifestyle, which stands in sharp contrast to crowded larval environments typical of ranid and bufonid frogs that produce much larger clutches (Duellman & Trueb, 1994; Wells, 2007). In these microhabitats, poison dart frog tadpoles use chemical cues in parent-offspring communication (Kam & Yang, 2002; Stynoski & Noble, 2012; Schulte & Mayer, 2017) and detect predators via kairomones (Surber-Cunningham et al., 2024; see also Szabo et al., 2021). *Ranitomeya sirensis*, for example, responds to conspecific but not heterospecific injury cues, hinting at highly specific chemical cue recognition (Lipkowski et al., 2023). Nonetheless, the broader ecological role of chemical communication in poison dart frog larvae remains poorly understood, underscoring the need for robust experimental designs.

On one hand, choosing a robust experimental design to obtain data from the receiver individuals is highly relevant (such as video tracking to help quantify complex behaviors in poison dart frog tadpoles; Szabo et al., 2021; Surber-Cunningham et al., 2024). On the other hand, the collection method of the chemical cues to be tested is just as important. While for some experiments chemical cues can be collected freshly without harming the donors (e.g., Schulte et al., 2015), for those experiments testing reactions to injury cues, donor tadpoles need to be harmed or even euthanized. The latter is a distinct pitfall (Achtymichuk et al., 2022; Lipkowski et al., 2025). Ethical guidelines require humane euthanasia (Leary et al., 2020), commonly administered with chemical anesthetics that suppress neural activity (Richards, 1972, 1973; Späth & Schweickert, 1977). Yet these compounds can persist in tissues (Scherpenisse & Bergwerff, 2007; Kiessling et al., 2009; Li et al., 2014; Xue et al., 2017) and can alter or diminish the efficacy of injury cues and novel cue-risk associations (Fraker et al., 2009; Achtymichuk et al., 2022; Lipkowski et al., 2025). Moreover, these effects may vary among species as evidence suggest anesthetic-contaminated injury cues increase the response activity in *Rana temporaria* (Ranidae) yet decrease it in *Bufo bufo* (Bufonidae; Lipkowski et al., 2025). Such interference and variability complicate within-species interpretations and cross-species comparisons and are particularly problematic for taxa whose chemical communication systems remain poorly characterized (Schulte et al., 2023). Understanding how euthanasia methods influence cue efficacy is therefore essential for robust experimental designs and interpretation, as well as for identifying alternative methods that minimize both animal distress and experimental artifacts (Leary et al., 2020; Lipkowski et al., 2025).

Building on previous studies investigating poison dart frog responses to injury cues (Lipkowski et al., 2023) and the influence of euthanasia methods on chemically mediated tadpole behavior (Achtymichuk et al., 2022; Lipkowski et al., 2025), we hypothesize that the method of euthanasia used to prepare conspecific extracts including injury cues will also affect behavioral responses (i.e., activity levels) in *Ranitomeya sirensis* tadpoles. Specifically, we compare extracts prepared after mechanical trauma (blunt force trauma, BFT), hypothermic euthanasia (liquid nitrogen, LN2), and chemical euthanasia (tricaine methanesulfonate, MS222), against a water control. Consistent with findings from Lipkowski et al. (2023), we predict that *R. sirensis* tadpoles will exhibit increased activity in response to injury cues derived from hypothermic or mechanically euthanized conspecifics, whereas no behavioral change is expected in response to cues generated via chemical euthanasia. We predict increased activity because *R. sirensis* tadpoles previously showed elevated movement to conspecific injury cues, consistent with escape/scan strategies in confined microhabitats (Lipkowski et al., 2023). By making the euthanasia method explicit and testing its effects on alarm-cue potency in a dendrobatid, this study provides methodological guidance for chemically mediated behavior research in small-volume larval systems.

2 Methods

This study builds on a previously published dataset by Lipkowski et al. (2023), which included behavioral responses of *Ranitomeya sirensis* tadpoles to control treatments and extracts derived from conspecifics euthanized via blunt force trauma (BFT). All bioassays, including those reported in the earlier study, were conducted concurrently between 2021 and 2022 under identical environmental and methodological conditions. For the current analysis, we incorporated this previously analyzed dataset and expanded it by conducting new trials using extracts prepared with two additional euthanasia methods: immersion in liquid nitrogen (LN2) and tricaine methanesulfonate (MS222). This design allowed us to assess whether the mode of euthanasia influences the behavioral potency of conspecific injury cues.

All procedures for animal housing, behavioral testing, and statistical evaluation followed the methodology described in Lipkowski et al. (2023), while euthanasia techniques and extract preparation adhered to the protocols detailed in Lipkowski et al. (2025).

2.1 Animals

Ranitomeya sirensis (Aichinger, 1991) is a dendrobatid frog native to the Serranía de Sira region in east-central Peru (Lötters et al., 2007; Fig. 1). *R. sirensis* lays two to four eggs and transports their predatory (potentially cannibalistic) tadpoles individually into

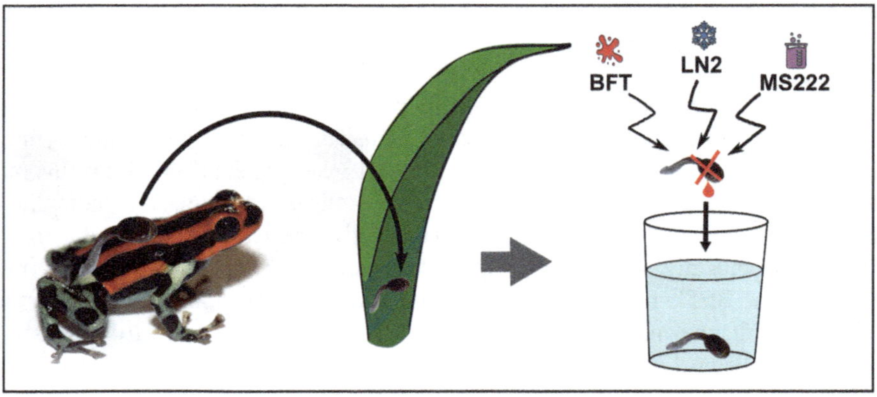

Fig. 1. Experimental overview. Adult frogs of the species *Ranitomeya sirensis* (left) transport their tadpoles individually into phytotelmata (middle). We divided these tadpoles into injury cue donors (right, upper part) and receivers (right, lower part). The donor tadpoles were euthanized in three different manners: mechanically (blunt force trauma, BFT), hypothermically (liquid nitrogen, LN2), or with chemical anesthesia (MS222). The chemical cues of the donors were presented to the receiver tadpoles and their activity was measured. (Picture of *R. sirensis* by Evan Twomey).

small phytotelmata (e.g., bromeliad leaf axils; Brown et al., 2011; Kahn et al., 2016). Tadpoles used in the current study originated from a laboratory colony maintained at Goethe University in Frankfurt am Main, Germany, since 2018. Adult frogs were housed in glass tanks measuring 60 × 30 × 30 cm at room temperatures between 23 and 25 °C and humidity levels of 70–90%. They were fed flightless fruit flies (*Drosophila melanogaster*) three times per week and occasionally supplemented with springtails. The tanks included a clay substrate, leaf litter, sticks, vines of *Epipremnum aureum*, and 50 mL Falcon tubes filled with demineralized water, which the adults used as nursery sites. These artificial nurseries mimic natural phytotelmata such as bromeliad leaf axils. On a regular basis, newly hatched tadpoles were moved from these tubes to individual tadpole housing units in the same room.

Each tadpole housing unit consisted of an 8.5-cm-diameter beaker with 4 cm water depth, demineralized water, and dried almond leaves (*Prunus dulcis*). Because tadpoles are solitary and exhibit cannibalistic tendencies, they were reared individually, with no conspecifics in the same container. Both adult and tadpole environments were maintained at a constant temperature of 23–25 °C under a 12 L:12D photoperiod. Tadpoles were fed nettle powder every second day and occasionally supplemented with rabbit chow. Water changes were not required during the study period (Lipkowski et al., 2023).

2.2 Euthanasia and Extraction of Injury Cues

To examine how different euthanasia procedures influence the behavioral activity of conspecific injury cues, donor tadpoles were randomly chosen from the rearing population and used to prepare stimulus solutions for the bioassays. To extract injury cues, each tadpole was rinsed with demineralized water prior to euthanasia, which was performed using one of three approaches: (a) mechanical trauma, (b) extreme cooling, or (c) chemical sedation (Fig. 1). For mechanical euthanasia, tadpoles were subjected to blunt force trauma (BFT) to the head, using a 1-cm-wide cylindrical tube and a pestle to rapidly destroy the central nervous system and ensure immediate death. In the hypothermic treatment, tadpoles were placed individually into 5 mL falcon tubes and submerged in liquid nitrogen (LN2; −196 °C) for 15 s. In the chemical condition, euthanasia was carried out by immersing tadpoles for 10 minutes in a buffered 0.03% solution of tricaine methanesulfonate (MS222; Trica Pharmaq® 1000 mg/g, Pharmaq Limited), adjusted to a neutral pH (7) with 1 M Tris(hydroxymethyl)-aminomethane. Throughout the exposure, tadpoles were monitored for the absence of swimming movements and loss of both positional and depth reflexes. Only individuals showing a complete absence of reflexes after the full exposure time were removed from the solution and rinsed to eliminate residual tricaine.

Immediately after each initial euthanasia method (mechanical, hypothermic, or chemical), animals were homogenized in 2 ml of water for 30 s using a handheld rotor-stator homogenizer (TissueRuptor II, Qiagen). The resulting crude extract was then centrifuged at 10,000 × *g* for 10 min at 4 °C to separate soluble from insoluble components. The supernatant was collected and diluted to a final volume of 5 mL per tadpole and loaded into syringes. Stimulus solutions were prepared immediately and brought to room temperature before each bioassay (within 30 min post-euthanasia) to minimize chemical degradation of biologically relevant compounds involved in chemically mediated antipredator mechanisms (van Buskirk et al., 2014; Crane et al., 2023). All extracts were handled using demineralized water to match the water chemistry of both housing and test conditions (see also Sect. 2.3).

One donor tadpole was euthanized for each recipient tadpole in cue treatments (BFT, LN2, MS222; total $N = 72$), while controls received water only. To estimate the concentration of biomass in the stimulus solutions while minimizing stress from handling, tadpoles were weighed every third day. For this, individuals were gently introduced into a beaker containing a known volume of water, and wet mass was determined by calculating the difference in total weight. Final extract concentrations used in trials ranged between 0.20 and 0.48 mg/mL, exceeding those reported in previous studies that successfully elicited behavioral responses in ranid and bufonid tadpoles (e.g., Hagman & Shine, 2008; Fraker et al., 2009; Lucon-Xiccato, 2019a, b; Crossland et al., 2019). Tadpoles used for extract preparation were between larval developmental stages 28 and 36 (Gosner, 1960).

2.3 Behavioral Assay

To assess how *Ranitomeya sirensis* tadpoles respond to conspecific injury cues generated through different euthanasia methods, we conducted a series of behavioral trials between April 2021 and May 2022. The experiment involved predator-naïve tadpoles ($N = 116$), each randomly exposed to one of four stimulus treatments: extracts from conspecifics euthanized via blunt force trauma (BFT), liquid nitrogen immersion (LN2), buffered tricaine solution (MS222), or demineralized water (control; Fig. 1).

In each trial, tadpoles were individually placed in plastic cups (8 cm in diameter) filled with 495 mL of demineralized water. Each behavioral trial was carried out in three sequential phases: (1) a pre-stimulus phase, (2) a stimulus phase, and (3) a post-stimulus phase, each lasting 5 min. After an acclimatization period (1 h), the trials started with the pre-stimulus phase, followed by the administration of one of the four stimuli mentioned. To introduce the stimuli, we gently placed the syringes on the cups' side without disturbing the animal. We then slowly dripped the content (5 mL) into the cups, which marked the start of the stimulus phase.

Preliminary tests using food dye confirmed that the added volume dispersed throughout the entire water column within 5 min. After the stimulus phase, the trial entered the poststimulus phase. Each trial ended after the poststimulus phase. Behavior was recorded using a Panasonic video camera positioned above the containers. A mirror, angled to provide a top-down view of the container interior, allowed the full area to be captured without the camera being positioned directly above the water surface. Prior pilot tests showed that observer presence near the mirror did not influence tadpole behavior (see also McIntyre et al., 2004; Carlson et al., 2015; Lipkowski et al., 2023). Therefore, stimuli were administered while standing briefly at the side of the mirror setup. To control for potential time-of-day effect-variation in tadpole responses, trials were distributed across five time slots between 10:00 and 17:00, with treatments rotated daily to avoid testing the same stimulus group at the same time on consecutive days. Each tadpole participated in only one trial. Developmental stages of all individuals tested ranged from stage 28–36 (Gosner, 1960).

2.4 Video Evaluation and Statistical Analysis

To evaluate changes in tadpole behavior following exposure to conspecific injury cues, we measured activity levels during the pre-stimulus (phase 1) and poststimulus (phase 3) periods. Video scoring was performed blind to treatment. Activity was defined as any visible movement, including swimming or tail motion, and recorded in seconds using CowLog software (version 3.0.2; Pastell, 2016). Brief tail movements that did not result in forward propulsion were also counted as activity, as dendrobatid tadpoles often inhabit small and spatially restricted aquatic environments (Brust, 1993; Caldwell & De Araújo, 1998; Summers, 1999; Poelman & Dicke, 2007). A response index was calculated for each individual by subtracting activity during the pre-stimulus phase from the poststimulus phase: [response activity = post - pre]. This variable reflected individual changes in movement following stimulus application in phase 2. Positive values indicated increased activity, while negative values represented reduced movement after stimulus exposure. Individuals that remained inactive in both phases were excluded from the analysis. With this, we excluded six data points from the control treatment due to zero activity in both phases, whereas no exclusions were necessary in the other treatments. We opted to use absolute rather than proportional changes as poison dart frog tadpoles typically exhibit low levels of spontaneous activity, often remaining inactive for extended periods. Such inactivity is a natural behavior in species adapted to small, nutrient-limited water bodies. Calculating proportional changes (e.g., (post-pre)/pre) would have necessitated excluding many biologically valid data points due to division by zero or near-zero values. Moreover, small absolute changes in behavior may appear disproportionately large when baseline activity is minimal, potentially distorting biological interpretations (Vickers, 2001). Because the activity data did not meet the assumptions of normality (as tested by the Kolmogorov-Smirnov test), we used a non-parametric Kruskal-Wallis test (k samples) to evaluate the effect of "treatment" (i.e., stimulus type) on the dependent variable "response activity." Upon detecting significant overall differences (see Sect. 3), we conducted pairwise post hoc comparisons using Mann-Whitney U tests, comparing each treatment group (BFT, LN2, MS222) against the control (water). A Bonferroni correction for multiple testing was applied, resulting in an adjusted alpha level of $P = 0.017$, above which comparisons were considered nonsignificant. Additionally, we reported effect sizes: epsilon-squared (ε^2) for the Kruskal-Wallis test and Rosenthal's r for the Mann-Whitney comparisons. Approximate 95% confidence intervals for r were obtained using Fisher's r-to-z transformation (based on the total sample size in each pairwise test), and the sign of r was assigned according to the direction of the group medians. All statistical analyses were carried out using SPSS version 20 (SPSS Inc., Chicago, IL, USA).

2.5 Ethical Note

All animal care and experimental procedures were conducted in accordance with national and institutional regulations in Germany (state of Hesse). The study received ethical approval from the *Regierungspräsidium Darmstadt* (approval numbers: V54-19c20/21I-FR Biologicum and V54-19c18-FR/1022) and the *Hessisches Landesamt für Naturschutz, Umwelt und Geologie* (reference: N2-103s0601/2020-3-06). All euthanasia procedures followed the guidelines set by the American Veterinary Medical Association (AVMA) for humane euthanasia in animals (Leary et al., 2020) and were carried out by trained personnel certified through a FELASA-accredited course in amphibian euthanasia (Federation of European Laboratory Animal Science Associations). The use of blunt force trauma (BFT) as a mechanical method ensured immediate and irreversible destruction of the central nervous system while avoiding the introduction of confounding chemical substances. When performed by qualified individuals, BFT is an accepted method in behavioral ecology for this reason. The hypothermic method, involving immersion in liquid nitrogen (LN2), was selected for its rapidity and lack of chemical by-products. This approach meets ethical standards for small-bodied aquatic organisms weighing less than 4 grams, a category that includes all tadpoles used in this study. Chemical euthanasia was applied using tricaine methanesulfonate (MS222) at a concentration of 0.03% (300 mg/L) for a 10-min exposure period. This dosage and duration follow established veterinary protocols at our institution and align with those used in comparable studies (e.g., Fraker et al., 2009; Achtymichuk et al., 2022; Lipkowski et al., 2025). Tricaine is widely accepted as a humane method to euthanize aquatic species (Leary et al., 2020). Following each euthanasia procedure (BFT, LN2, or MS222), individuals were immediately processed for extract preparation (see Sect. 2.2). This process ensured irreversible death and marked the completion of the euthanasia procedure, classifying each method as a two-stage process (Leary et al., 2020).

3 Results

Building on the previously published dataset from Lipkowski et al. (2023), which included tadpole responses to control stimuli and extracts from conspecifics euthanized via blunt force trauma (BFT), we expanded the analysis by incorporating two additional euthanasia methods: immersion in liquid nitrogen (LN2) and tricaine methanesulfonate (MS222). This resulted in a total sample size of $N = 110$ individuals tested in behavioral bioassays, distributed across four treatment groups: control ($N = 38$) and BFT ($N = 24$) from Lipkowski et al. (2023) and LN2 ($N = 24$) and MS222 ($N = 24$), additionally analyzed for the current study.

Median response activity in the control group was 0.29 s (IQR: −4.55 to 9.43), indicating minimal change from baseline behavior (i.e., activity before the introduction of stimuli). In contrast, tadpoles exposed to BFT extracts showed a median activity of 19.73 s (IQR: −9.16 to 33.21), those exposed to LN2 extracts exhibited the strongest response with a median of 23.42 s (IQR: 16.57–61.87), and MS222 extracts elicited an intermediate response of 10.25 s (IQR: −15.89 to 49.70).

A Kruskal-Wallis test revealed a significant difference in median response activity among the four treatments ($H = 9.53$, df = 3, $P < 0.050$, $\varepsilon^2 = 0.06$, small to moderate), indicating treatment-related differences in tadpole responses. Follow-up pairwise comparisons using Mann-Whitney U tests showed that tadpoles exhibited significantly greater response activity to extracts prepared via BFT ($Z = -2.59$, $P = 0.010$, $r = 0.33$, 95% CI 0.09–0.53; medium) and LN2 ($Z = -4.71$, $P < 0.001$, $r = 0.60$, 95% CI 0.41–0.74; large) compared to the control. In contrast, no significant difference in response activity was found between the control and the MS222 group ($Z = -1.44$, $P = 0.150$, $r = 0.18$, 95% CI −0.07–0.41; small), consistent with the broader spread of individual responses in that group (Fig. 2).

4 Discussion

We tested whether the method of euthanasia used to prepare conspecific injury cues alters behavioral responses in *Ranitomeya sirensis* tadpoles. Building on responses to water controls and blunt force trauma (BFT) reported by Lipkowski et al. (2023), we added hypothermic euthanasia via liquid nitrogen (LN2) and chemical euthanasia with tricaine methanesulfonate (MS222). Tadpoles increased activity to BFT- and LN2-derived cues but showed no consistent change to MS222-derived cues. Effect sizes mirrored this pattern, with medium effects for BFT, large for LN2, and small for MS222, indicating that euthanasia method is a key, controllable source of variation in apparent injury cue potency (Fig. 2).

These results align with prior work showing that chemical euthanasia can impair the efficacy of injury cues and dampen antipredator responses in gregarious tadpoles (Fraker et al., 2009; Achtymichuk et al., 2022; Lipkowski et al., 2025). Comparable disruptions are documented in fishes, where exposure to anesthetics alters thermoregulatory behavior and responses to injury cues at subeffective concentrations (Goddard

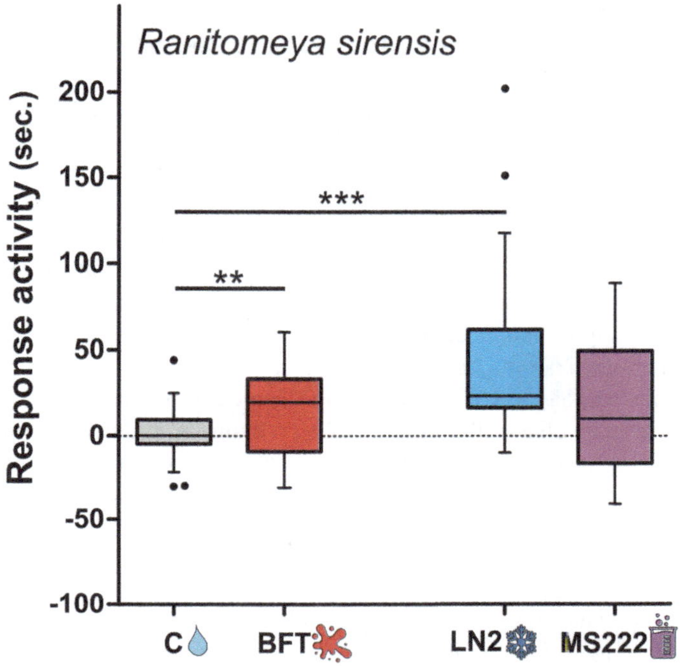

Fig. 2. Short-term changes in activity (response activity) in *Ranitomeya sirensis* tadpoles following exposure to conspecific injury cue extracts generated via mechanical (BFT; $N = 24$), hypothermic (LN2; $N = 24$), or chemical (MS222; $N = 24$) euthanasia, as well as a water control (C; $N = 38$). Boxplots display raw data, showing the median (internal line), interquartile range (box edges), most data points' range (whiskers; up to 1.5 IQR), and outliers (black dots). The dashed line represents equal pre- and post-activity. Datasets previously published as part of Lipkowski et al. (2023) are shown on the left and include the control and injury cue extracts from mechanically euthanized conspecifics (BFT). Trials additionally analyzed as part of this study (LN2 and MS222) are shown on the right. Asterisks denote significance levels in Mann-Whitney U post hoc comparisons against the control group: *** $P < 0.001$, ** $P < 0.01$. Alpha level adjusted for multiple comparisons (Bonferroni correction: $P = 0.0167$).

et al., 1974; Losey & Hugie, 1994; Readman et al., 2013; Wong et al., 2014). Together, these studies suggest that even low residual anesthetic can interfere with ecologically relevant, chemically mediated behaviors.

Several mechanisms could explain the weak MS222 response. Residual anesthetic might directly alter cue molecules or water chemistry (Achtymichuk et al., 2022). However, in control assays with different concentrations of MS222, including trials with the putative injury cue arginine, point instead to receiver-side interference in which anesthetic residues disrupt chemosensory reception and/or neural processing rather than altering the cue itself (Lipkowski et al., 2025). Such effects also appear species-specific (Lipkowski et al. 2025), reflecting distinct pharmacokinetic or pharmacodynamic profiles between species (Zahl et al., 2012). For instance, while *Rana temporaria* tadpoles exhibited an increase in activity when exposed to MS222 controls, *Bufo bufo* tadpoles displayed a marked reduction (Lipkowski et al., 2025). In *R. sirensis*, we found neither a clear increase nor decrease but rather a broad spread of responses to MS222 extracts. This is reflected in the low effect size, suggesting interindividual variation in sensitivity or response thresholds.

By contrast, LN2-derived extracts elicited strong, consistent responses, suggesting that *R. sirensis* injury cues are not degraded by extreme cold. This matches earlier work indicating that ecologically relevant cue components in tadpoles are thermally stable (Fraker et al., 2009; Austin et al., 2018; Raices et al., 2020; Achtymichuk et al., 2022). The large effect size for LN2 (i.e., the strength and consistency of LN2 responses) indicates that hypothermic euthanasia can at least match mechanical euthanasia in yielding behaviorally effective extracts while avoiding external chemical interference. In fact, flash freezing likely promotes cellular lysis, facilitating release of injury cues during subsequent homogenization (Austin et al., 2018).

When donor tadpoles were euthanized mechanically or hypothermically (i.e., no anesthetics where involved), the usually solitary living *R. sirensis* tadpoles increased their activity after exposure to conspecific extracts. The ecological significance of this response remains opaque. Lipkowski et al. (2023) noted specificity to conspecific injury cues and proposed specialized recognition that could be directed at conspecific competitors. In small phytotelmata, conspecific injury cues may indicate not only predation risk but also nutrient opportunities and increase the likelihood of competitive/cannibalistic interactions. Thus, increased activity could reflect escape, foraging, or aggressive behavior. Disentangling these alternatives will require assays that administer different cue types and manipulate social context like the presence or absence of conspecifics and cross injury cues with predator kairomones (Lipkowski et al., 2023). Surber-Cunningham et al. (2024) exposed cannibalistic *Dendrobates tinctorius* tadpoles to conspecific injury, predator, and food cues and found that injury cues increased exploration relative to baseline and predator cues but produced less overall movement than food cues. Their study is an important step toward disentangling responses to different cue types. Because their injury cue extracts were prepared using MS222, and our study shows that residual anesthetic can modulate behavior not only in gregarious but also in solitary/cannibalistic tadpoles, a contribution of anesthetic residues cannot be ruled out. Replication with nonchemical euthanasia, or inclusion of anesthetic-only controls, would therefore help confirm that the pattern reflects cue identity rather than interactions between the cue and the anesthetic.

Our study shows that euthanasia method is not a trivial procedural choice in chemically mediated behavior studies (see also Achtymichuk et al. 2022 and Lipkowski et al. 2025). Many aquatic studies therefore use nonchemical euthanasia to ensure immediate and irreversible destruction of the central nervous system while avoiding confounding residues (e.g., Meuthen et al., 2018; Lucon-Xiccato, 2019a, b; Lipkowski et al., 2023), as alternatives are permissible where chemical anesthesia compromises experimental integrity (Leary et al., 2020). For dendrobatids in small-volume habitats, such artifacts risk misinterpreting cue function and the processes shaping antipredator or competitive strategies. In practical terms, nonchemical options (LN2 or BFT) appear preferable for studies involving chemically mediated responses in *R. sirensis*. Where MS222 is mandated, studies should, at minimum, include anesthetic-only controls and report wash steps explicitly.

5 Conclusion

Euthanasia method is a controllable source of variation in tadpole alarm-cue assays. In *Ranitomeya sirensis*, nonchemical approaches preserved the behavioral potency of conspecific extracts, whereas anesthetic-derived euthanasia produced no consistent response and greater interindividual variability. We therefore recommend avoiding the use of anesthetics during injury cue studies in solitary and gregarious tadpoles alike. Future work should focus on the identity of active cue components under different preparation methods as well as further comparisons across solitary and gregarious taxa. Lastly, reporting effect sizes facilitates cross-study comparisons and helps assess the practical significance of findings. Adopting these practices will improve the reliability and comparability of amphibian chemical-ecology assays, including those developing in small, resource-limited nurseries.

Acknowledgments. We gratefully acknowledge Diana Abondano Almeida, Carmen Meyer, Evan Twomey, and the animal care team at the Biologicum, Goethe University Frankfurt, for support in the animal room. This work was supported by the Wilhelm-Peters-Fonds of the Deutsche Gesellschaft für Herpetologie und Terrarienkunde (DGHT).

References

Achtymichuk, G. H., Crane, A. L., Simko, O. M., Stevens, H. E. F., & Ferrari, M. C. O. (2022). The choice of euthanasia techniques can affect experimental results in aquatic behavioural studies. *Animal Behaviour, 185*, 1–8.

Austin, C. E., March, R. E., Stock, N. L., & Murray, D. L. (2018). The origin and ecological function of an ion inducing anti-predator behavior in *Lithobates* tadpoles. *Journal of Chemical Ecology, 44*, 178–188.

Bradbury, J. W., & Vehrencamp, S. L. (2011). *Principles of animal communication* (2nd ed.). Oxford University Press.

Brown, J. L., Twomey, E. M., Amézquita, A., de Souza, M. B., Caldwell, J. P., Lötters, S., von May, R., Melo-Sampaio, P. R., Mejía-Vargas, D., Pérez-Peña, P. E., Pepper, M., Poelman, E. H., Sánchez-Rodríguez, M., & Summers, K. (2011). A taxonomic revision of the Neotropical poison frog genus *Ranitomeya* (Amphibia: Dendrobatidae). *Zootaxa, 3083*, 1–120.

Brust, D. G. (1993). Maternal brood care by *Dendrobates pumilio*: A frog that feeds its young. *Journal of Herpetology, 27*, 96–98.

Caldwell, J. P., & de Araújo, M. C. (1998). Cannibalistic interactions resulting from indiscriminate predatory behavior in tadpoles of poison frogs (Anura: Dendrobatidae). *Biotropica, 30*, 92–103.

Carlson, B. E., Newman, J. C., & Langkilde, T. (2015). Food or fear: Hunger modifies responses to injured conspecifics in tadpoles. *Hydrobiologia, 743*, 299–308.

Crane, A. L., Bairos-Novak, K. R., Goldman, J. A., & Brown, G. E. (2022). Chemical disturbance cues in aquatic systems: A review and prospectus. *Ecological Monographs, 92*, 1–18.

Crane, A. L., Achtymichuk, G. H., & Rivera-Hernández, I. A. E. (2023). Uncertainty about old information results in differential predator memory in tadpoles. *Proceedings of the Royal Society B: Biological Sciences, 290*, 20230746.

Crossland, M. R., Salim, A. A., Capon, R. J., & Shine, R. (2019). The effects of conspecific alarm cues on larval cane toads (*Rhinella marina*). *Journal of Chemical Ecology, 45*, 838–848.

Duellman, W. E., & Trueb, L. (1994). *Biology of amphibians*. The Johns Hopkins University Press.

Ferrari, M. C. O., Wisenden, B. D., & Chivers, D. P. (2010). Chemical ecology of predator–prey interactions in aquatic ecosystems: A review and prospectus. *Canadian Journal of Zoology, 88*, 698–724.

Fraker, M. E., Hu, F., Cuddapah, V., McCollum, S. A., Relyea, R. A., Hempel, J., & Denver, R. J. (2009). Characterization of an alarm pheromone secreted by amphibian tadpoles that induces behavioral inhibition and suppression of the neuroendocrine stress axis. *Hormones and Behavior, 55*, 520–529.

Gazzola, A., Brandalise, F., Rubolini, D., Rossi, P., & Galeotti, P. (2015). Fear is the mother of invention: Anuran embryos exposed to predator cues alter life-history traits, post-hatching behaviour and neuronal activity patterns. *Journal of Experimental Biology, 218*, 3919–3930.

Goddard, C. I., Lilley, J. W., & Tait, J. S. (1974). Effects of MS222 anesthetization on temperature selection in lake trout, *Salvelinus namaycush*. *Journal of the Fisheries Research Board of Canada, 31*, 100–103.

Gosner, K. L. (1960). A simplified table for staging anuran embryos and larvae with notes on identification. *Herpetologica, 16*, 183–190.

Hagman, M., & Shine, R. (2008). Understanding the toad code: Behavioural responses of cane toad (*Chaunus marinus*) larvae and metamorphs to chemical cues. *Austral Ecology, 33*, 37–44.

Hettyey, A., Tóth, Z., Thonhauser, K. E., Frommen, J. G., Penn, D. J., & van Buskirk, J. (2015). The relative importance of prey-borne and predator-borne chemical cues for inducible antipredator responses in tadpoles. *Oecologia, 179*, 699–710.

Hossie, T., Landolt, K., & Murray, D. L. (2016). Determinants and co-expression of anti-predator responses in amphibian tadpoles: A meta-analysis. *Oikos, 126*, 173–184.

Kahn, T. R., Lötters, S., La Marca, E., Brown, J. L., Twomey, E., & Amézquita, A. (2016). *Aposematic poison frogs (Dendrobatidae) of the Andean countries: Bolivia, Colombia, Ecuador, Perú and Venezuela*. Conservation International.

Kam, Y.-C., & Yang, H.-W. (2002). Female–offspring communication in a Taiwanese tree frog, *Chirixalus eiffingeri* (Anura: Rhacophoridae). *Animal Behaviour, 64*, 881–886.

Kats, L. B., & Dill, L. M. (1998). The scent of death: Chemosensory assessment of predation risk by prey animals. *Écoscience, 5*, 361–394.

Kiessling, A., Johansson, D., Zahl, I. H., & Samuelsen, O. B. (2009). Pharmacokinetics, plasma cortisol and effectiveness of benzocaine, MS-222 and isoeugenol measured in individual dorsal aorta-cannulated Atlantic salmon (*Salmo salar*) following bath administration. *Aquaculture, 286*, 301–308.

Leary, S., Underwood, W., Anthony, R., Cartner, S., Grandin, T., Greenacre, C., Gwaltney-Brant, S., McCrackin, M. A., Meyer, R., Miller, D., Shearer, J., Turner, T., & Yanong, R. (2020). AVMA guidelines for the euthanasia of animals: 2020 edition. American Veterinary Medical Association.

Li, J., Liu, H., Yu, M., Wu, L., Wang, Q., Lv, H., Ma, B., & Song, Y. (2014). Rapid determination of tricaine mesylate residues in fish samples using modified QuEChERS and high-performance liquid chromatography–tandem mass spectrometry. *Analytical Methods, 6*, 9124–9128.

Lipkowski, K., Almeida, D. A., & Schulte, L. M. (2023). Perception of con- and heterospecific injury cues in tadpoles of dendrobatid, ranid and bufonid frogs (Anura). *Evolutionary Ecology, 38*, 37–59.

Lipkowski, K., Almeida, D., Wenzel, D., & Schulte, L. M. (2025). Alarm cue properties and euthanasia impact on tadpole antipredator behaviour in *Rana temporaria* and *Bufo bufo*. *Animal Behaviour, 222*, 123109.

Losey, G. S., & Hugie, D. M. (1994). Prior anesthesia impairs a chemically mediated fright response in a gobiid fish. *Journal of Chemical Ecology, 20*, 1877–1883.

Lötters, S., Jungfer, K.-H., Henkel, F.-W., & Schmidt, W. (2007). *Pfeilgiftfrösche: Biologie, Haltung, Arten* (1st ed.). Chimaira.

Lucon-Xiccato, T. (2019a). Chemical alarm cues allow prey to adjust their defensive behaviour to cover abundance. *Behavioural Processes, 162*, 86–89.

Lucon-Xiccato, T. (2019b). Tadpoles modulate antipredator responses according to the abundance of vegetation experienced during the embryonic stage. *Journal of Zoology, 308*, 259–265.

McDiarmid, R. W., & Altig, R. (1999). *Tadpoles: The biology of anuran larvae*. University of Chicago Press.

McIntyre, P. B., Baldwin, S., & Flecker, A. S. (2004). Effects of behavioral and morphological plasticity on risk of predation in a Neotropical tadpole. *Oecologia, 141*, 130–138.

Meuthen, D., Flege, P., Brandt, R., & Thünken, T. (2018). The location of damage-released alarm cues in a cichlid fish. *Evolutionary Ecology Research, 19*, 529–546.

Mitchell, M. D., Bairos-Novak, K. R., & Ferrari, M. C. O. (2017). Mechanisms underlying the control of responses to predator odours in aquatic prey. *Journal of Experimental Biology, 220*, 1937–1946.

Pastell, M. (2016). CowLog: Cross-platform application for coding behaviours from video. *Journal of Open Research Software, 4*, 15.

Poelman, E. H., & Dicke, M. (2007). Offering offspring as food to cannibals: Oviposition strategies of Amazonian poison frogs (*Dendrobates ventrimaculatus*). *Evolutionary Ecology, 21*, 215–227.

Raices, M., Jungblut, L. D., & Pozzi, A. G. (2020). Evidence of the peptide identity of the epidermal alarm cue in tadpoles of the toad *Rhinella arenarum*. *Herpetological Journal, 30*, 230–233.

Readman, G. D., Owen, S. F., Murrell, J. C., & Knowles, T. G. (2013). Do fish perceive anaesthetics as aversive? *PLoS One, 8*, e73773.

Relyea, R. A. (2001). Morphological and behavioral plasticity of larval anurans in response to different predators. *Ecology, 82*, 523–540.

Richards, C. D. (1972). On the mechanism of barbiturate anaesthesia. *The Journal of Physiology, 227*, 749–767.

Richards, C. D. (1973). On the mechanism of halothane anaesthesia. *The Journal of Physiology, 233*, 439–456.

Rödin-Mörch, P., Forsman, A., & Hagman, M. (2011). Taxonomic patterns of tadpole behavioral responses to alarm cues. In J. L. Murray (Ed.), *Frogs: Biology, ecology and uses* (pp. 123–140). Nova Science Publishers.

Scherpenisse, P., & Bergwerff, A. A. (2007). Determination of residues of tricaine in fish using liquid chromatography–tandem mass spectrometry. *Analytica Chimica Acta, 586*, 407–410.

Schulte, L. M., & Mayer, M. (2017). Poison frog tadpoles seek parental transportation to escape their cannibalistic siblings. *Journal of Zoology, 303*, 83–89.

Schulte, L. M., Schulte, R., Rödder, D., & Lötters, S. (2010). Preference and competition for breeding plants in coexisting *Ranitomeya* species (Dendrobatidae): Does height play a role? *Salamandra, 46*, 180–184.

Schulte, L. M., Krauss, M., Lötters, S., Schulze, T., & Brack, W. (2015). Decoding and discrimination of chemical cues and signals: Avoidance of predation and competition during parental care behaviour in sympatric poison frogs. *PLoS One, 10*, e0129929.

Schulte, L. M., Lipkowski, K., & Abondano Almeida, D. (2023). Chemical communication and semiochemical recognition in frogs – From eggs to adults. In B. Schaal, D. Rekow, M. Keller, & F. Damon (Eds.), *Chemical signals in vertebrates, 15* (pp. 75–111). Springer Nature.

Späth, M., & Schweickert, W. (1977). The effect of metacaine (MS-222) on the activity of the efferent and afferent nerves in the teleost lateral-line system. *Naunyn-Schmiedeberg's Archives of Pharmacology, 297*, 9–16.

Stynoski, J. L., & Noble, V. R. (2012). To beg or to freeze: Multimodal sensory integration directs behavior in a tadpole. *Behavioral Ecology and Sociobiology, 66*, 191–199.

Summers, K. (1999). The effects of cannibalism on Amazonian poison frog egg and tadpole deposition and survivorship in *Heliconia* axil pools. *Oecologia, 119*, 557–564.

Surber-Cunningham, L. L., Oza, S. S., & Fischer, E. K. (2024). Chemical cues of conspecific predation elicit distinct behavioural responses in cannibalistic poison frog tadpoles. *Animal Behaviour, 208*, 79–89.

Szabo, B., Mangione, R., Rath, M., Pasukonis, A., Reber, S. A., Oh, J., Ringler, M., & Ringler, E. (2021). Naive poison frog tadpoles use bi-modal cues to avoid insect predators but not heterospecific predatory tadpoles. *Journal of Experimental Biology, 224*, 243647.

van Buskirk, J., & McCollum, S. A. (2000). Functional mechanisms of an inducible defence in tadpoles: Morphology and behaviour influence mortality risk from predation. *Journal of Evolutionary Biology, 13*, 336–347.

van Buskirk, J., Krügel, A., Kunz, J., Miss, F., & Stamm, A. (2014). The rate of degradation of chemical cues indicating predation risk: An experiment and review. *Ethology, 120*, 942–949.

Vickers, A. J. (2001). The use of percentage change from baseline as an outcome in a controlled trial is statistically inefficient: A simulation study. *BMC Medical Research Methodology, 1*, 6.

Wells, K. D. (2007). *The ecology and behavior of amphibians* (1st ed.). University of Chicago Press.

Wong, D., von Keyserlingk, M. A. G., Richards, J. G., & Weary, D. M. (2014). Conditioned place avoidance of zebrafish (*Danio rerio*) to three chemicals used for euthanasia and anaesthesia. *PLoS One, 9*, e88030.

Xue, Y. J., Chang, C. C., Lai, J. M., & Wang, J. H. (2017). Determining the tranquilization dose and residue of tricaine methanesulfonate (MS-222) in sea bass *Lates calcarifer* tissue. *Fisheries Science, 83*, 625–633.

Zahl, I. H., Samuelsen, O. B., & Kiessling, A. (2012). Anaesthesia of farmed fish: Implications for welfare. *Fish Physiology and Biochemistry, 38*, 201–218.

The Use of Habituation-Discrimination Tests in Chemical Communication Studies: Evaluation of the Benefit and Limitations on Model Rodent Species, Djungarian and Golden Hamsters

Nina Y. Vasilieva[1](✉), Anton P. Zakharov[1], Polina D. Semenova[1], Irina V. Kolesnikova[1], and Anastasia M. Khrushchova[1]

[1] Severtsov Institute of Ecology and Evolution RAS, Moscow, Russia

Abstract. This study was conducted to show the possibilities of expanding the scope of the habituation-dishabituation/discrimination (H-D) test that is widely used to study social memory, ability for individual and kin discrimination, over-marking phenomenon, etc. The H-D test is sometimes used to determine the ability of animals to distinguish the sex of individuals in cases where this cannot be detected with preference tests (e.g., in castrated subjects). In such experiments, stimuli of different sex individuals are presented in habituation trials followed by a dishabituation test. It is obvious that such stimuli combine features belonging to different categories—gender and individuality. Therefore, the question remains whether one is dealing with gender or individual recognition. To solve this problem, we propose using gonadectomized males treated with sex hormones, testosterone (T) or estradiol (E), as chemosignal donors. Djungarian hamsters served as a model species, and intact, sham-operated and castrated males were used as test subjects. Unlike intact and sham-operated males, castrated male subjects did not show a preference between female and male urine, nor between the urine of gonadectomized male donors treated with E vs. castrated males treated with T. However, in H-D tests, castrated male subjects were able to discriminate subtle differences in urine. This was likely related to the appearance/absence of sex steroids (or their metabolites) in samples of a castrated male donor that undergone therapy with T or E. This suggests that such discrimination may represent the basis for the ability of castrated males to determine the sex of a urine donor. In its turn, the results indicate that the lack of a female preference exhibited by castrated subjects is not due to the inability to perceive and/or process sex related information via urine, but rather to a lack of motivation. The results obtained allow us to propose an experimental model that can help to cope with seemingly unsolvable stimulus problems, namely, a combination of categories of gender and individuality in stimuli belonging to donors of different sexes. We propose that castrated individuals treated first with T and then with E and vice versa can be used as experimental models to simulate males and females in future research. Another experiment was conducted to reveal the possible ability of Djungarian and golden hamster males to process binary mixtures of urine samples. The results demonstrated that Djungarian and golden hamster males perceive binary urine mixtures elementally, meaning that they are able to discriminate between individual

L. M. Schulte (ed.), *Chemical Signals in Vertebrates 16*,
https://doi.org/10.1007/978-3-032-30752-1_12

components of the mixture. Thus, our data indicate that the H-D protocol can be easily adapted to also study mixtures of secretions.

Keywords: Habituation-dishabituation test · Odor preference · Sex discrimination · Individual discrimination · Replacement therapy · Binary urine mixture · Djungarian hamster · Golden hamster

1 Introduction

Behavioral tests based on the detection and recognition of odors remain important tools for analyzing and verifying experimental data obtained by applying modern technologies (Friedrich, 2006; Gabor et al., 2012; Aspesi & Choleris, 2022). Using genetic engineering methods allows creating animal models with specified characteristics that may represent useful tools to study the genesis and treatment of diseases accompanied by olfactory dysfunctions (e.g., Alzheimer's, Huntington's, and Parkinson's diseases). Although the predictive value of the results obtained using such models has certain limitations, their use is becoming more widespread. Obviously, choosing the right behavioral experimental procedure and the appropriate test is very important for the correct interpretation of the results of relevant physiological data, so careful selection and adequate application of behavioral testing procedures in this area come to the fore (Friedrich, 2006; Sukoff Rizzo & Crawley, 2017; Papes et al., 2018).

Various kinds of applied research, e.g., behavioral phenotyping of new lines of knockout animals, testing of pharmacological preparations, and testing the effects of various experimental procedures, require standardization of the methods used (Bailey et al., 2006; Yang & Crawley, 2009; Gabor et al., 2012). The most popular, informative, and easy to conduct experimental procedures are pairwise preference tests and so-called habituation-dishabituation/discrimination (H-D) tests (Gheusi et al., 1994; Halpin, 1974; Thom & Hurst, 2004; Yang & Crawley, 2009; Gabor et al., 2012). The H-D test is based on a fundamental feature of the nervous system common to all living organisms—habituation to repetitive and unrewarded stimuli and dishabituation to stimuli that differ from the previously presented (for historical reviews, see Berlyne, 1950; Thompson, 1986, 2009). During the habituation trial of the H-D test, the same stimuli are presented to a subject animal several times at regular time intervals. Such repetitive presentations lead to a decline of interest to the stimulus, indicating its discrimination. When a new stimulus that differs from the one used in the habituation trial is presented, dishabituation expressed in the restoration of interest toward this new stimulus is considered as discrimination (for more details, see experiments 2–4). The number of modifications of the H-D test is extremely large and varies depending on the tasks, conditions, and species (e.g., Halpin, 1974, 1986; Brown et al., 1987; Cleland et al., 2002; Ferguson et al., 2002; Johnston, 2008; McNamara et al., 2008; Schellinck et al., 2008). Both H-D and pairwise preference tests were the main tools to show that odor cues are involved in all aspects of animals' live, such as species and strain discrimination, sex and reproductive state discrimination, mate choice and individual discrimination, etc. (e.g., Brown,

1985; Brennan & Kendrick, 2006; Johnston & delBarco-Trillo, 2009; Wyatt, 2014; Ferkin, 2018).

One of the most frequent potential confounds of the pairwise odor preference test, when two stimuli are presented simultaneously, is the problem with the interpretation of the experimental results, which arises from a situation when subjects exhibit no preference for any of the stimuli. Indeed, a lack of asymmetry in the response to two stimuli indicates an absence of preference, but does not necessarily indicate a lack of discrimination. Ignoring this possibility can be a source of errors in the interpretation of experimental results. To cope with this possible problem, other methods, including operant conditioning and H-D tests, may be applied. Some research asserts that social preference tests are more sensitive than H-D tests (Gabor et al., 2012). However, each of the tests has its own benefits and restrictions (also see experiments 3–4). They have different resolutions and are often used independently of each other. Here we will focus only on research that has used both methodological approaches to study the effect of various experimental manipulations on the ability of animals to recognize sex or reproductive condition of conspecifics by odor cues. We will also try to analyze the results described in terms of the resolution of these methods, draw attention to the related possible problems that arise when interpreting the results, and propose a possible experimental solution to avoid "pitfalls" when using the H-D test in order to determine the ability of sex discrimination by odor cues.

In many mammalian species and, particularly, in rodents (e.g., rats, mice, hamsters, meadow voles), it is well documented that the sex steroids modulate the mechanisms of perception and motivation and affect the social behavior of individuals of both sexes. Thus, gonadectomy leads to a general decrease in interest in stimuli and eliminates the preference of individuals of the opposite sex that usually may be assessed with the use of different modifications of preference test (e.g., Carr et al., 1966; Thor, 1980; Thor et al., 1982; Ferkin, 1992; Hughes et al., 1990; Oliveira, 2004; Ballard & Wood, 2007; Petrulis, 2013; Achiraman et al., 2014; Karlsson et al., 2015; Le Moëne & Ågmo, 2018; Kondo & Hayashi, 2021). Such an effect of gonadectomy is usually interpreted as a consequence of a general decrease in sexual motivation, which still seems to be a largely speculative concept, since there are no clear criteria defining it (Sachs, 2007; Harding & Velotta, 2011; Karlsson et al., 2015). The question of what is behind the disappearance of interest in opposite sex stimuli—the loss of the ability to perceive and process sex related information or a general decrease in interest in social stimuli—still remains relevant. In a brilliant experimental study, which can rightfully be considered a textbook example, W. J. Carr and W. F. Caul used operant conditioning to show that gonadectomized male and female rats are able to discriminate sex-associated odors at the extend that is found in intact subjects (Carr & Caul, 1962). This may indicate that the impairment of sexual preferences in castrated subjects is determined by reduced motivation, but not the inability to discriminate sex related odors. However, the operant discrimination learning method is very laborious and has limitations regarding the use of species, which limits its wide application. Thus, to address similar issues, some studies have been conducted using the H-D paradigm.

For example, in ferrets, volatile compounds of anal gland secretions are sexually dimorphic, and in choice test, they are more attractive to opposite sex subjects (Clapperton et al., 1988). Castrated ferrets of both sexes did not show a preference for volatiles of body odor of opposite sex donors; however, a sex-specific preference was observed after hormonal replacement therapy (Kelliher & Baum, 2002). Based on these findings, a study aimed at clarifying the influence of gonadectomy and replacement therapy on sex discrimination via volatiles of anal gland secretions was conducted using an H-D test (Woodley & Baum, 2003). Importantly, during the habituation trial as well as the dishabituation test, secretions from donors of different sexes were used. After habituation toward one sex donor, male and female gonadectomized subjects showed dishabituation to the odor of the other sex, indicating discrimination. However, any stimuli from male and female donors ultimately possess at least two independent variables that may determine a reaction to them by recipients, namely, gender and individuality. Because of this, in the aforementioned experiment, subjects could exhibit not sex but rather individual discrimination. Based on the results of another experiment, in which the animals did not demonstrate individual discrimination in H-D test, the authors concluded that gonadectomized male and female ferrets are able to discriminate rather sex of donor samples but not individual via volatiles of anal gland secretions (Woodley & Baum, 2003).

Once applied, this modification of the H-D method was repeatedly used in other ferrets and mice studies. In experiments that tried to evaluate the possible role of the additional olfactory system (AOS) in sex discrimination, it was shown that ferrets and mice with removed vomeronasal organs (VNO) failed to discriminate sex via odor cues in choice preference tests but exhibited sex discrimination in the habituation and dishabituation phases of H-D tests, when urine of opposite sex donors was presented (Pankevich et al., 2004; Woodley, 2004; Jakupovic et al., 2008; Liu et al., 2010). These findings allowed authors to conclude that the VNO is not necessarily involved in sex discrimination (Pankevich et al., 2004; Woodley, 2004; Jakupovic et al., 2008; Liu et al., 2010). However, in these studies, dishabituation to volatiles from a new donor of the other sex/state indicated rather individual but not sex/state discrimination. This is consistent with the numerous data that, in spite of partial functional overlapping of main and accessory olfactory systems, gender recognition is subserved by VNO/AOS. AOS perceives mainly nonvolatile compounds, while the MOS, which perceives volatile substances, is responsible for individual discrimination (the Bruce effect is an exception and is realized through a special mechanism) (Wysocki et al., 1982; Stowers et al., 2002; Luo et al., 2003; Brennan, 2004; Brennan & Kendrick, 2006; Spehr et al., 2006; Becker & Hurst, 2009; Hurst, 2009; Muroi et al., 2017).

The same experimental procedure—presentation of odors from different sex donors during the habituation trial and the dishabituation phase of the H-D test—has been applied when the involvement of different brain structures in the regulation of sexual behavior and social preferences was studied. Based on the results of experiments showing that lesions of the ventromedial hypothalamic nucleus (VMH) in mice and posteromedial cortical amygdale (PMCo) in male golden hamsters do not disrupt sex discrimination by odors, the authors concluded that VMH and PMCo are not required for sex/state discrimination in these species (Robarts & Baum, 2007; Maras & Petrulis, 2008).

In all the studies mentioned above, odors of different sex donors were presented during habituation and dishabituation phases of H-D tests. Obviously, stimuli of male and female donors invariantly possess features determined both by sex and individuality (for more discussion, see experiment 3; for justification, Gheusi et al., 1994). Thus, in all these studies, the dishabituation could be due to discrimination of individuals, not sex. This circumstance requires a very careful and critical evaluation of the results and conclusions of the above studies, which, due to their fundamental nature, need additional experimental verification.

Are there any experimental solutions to solve the problem of using stimuli from donors of both sexes, which invariably carry signs of gender and individuality, in addition to the already mentioned operant conditioning paradigm? Our work is devoted to an attempt to develop such an experimental model with the use of H-D tests (experiments 1–3), as well as to show additional possibilities of using it to solve one of the key issues of chemical communication—the analysis of mixtures of natural excreta, particularly urine (experiment 4). The work was carried out on model species, the Djungarian (*Phodopus sungorus*) and golden (*Mesocricetus auratus*) hamsters. These species have been repeatedly used to study the role of odor cues in individual and kin recognition, discrimination of sex and female reproductive state, species preference, over-marking phenomenon, as well as the involvement of VNO and brain structures in processing of chemosensory information and sociosexual behaviors (e.g., Johnston, 1993, 2008; Johnston et al., 1993; Lai & Johnston, 1994; Sokolov et al., 1994; Lai et al., 1996; Vasilieva & Johnston, 1996; Petrulis et al., 1999, 2000; Petrulis & Johnston, 1999; Vasilieva et al., 2001; Litvinova, 2003; Petrulis & Eichenbaum, 2003; Litvinova & Vasilieva, 2004a, b; Maras & Petrulis, 2008).

1.1 Influence of Castration, Testosterone Therapy, and Hormonal State of the Urine Donors on Sex Discrimination in Djungarian Hamster Males: Development of a Model for Easy Assessment

We designed our experiments based on generally accepted notion concerning a leading role of sex steroids in the determination of sex-specific odor properties (Leonard & Ferkin, 1999; Stowers & Logan, 2010; Petrulis, 2013; Baum & Bakker, 2013; Ervin et al., 2015; Aspesi & Choleris, 2022). Since the odor of males is determined primarily by testosterone and by-products of its biotransformation (such chemical derivatives as dihydrotestosterone and estradiol and other metabolic by-products) and female odors and their attractiveness depend on the estradiol level (Baum & Bakker, 2013; Ervin et al., 2015), we used these two hormones to develop an experimental model that may allow to solve the problem of the interference of individuality and gender categories described above for the H-D test. As a first step (experiment 1), we conducted a set of paired odor preference tests in which castrated males and males with intact reproductive system were used as subjects. Intact and castrated males and females, as well as gonadectomized individuals receiving hormone replacement therapy, were used as urine donors. Then (experiment 2) we checked whether castrated males are able to discriminate individuals and sex of intact conspecifics via urine with the use of habituation-dishabituation (H-D) tests. Experiment 3 was devoted to develop an experimental model that may allow assessing an ability to discriminate sex of urine donors using the H-D test and to cope with the problem arising from the combination of features belonging to different stimuli categories—gender and individuality.

2 General Methods

2.1 Animals

The Djungarian hamsters used in our experiments were from a laboratory colony that has been established from animals captured in the Novosibirsk region in Russia and maintained since the early 1980s. Hamsters were housed in family groups consisting of a breeding pair and their litters in 38 x 27 x 15 cm plastic cages, covered with a metal grid and filled with wood shavings. Golden hamsters came from the Stolbovaya Laboratory Animal Nursery (Moscow region). Male hamsters were kept individually in 38 x 27 x 15 cm plastic cages and tested in their home cages. Water and food (rodent chow, oat, bread, and fresh vegetables) were fed ad libitum. All animals were kept in ventilated rooms at 22 ± 1 °C and 16 L:8D, with an inverted photoperiod where the light turned off at 8 a.m. The number of animals used in the different experiments is indicated below.

Animal care and experimental manipulations were conducted in accordance with the Council Directive 86/609/EEC of November 24, 1986, (https://faolex.fao.org/docs/pdf/eur36598.pdf) and approved by the Bioethics Committee of IPEE RAS (protocols #25a-2023, #74–2023).

2.2 Surgery and Replacement Hormonal Therapy

We used gonadectomy and replacement hormonal therapy as tools to manipulate the hormonal state of urine donors and subject males. Castration of males was performed with the use of standard procedure, where the testes and epididymis were removed through bilateral scrotal dissections. Ovariectomy was conducted with a standard procedure through two incisions about 5 mm long above the location of the ovaries on the dorsal side of the body. Castration or sham operation and ovariectomy were conducted under inhalation of isofluran anesthesia with postoperation subcutaneous lidocaine analgesia.

For testosterone replacement therapy, we used a single injection of "Omnadren" (Pharmaceutical Works JELFA, SA, Poland)—a mixture of four ethers of testosterone that is used for ensuring the maintenance of stable hormone levels for at least 3 weeks from the moment of injection. 1 ml of Omnadren was diluted in 20 ml of highly refined olive oil and injected at the rate of 0.3 ml of solution/kg of body weight. For estradiol treatment of castrated males, we used "Oestrogel" (Besins Manufacturing Belgium, SA, Belgium), which is a widely accepted medicine for transdermal delivery of estradiol during a course of hormone replacement therapy. The gel was applied with a glass spatula to a shaved skin area about 5 x 5 mm in size in the middle of the abdomen next to the midventral gland. The gel is absorbed within 10–15 min. In order to prevent possible efforts of a hamster to remove the gel (e.g., by grooming and/or belly rubbing) at the time around application, sunflower seeds were poured into the home cage. This immediately caused a switch of the animal's activity to filling its cheek pouches and thereby preventing an attempt to remove the gel.

2.3 Urine Samples

For urine collection, the donor hamster was placed into a metal cylinder (D = 10, h = 12 cm) and provided with a small piece of cabbage. The cylinder was placed on fine-mesh grid stand preventing feces from entering the tray. The donor was kept in the cylinder 30–60 min until urination, and samples were collected from a tray and put in a freezer until use (for further details, see the method description of the experiments).

2.4 Experimental Procedures

We presented urine samples in glass disposable Pasteur pipettes (1.0 mm tip diameter). Urine was drawn into the pipette by capillary action to a height of about 1 cm. Hamsters never consumed all the urine in the pipette.

2.4.1 The Pairwise Preference Test. Two odors were presented simultaneously in glass Pasteur pipettes in the experimental hamsters' home cages. The pipettes were inserted to a depth of about 2 cm through the wire mesh cage grate at a distance of 5 cm from each other and fixed with clamps. The time that hamsters sniffed and licked the samples during the 3-min test period was recorded by two stopwatches. All the experiments were conducted blind. If the subject investigated only one stimulus, the test was not taken into account, and the animal was tested later with a similar pair of stimuli again.

2.4.2 Habituation-Dishabituation Test. We used a modification of the H-D test, which was successfully used in experiments on individual discrimination in golden and Djungarian hamsters (Lai & Johnston, 1994; Sokolov et al., 1994; Litvinova & Vasilieva, 2004a,b; Johnston, 2008). Following four consecutive presentations of identical urine samples (first odor in a habituation trial: presentations 1–4), a new urine sample that differed from the previously presented by any studied parameter (test: presentation 5) was presented. In all experiments where dishabituation was not observed, we used additional control presentation of a stimulus analogous to that presented in the test (for details, see the relevant experiments). As a lack of dishabituation in a test may be not necessary due to test specificity (see for possible pitfalls of interpretations—Schellinck et al., 2008), but because an animal simply refused to "work" due to fatigue, we introduced the control stimulus presentation just in order to avoid errors in the interpretation of the results. If a subject did not show a dishabituation toward the control stimulus, the experiment was not included in the analysis.

If the subjects were used more than once, the interval between experiments was at least 3 days. All experiments, including pairwise and H-D tests, were conducted during the first hours after the onset of the dark phase of the light cycle.

2.5 Statistics

Statistical analyses were conducted using R version 4.4.1 (R Core Team, 2024) with the "rstatix" (Kassambara, 2023) and "tidyverse" (Wickham et al., 2019) packages. All data were tested for normality and analyzed, depending on the type of distribution using parametric or nonparametric methods for pairing comparisons, which is reflected in the criteria t (paired t-test) or V (Wilcoxon signed-rank test). All data shown in the figures are presented as mean ± standard error to make them comparable.

3 Experiment 1: Sex Discrimination via Urine in Djungarian Hamster Males Using Odor Preference Tests

The aims of this experiment were (1) to reveal the consequences of gonadectomy followed by replacement testosterone therapy on the abilities of castrated Djungarian hamster males to discriminate sex via urine in choice preference test and (2) to assess the influence of hormonal state of urine donors on sex discrimination performance in these subjects.

3.1 Animals and Methods Used in Experiment 1

As a first step, we conducted a set of sex odor preference tests in which castrated males and males with intact reproductive systems were used as subjects. Intact and castrated males and females, as well as gonadectomized individuals receiving hormone replacement therapy, were used as urine donors.

Eleven intact 4–6-month-old males randomly chosen from the breeding colony were used as test subjects. In addition, we used 12 pairs of sibling males from 9 families that lived with their parents and other siblings until the age of 45–50 days. The pairs of brothers were put together in separate cages, where they were kept until they were separated into individual cages for the entire testing period. The pairs were chosen in such a way that the brothers' body weight did not differ. At the age of 60–65 days, one male in a pair was castrated (CM) and his brother was sham operated (ShM). About 1 month later, ShM and CM were used in behavioral tests. After all experiments with the castrated males were completed (including experiments 1–3), they received testosterone replacement therapy being injected with "Omnadren" (CM + T group) and then after 10–14 days were used in further experiments, described as a part of experiment 1.

As urine donors, the following animals were used:

1. Three intact males and three intact diestrous females at the age of 4–5 months from the breeding colony were used as intact urine donors. Females' physiological state was determined by a routine vaginal cytology.
2. Another three females and three males at the age of 4–5 months were gonadectomized, and after getting their urine samples following about 1 month after surgery, these hamsters received replacement hormonal therapy with E (CF + E) or T (CM + T) according to sex.
3. Four more castrated males of the same age were used as urine donors in experiments with ShM and CM subjects. Two of CM males received T injection (CM + T) and two were treated with estradiol (CM + E). Urine samples were collected 7 and 10 days after E and T treatment, respectively.

Intact male subjects were presented with pairs of urine samples from (1) intact male (M) vs. diestrous female (F), (2) castrated male (CM) vs. ovariectomized female (CF), and (3) castrated male treated with testosterone (CM + T) vs. ovariectomized female treated with estradiol (CF + E). Sham-operated males (SmM) and their castrated siblings (CM; before and after T treatment) were allowed to investigate a set of urine samples from (1) intact male (M) vs. diestrous female (F) and (2) castrated male treated with testosterone (CM + T) vs. castrated male treated with estradiol (CM + E).

3.2 Results and Discussion of Experiment 1

Intact and sham-operated Djungarian hamster males in preference tests investigated female urine significantly longer than that of male donors ($t = 5.69$, $df = 9$, $p < 0.001$, Fig. 1a; $t = 8.82$, $p < 0.001$, Fig. 2a). Castrated subjects did not show any preference ($V = 55$, $p = 0.054$; Fig. 2a). Castration of male and female donors led to the disappearance of preference for female urine by intact males ($t = 0.84$, $df = 10$, $p = 0.42$; Fig. 1b), but treatment of gonadectomized male and female donors with testosterone and estradiol, respectively, resulted in restoration of preference for female stimuli by intact males ($t = 4.54$, $df = 9$, $p < 0.001$; Fig. 1c). When urine of castrated male donors that received T (CM + T) and E (CM + E) therapy were presented as stimuli, sham-operated subjects investigated CM + E urine significantly longer, but castrated subject males demonstrated no preference ($V = 0$, $p = 0.004$, and $V = 43$, $p = 0.41$, respectively; Fig. 2b). After testosterone treatment, castrated male subjects investigated much longer both urine of intact diestrous female vs. urine of intact male and urine of estradiol

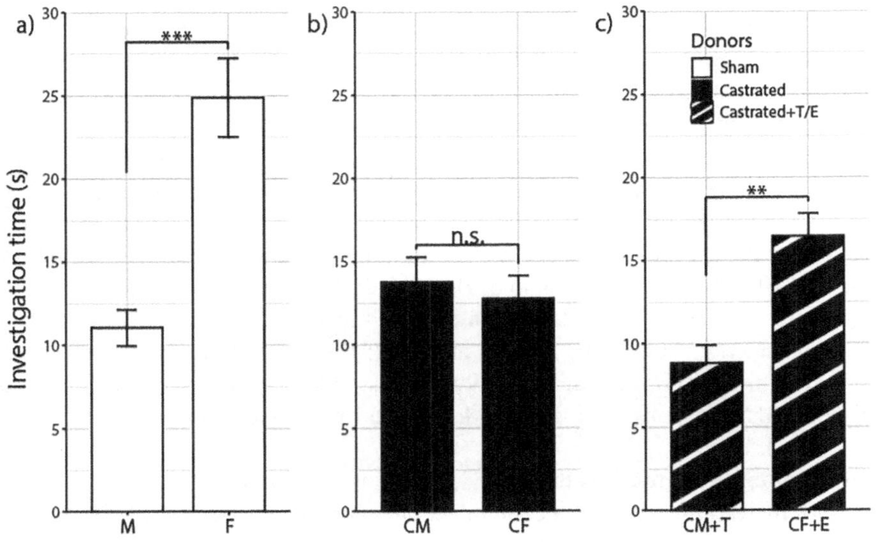

Fig. 1. Time (s; m ± SE) that intact Djungarian hamster males investigated urine samples in preference tests: male (M) vs. female (F) (**a**, white bars), castrated male (CM) vs. gonadectomized female (CF) (**b**; black bars), castrated testosterone treated male (CM + T) vs. castrated estradiol treated female (CF + E) (**c**, black background white striped bars). Statistical significance of diefferences between the corresponding values are marked with asterisks (** – p<0.01, *** – p<0.001).

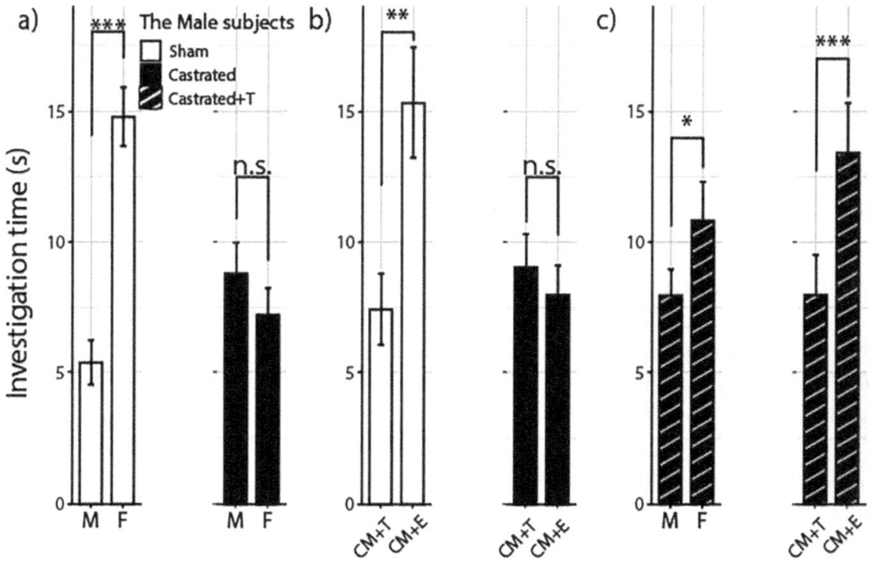

Fig. 2. Time (s; m ± SE) that sham-operated (white bars), castrated (black bars), (**a**, **b**) and castrated testosterone treated (black background white striped bars) (**c**) Djungarian hamster males investigated conspecifics' urine samples in two choice test: male (M) vs. female (F) and castrated testosterone treated male (CM + T) vs. castrated estradiol treated male (CM + E). Statistical significance of differences between the corresponding values are marked with asterisks (* – $p<0.05$, ** – $p<0.01$, *** – $p< 0.001$).

treated castrated male (CM + E) vs. castrated testosterone treaded male (CM + T) donors ($t = 3.06$, $df = 10$, $p < 0.001$, and $t = 6.96$, $df = 10$, $p < 0.001$, respectively; Fig. 2c).

These results indicate that castrated males do not show a sex-specific sexual odor preference. However, it can be restored by testosterone administration. Such a result is expected. Indeed, it is well documented that after castration, male sexual motivation decreases, which is manifested in the disappearance of sexual behavior and preference for female odors, which are restored after hormone therapy (Ågmo, 2003; Stowers & Logan, 2010; Petrulis, 2013; Kondo & Hayashi, 2021). In particular, testosterone therapy restores interest in female scents in castrated male rats, mice, meadow voles, and golden hamsters (Thor, 1980; Thor et al., 1982; Powers & Bergondy, 1983; Merkx, 1984; Ferkin, 1992; Havens & Rose, 1992; Ågmo, 2003; Achiraman et al., 2014;) and enhances olfactory sensitivity in gonadectomized male and female mice (Pankevich et al., 2003).

Another expected result is that gonadectomy of urine donors caused the disappearance of females' biased preference in intact Djungarian hamster males (Fig. 1b). Such phenomenon has been repeatedly shown in other mammalian species (Ferkin & Johnston, 1993; Ågmo, 2003; Ervin et al., 2015).

It should be stressed once more that all male subjects (intact, sham-operated and castrated T treated males) investigated the urine of CM + E donors longer than that of CM + T donors, just as they examined the urine of intact males and females (Fig. 1a, c;

Fig. 2a, b, c). Similarly, estradiol (+progesterone) treated castrated male rats became attractive for intact adult males (Merkx, 1984; Merkx et al., 1987). It has to be stressed here that we used urine from CM + E males that received E transdermally within only 7 days. We cannot rule out that long-lasting chronic application of E could cause an opposite influence on attractiveness of CM + E males. For example, it has been shown that different doses and duration of estradiol therapy in castrated male rats have different effects on their attractiveness (Xiao et al., 2004, 2015).

Taken together, all the results of this part of the study indicate that the sex-specific bias toward female odors in male Djungarian hamsters to a greater extent depends on circulating estradiol.

3.3 Conclusive Remarks of Experiment 1

The data obtained indicate that 1) attractiveness of sex related urine stimuli depends on sex steroids and females' biased preference of urine is determined by estradiol and 2) castrated Djungarian hamster males did not demonstrate masculine sexual preference for female urine, but this was restored after replacement testosterone therapy. The question arising from this finding is why castrated males do not show male specific preferences toward female urine? Is it because of low sexual motivation or inability to perceive/process sex related information via urine? In the section below we will try to answer these questions.

4 The Ability of Castrated Djungarian Hamster Males to Discriminate Sex and Individual Revealed by the H-D Test: The Development of an Experimental Model

The experiments described below were performed to answer the question that has been mentioned in the introduction and follow the results of experiment 1. There are two possibilities to show that castrated males are able/not able to discriminate between odors of different sex conspecifics—to use an operant conditioning procedure or a habituation-dishabituation test. The restrictions of the operant conditioning paradigm are obvious—it is a time-consuming procedure demanding a lot of trials to reach the acquisition-performance criterion (Penn & Potts, 1998; Schellinck et al., 2008). Moreover water/food restriction/reward procedure is not suitable for all species, including hamsters. The habituation-dishabituation paradigm is a robust tool to evaluate the ability of subjects to discriminate between two stimuli but has serious restrictions related to results interpretation when stimuli carry attributes belonging to different categories. The following experiment 2 is to some extent a prelim but necessary stage for conducting and reasonably discussing the results of further experiments (experiment 3). Its goal is to determine the ability of castrated males to distinguish between sex and individuals by urine using the H-D procedure.

5 Experiment 2: Sex Discrimination and Individual Discrimination in Castrated Djungarian Hamster Males by the Means H-D Test

In this section, we once again wanted to highlight the problem of ambiguity in the interpretation of the results of experiments conducted to determine the ability of animals to discriminate the sex of individuals via chemosignals using the H-D test, when the stimuli presented during the habituation trial and in the dishabituation/discrimination test belong to donors of different sexes (e.g., Woodley & Baum, 2003; Pankevich et al., 2004, 2006; Jakupovic et al., 2008).

5.1 Animals and Methods Used in Experiment 2

Castrated subject males were the same that were used in experiment 1. Seven new intact males and three intact diestrous females at the age of 6–8 months from the breeding colony were used as urine donors. The standard H-D procedure was applied. In sex discrimination experiments, female urine was presented in habituation trials and male urine in the test trials, and vice versa. In individual discrimination experiments, urine of one male donor was presented in the habituation trial and a new male in the test trial. None of the individually derived samples were presented to the experimental animals twice.

5.2 Results and Discussion of Experiment 2

During habituation trials, castrated male subjects demonstrated a decrease in investigation time of male ($t = 13.65$, $df = 9$, $p < 0.001$; see presentations 1–4 in Fig. 3a) and female urine ($t = 11.80$, $df = 9$, $p < 0.001$; see presentations 1–4 in Fig. 3b) that indicates habituation due to identification of the stimuli. Investigation time significantly increased when urine from the opposite sex donor was presented in the test ($t = 16.79$, $df = 9$, $p < 0.001$ and $t = 10.93$, $df = 9$, $p < 0.001$; see presentations 4–5 in Fig. 3a, b), indicating that castrated males discriminated between urine of male and female donors. Castrated males also showed discrimination of individual male urine samples: when being habituated to urine of one male ($V = 66$, $p = 0.01$; see presentations 1–4 in Fig. 3c), they investigated the sample of a new male donor longer ($V = 2$, $p = 0.021$; see presentations 4–5 in Fig. 3c).

Thus, castrated Djungarian hamster males as well as intact males (Litvinova, 2003; Fig. 3c) are able to discriminate between two individual urine samples. This finding allows using castrates in further experiments as a category of subjects. Although in preference choice tests castrated males did not show differences in investigation of simultaneously presented male and female urine (experiment 1), H-D tests revealed a discrimination. However, based on these results, we cannot rule out that castrated subjects discriminated individually distinctive but not gender-specific substances in male and female urine.

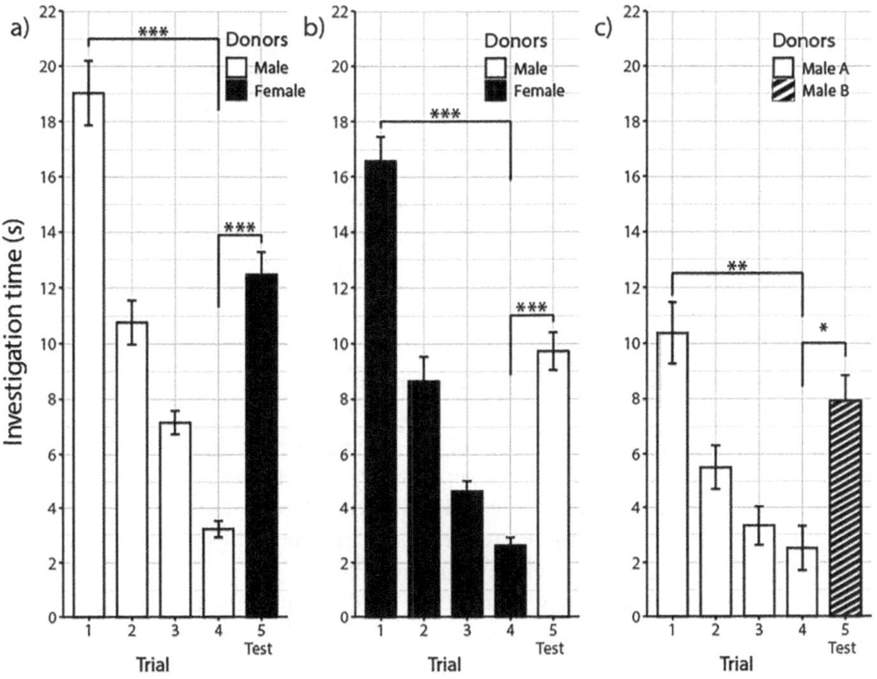

Fig. 3. Time (s; m ± SE) that castrated Djungarian hamster males investigated urine samples from intact male and female donors ("sex discrimination"): **a**) male urine in a habituation trial (presentations 1–4) and female urine in a test (presentation 5) and **b**) vice versa; **c**) male A urine in a habituation trial (presentations 1–4) and male B urine in a test (presentation 5) ("individual discrimination"). Statistical significance of differences between the corresponding values are marked with asterisks (* – $p<0.05$, ** – $p<0.01$, *** – $p<0.001$).

To clarify this issue, we needed to create a model that would allow us to manipulate the category of gender without affecting the category of individuality. In other words, we had to develop model stimuli that carry hormone-dependent sex characteristics, in the absence of differences determined by individuality. The results of the experiments described below give us hope that this is possible.

6 Experiment 3: Discrimination of Changes in the Urine of Castrated Individual Male Donors Caused by Hormone Replacement Therapy by Gonadectomized Male Djungarian Hamsters with the Use of the H-D Test

Based on the results of experiment 1, indicating that Djungarian hamster males in preference choice test investigated the urine of CM + E longer than the urine of CM + T and that such a bias is similar to that seen between native female and male urine, we propose an experimental model that allows to solve a seemingly unsolvable stimulus problem, namely, combining categories of gender and individuality in stimuli from donors of different sexes.

We suggested that castrated E treated males (CM + E) may represent, at least to some extent, a simulated female model. Based on this assumption, we used castrated males treated with E (CM + E) as simulated female models ("hormonal females") and castrated males treated with T (CM + T) as simulated male models ("hormonal males").

Thus, in the following set of experiments, we tried to reveal whether castrated male subjects are able to discriminate changes in urine of one particular gonadectomized individual male donor caused by either testosterone or estradiol administration using H-D tests.

6.1 Animals and Methods Used in Experiment 3

In this part of the study, we used urine of the same castrated male donors collected at different times—before and after hormonotherapy—as odor source. In order to be sure that the age of the samples (shelf life in the freezer) does not affect the properties of urine, which can be detected by the tested castrated male subjects in the H-D test, an additional auxiliary experiment was conducted.

For this experiment, four new Djungarian hamster males castrated at the age of about 4 months were used as urine donors. Urine samples from each donor were taken twice: first, about 1 month after castration (represents "old" urine) and then 1.5–2 months later, 5–7 days before use in the H-D test ("fresh" urine). "Old" or "fresh" urine from one individual donor was presented in habituation trial, and then the different age sample of the same donor was presented in the test ("fresh" or "old," respectively). Urine of the same age as presented in the test from different castrated donors was used as control sample (presentation 6).

In the main part of the experiments, 12 castrated males aged 4–5.5 months (CM) that have not been used in other experiments were urine donors. One month after castration, six males were injected intraperitoneally with a microsyringe with purified olive oil which was used to prepare a testosterone solution for injection, and then 10 days later, urine samples from these hamsters were collected. Then 45–50 days later, these CM received one injection of "Omnadren" (CM + T), and their urine was collected 10–12 days after. Urine samples from another six castrated males also were collected twice. The first samples were taken 45–50 days before estradiol administration. These castrated males were treated daily with "Estrogel" (CM + E) within 7 days, and then their urine was collected. All the experiments with the use of these samples (CM + E) were conducted within 3–10 days.

6.2 Results and Discussion of Experiment 3

In the first experiment, when old (stored in a freezer for 1.5–2 months) urine of a castrated male was presented in the habituation trial and fresh (stored in a freezer for 3–5 days) urine of the same donor was used in the test (Fig. 4a) and vice versa (Fig. 4b), subject males demonstrated a typical decrease of investigation time of the samples ($V = 55$, $p = 0.002$, $n = 10$, for both "a" and "b"). There was no increase of interest

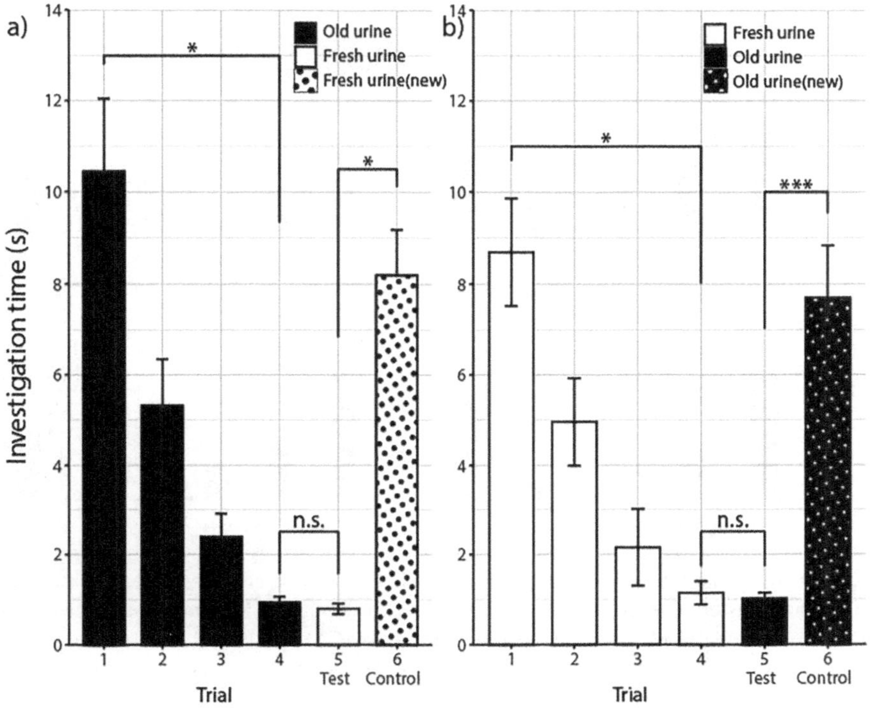

Fig. 4. Time (s; m ± SE) that Djungarian hamster males investigated "old" and "fresh" urine samples of the same castrated male donor in a habituation trial (presentations 1–4) and in a test (presentation 5) (**a**) and vice versa (**b**) in a dishabituation test. Control (presentation 6)—"fresh" (**a**) or "old" (**b**) urine of a new castrated male. Statistical significance of differences between the corresponding values are marked with asterisks (* – p<0.05, *** – p<0.001).

toward stimuli of a different age in the test ($V = 35$, $p = 0.492$, $n = 10$, and $V = 31.5$, $p = 0.721$, $n = 10$, respectively). However, when a urine sample of the same age but from a new castrated male donor was presented (control, presentation 6), pronounced dishabituation was observed ($V = 0$, $p = 0.002$, $n = 10$ and $t = 6.25$, $df = 9$, $p < 0.001$, respectively; Fig. 4a, b). The lack of increase of interest toward the stimuli of a different age from the same castrated donor and restoration of interest to urine of a new castrated donor indicated that hamsters easily discriminated between urine of different ages from the same donor and treated them as equally familiar. These results indicate that storage in a freezer does not influence individually distinctive urine quality. This allowed us to collect urine from a particular donor at different times and be sure that the age of the sample would not affect the result. The following experiments were conducted to find out whether castrated male subjects are able to detect differences determined by the presence or absence of hormonal derivatives in the urine of a particular castrated male donor. In habituation trials and dishabituation tests, urine stimuli from the same castrated male donors collected before and after replacement therapy with T or E (and vice versa) were presented to castrated male subjects.

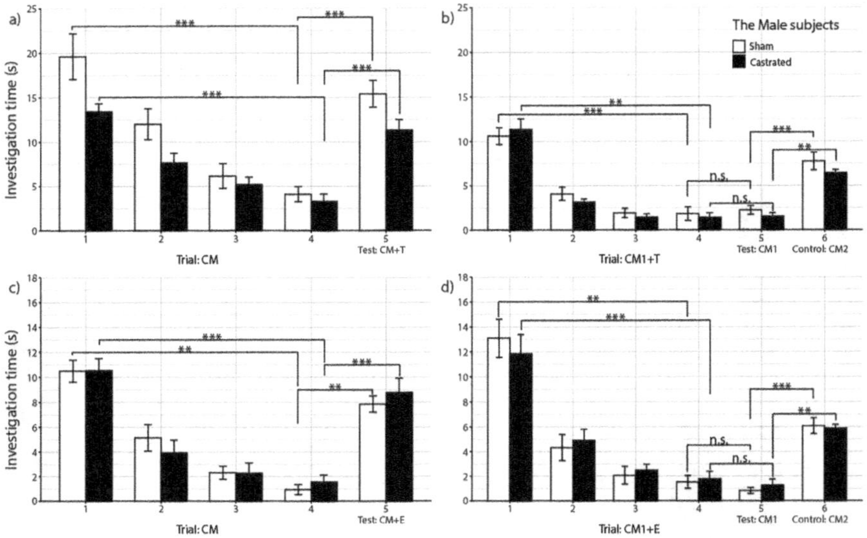

Fig. 5. Time (s; m ± SE) that sham-operated (white columns) and castrated (black columns) Djungarian hamster males investigated urine of a castrated male (CM) and the same male treated with testosterone (CM + T) (top charts: **a**, **b**) or estradiol (CM + E) (lower charts: **c**, **d**). **a**) habituation trial: CM urine (presentations 1–4), test: CM + T urine (presentation 5); **b**) habituation trial: CM1 + T urine (presentations 1–4), test: CM1 urine (presentation 5), control: new male donor, CM2 (presentation 6); **c**) habituation trial: CM urine (presentations 1–4), test: CM + E urine (presentation 5); **d**) habituation trial: CM1 + E urine (presentations 1–4), test: CM1 urine (presentation 5), control: new donor, CM2 (presentation 6). The designations C1 and C2 emphasize that in control (presentation 6) urine of a new donor was used. Statistical significance of differences between the corresponding values are marked with asterisks (** – p<0.01, *** – p< 0.001).

Both ShM and CM male subjects exhibited habituation when they were presented four times with urine of the castrated donor (CM) in T treatment (Fig. 5a; $t = 7.85$, $df = 7$, $p < 0.001$, and $t = 12.8$, $df = 8$, $p < 0.001$, respectively) or E treatment experiments (Fig. 5c; $V = 78$, $p = 0.0025$, and $V = 78$, $p < 0.001$, respectively). All subjects, ShM and CM, demonstrated pronounced dishabituation when urine from the same donors after undergoing T (Fig. 5a; $t = 10.24$, $df = 7$, $p < 0.001$, and $t = 6.96$, $df = 8$, $p < 0.001$, respectively) or E administration (Fig. 5c; $V = 0$, $p = 0.0025$, and $V = 0$, $p < 0.001$, respectively) was presented in a test. When urine from CM + T and CM + E donors was presented in the habituation trials, all subjects, ShM and CM, demonstrated a decrease in investigation time, too (CM + T: Fig. 5b; $V = 55$, $p = 0.0059$ and $t = 10.80$, $df = 9$, $p < 0.001$, respectively; CM + E: Fig. 5d; $V = 55$, $p = 0.002$ and $t = 7.79$, $df = 9$, $p < 0.001$, respectively). There were no dishabituation observed when ShM and CM subjects were presented with urine from the same CM donors (urine collected before replacement hormonal therapy) after being habituated to CM + T (Fig. 5b; $V = 20$, $p = 0.475$, and $V = 22$, $p = 0.61$, respectively) or CM + E urine (Fig. 5d; $V = 38$, $p = 0.076$, and $V = 24$, $p = 0.44$, respectively). Urine from a new castrated donor caused dishabituation in ShM and CM hamsters in T treatment (Fig. 5b; $t = 6.55$, $df = 9$, $p < 0.001$ and $V = 0$, $p < 0.001$, respectively) and E treatment experiments (Fig. 5d; $t = 7.79$, $df = 9$, $p = 0.0025$ and $V = 0$, $p = 0.004$, respectively).

Both ShM and CM male subjects exhibited classical habituation toward all the stimuli, indicating that they perceived them as familiar. All subjects exhibited dishabituation when they were presented in a test with urine of the same castrated donor as in habituation trial but after it undergone either T or E treatment (Fig. 5a, c). According to the logic of H-D paradigm, this indicates that they were able to discriminate between two samples. On the other hand, when the presentation sequence was reversed (samples from CM + T and CM + T were presented in the habituation trials followed by the presentation of CM urine samples), none of the subjects responded to CM urine by increasing investigation time. At that, they dishabituated to control samples (presentation 6) from the same type of donor (Fig. 5b, d). Thus, a lack of dishabituation indicates that subjects reacted to CM urine as familiar to that presented in the habituation trial, in spite of the obvious difference in the stimuli. These results suggest that ShM and CM male subjects seem to be able to recognize a familiar component of the stimulus—independent of sex steroids—and respond to it as they would to a familiar compound present in CM + E or CM + T urine to which they have been previously habituated. As urine samples presented in habituation trials and in the tests belonged to the same individual donors, the differences between them could be due only to some changes caused by hormone treatment. Such a response cannot be due to different ages of the samples, as we have shown that the freezer shelf life in the used time interval does not influence the sample qualities.

The dishabituation in a test using urine of CM + T and CM + T donors (after being habituated to urine of the same donors before T and E treatment) exhibited by both ShM and CM subjects may indicate that they discriminated some substances that appeared in urine after hormone treatment (hormones and/or their metabolites). Some studies provide examples of this kind. It was shown that gonadectomy leads to disappearance of gender-specific substances in male and female mice urine that can be restored by hormone replacement therapy (e.g., Jemiolo et al., 1986; Boyer et al., 1989; Harvey et al., 1989). Our results are in line with studies conducted with rats and mice indicating subjects' abilities to recognize and respond to even subtle changes in composition of urine (Janssen et al., 2001; Thom & Hurst, 2004). In its turn, our findings support the notion that the H-D test is a sensitive tool that is able to reveal any novelty in tested samples (Thom & Hurst, 2004).

The results of the here presented experiments also indicate that castrated males in the H-D tests behave identical to their sham-operated siblings and are able to discriminate urine samples based on sex steroid changes. This finding is in line with the results of experiments conducted with rats and mice demonstrating that castrated males as intact subjects exhibited individual discrimination performance in H-D tests (Ferguson et al., 2002; Karlsson et al., 2015).

Our results are only the first step toward solving the question mentioned above concerning the possible reason for the lack of preference for female urine in castrated males. It has been shown that they can discriminate sex-specific hormone-dependent characteristics of urine, differentiating between urine samples of gonadectomized donors and urine of the same individuals after they underwent treatment with T and E, thus representing simulated males and females. Although this cannot yet serve as an absolute proof of the ability of castrated males to discriminate the sex of urine donors, it nevertheless provides tools for setting up further experiments. One of the possible next steps could be the use of urine samples from the same donor, when it first represents a simulated female and then a simulated male (and other experimental manipulations are possible). These experiments will clarify the question of the ability of castrated males to discriminate sex of conspecific individuals and, consequently, to answer the original question about the reasons of the lack of preference for female vs. male urine in the sexual preference test. Indeed, if this possibility is confirmed, it can be argued that in castrated males, the lack of preference for female urine is a consequence of the lack of motivation to respond to sex-associated stimuli like it was shown with the use of operant conditioning paradigm in rats (Carr & Caul, 1962).

6.3 Conclusive Remarks of Experiment 3

Our results indicate that Djungarian hamster males are able to discriminate even subtle hormone-dependent differences in the urine samples from one particular individual and that gonadectomy does not influence this ability. The most likely candidates influencing these differences in urine samples are sex-specific steroids and their metabolites. The data allow suggesting that subject males are able to recognize basic hormone-independent constituents in the urine of hormone treated castrated individuals, identify them, and therefore respond to them as to familiar. This obviously represents an ability to analyze and retrieve information from urine.

Finally, the proposed model using simulated females and simulated males can be applied in other studies in which it is necessary to eliminate the interference of gender and individuality traits, which is inevitable when using donors of different sexes.

7 Experiment 4: The Use of H-D Tests as a Tool to Assess the Ability of Hamsters to Analyze Binary Urine Mixtures

Animals and humans live in a world filled with odors, which are mixtures of various chemicals, like different social odors, smell of food or perfumes, gasoline, etc. However, there is still a gap in our knowledge concerning mechanisms that underline odor mixture perception. Even studies of the simplest binary mixtures of distinct chemicals provide controversial results. In spite of the growing interest to the field of binary odor mixture perception and analysis (according the PubMed database, a number of papers published during the last four decades increased by about 800 times), there is no clear understanding of animal and human abilities to process information from complex odor mixtures (Pannunzi & Nowotny, 2019; Young et al., 2020). This is due to the fact that the perception of odor mixtures depends on many factors, such as the chemical properties of constituents, their concentration ratio, experience of a subject with chemicals, and experimental design (Wiltrout et al., 2003; Kay et al., 2005; McNamara et al., 2008; Auffarth, 2013; Young et al., 2020; Coureaud et al., 2022).

Even more pronounced problems arise when analyzing how animals deal with the perception of mixtures of natural excreta, which themselves are very complex mixtures (e.g., like urine, glandular secretions, saliva, feces, etc.). We suggest, however, that the perception of such mixtures may share the same mechanisms that interline perception of binary/complex scent marks and blends of simple chemicals. The following scenarios are possible: (1) the forming mixture is homogenous and perceived configurally or synthetically as a substance with a new quality, different from its constituents; (2) one of the constituents of the homogenous mixture overshadows or masks the other component(s); and (3) the components are perceived elementally (each component is discriminated) within a heterogeneous mixture that represents the analytical processing of olfactory information (Laing & Francis, 1989; Johnston et al., 1994; Wilson & Stevenson, 2003; Thomas-Danguin et al., 2014).

There are just a few studies devoted to evaluate animals' abilities to perceive mixtures of natural excreta. In operant conditioning experiments, it was shown that rat subjects were able to recognize one of the individual urine samples in a mixture consisting of urine samples from two to seven donors. However, as the number of odor samples in a mixture increased, the number of recognizing individuals decreased dramatically (Voznessenskaya et al., 1992). Specially trained Pomeranian dogs were able to easily identify an individual rat urine sample in a urine mixture consisting of up to 21 individual urine samples, after which, as the mixture became more complex, the number of incorrect answers increased dramatically (Krutova & Zinkevich, 2003). Results similar to those shown in the rat study (Voznessenskaya et al., 1992) were obtained using the habituation-discrimination paradigm in our pilot experiments conducted on male golden and Djungarian hamsters (Vasilieva & Johnston, 1996; Litvinova & Vasilieva, 2001; Litvinova, 2003).

Since the effectiveness of research is to a great extent determined by the methodology, choosing a reliable method is an important step of any study. The aim of this experiment was to show that the H-D test can be used as a convenient and robust behavioral method for studying the perception of binary mixtures of natural excretions (urine). Here we present the results of H-D experiments with Djungarian and golden hamster males when (1) urine of one male donor was presented during the habituation trial and then, in discrimination tests, subjects were introduced to a binary mixture consisting of a urine sample presented during the habituation trial and urine of a new individual donor and (2) during the habituation trials, binary mixtures consisting of urine samples from two donor males were presented, and one of the urine samples from this mixture was later presented in the test.

7.1 Animals and Methods Used in Experiment 4

Sixteen Djungarian and 15 golden hamster males that did not participate in the experiments described above were used as test subjects. In addition, 13 Djungarian and 12 golden hamster males were used as urine donors. All the conditions were as described under General Methods. The standard H-D procedure was applied in two sets of experiments: (1) in habituation trials, urine from one donor male (male I) was used, and during the discrimination test, a mixture consisting of urine samples of male I and urine of a new donor, male II, in a ratio 1:1 was presented (Fig. 6a, c); (2) in habituation trials, a binary mixture consisting of urine samples from two donors (male I + male II), and in the test one of the mixture' constituents (urine of male I or male II) was presented

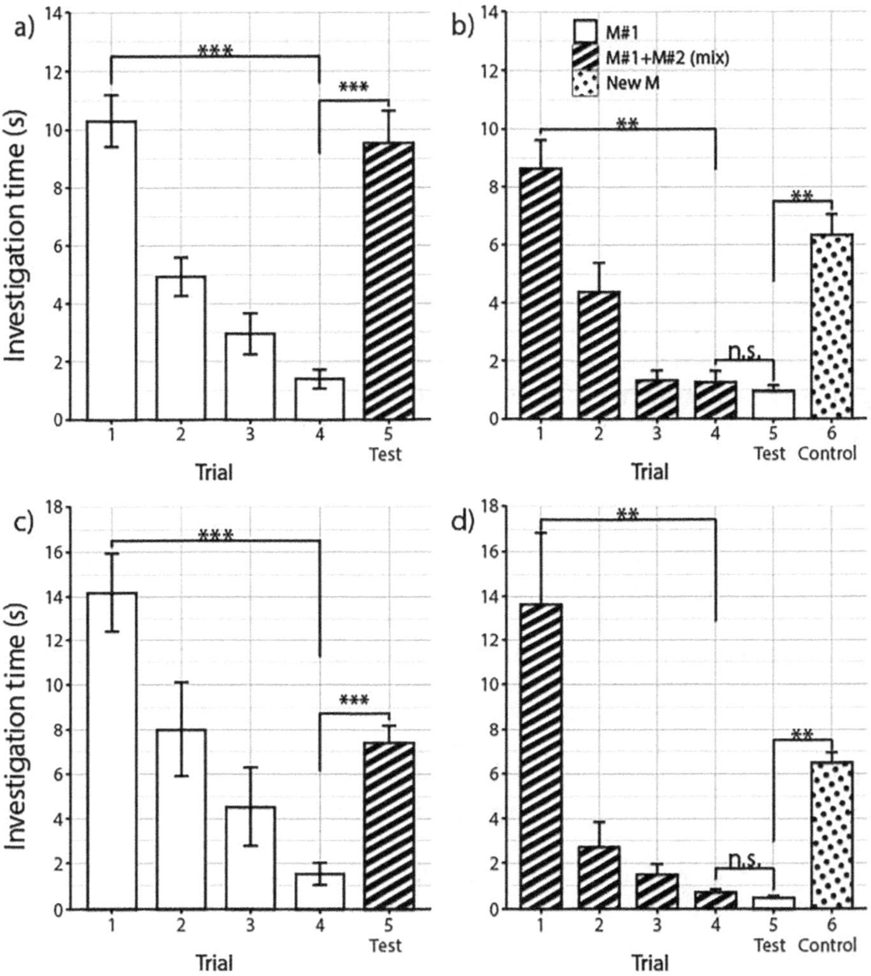

Fig. 6. Time (s; m ± SE) that Djungarian hamster males (top charts: **a**, **b**) and golden hamster males (lower charts: **c**, **d**) investigated: urine of a male #1 in the habituation trials (presentations 1–4) and a mixture of urine samples from the same male (#1) + urine of another male (#2) in the test (presentation 5) (**a**, **c**) and a mixture of urine samples from two males—male #1 + male #2 (mix) in the habituation trials (presentations 1–4), then one of the urine samples from this mixture (either male #1 or male #2; presentation 5), and urine from a new male (control presentation 6) (**b**, **d**). Statistical significance of differences between the corresponding values are marked with asterisks (** – P<0.01, *** – P<0.001).

(Fig. 6b, d). As in the last case animals usually did not demonstrate dishabituation, we used a presentation of a control stimuli (urine from a new donor male). Urine from none of the donors was presented to the male subjects twice.

7.2 Results and Discussion of Experiment 4

Djungarian and golden hamster males exhibited a classical decrease in interest toward urine of one male donor (male I) during the habituation trials ($V = 136$, $p < 0.001$ and $t = 9.03$, $df = 14$, $p < 0.001$, respectively; Fig. 6a, c,). At that, they demonstrated an increase in investigation toward the mixture consisting of the habituated urine stimulus and the urine of a new individual donor ($V = 0$, $p < 0.001$ and $t = 7.18$, $df = 14$, $p < 0.001$, respectively; Fig. 6a, c). When a binary mixture consisting of male urine samples was presented during the habituation trials, subjects of both species also showed a decreased interest toward this stimulus ($V = 66$, $p = 0.006$ and $t = 4.09$, $df = 14$, $p = 0.007$, respectively; Fig. 6b, d). In the following test, male subjects did not exhibit dishabituation toward the urine sample that was a part of the habituated mixture ($V = 41$, $p = 0.52$ and $V = 58$, $p = 0.146$, respectively; Fig. 6b, d). However, both Djungarian and golden hamster males investigated the control urine from a new individual male significantly longer ($V = 0$, $p = 0.006$ and $V = 0$, $p = 0.003$, respectively; Fig. 6b, d).

The results of those experiments where urine of a new donor was added to the habituated urine sample (Fig. 6a, c) indicate that subject males of both species reacted to the mixture as a stimulus that differed from that presented in the habituation trials. On one hand, such a behavioral response may indicate that when two samples are mixed, a complex mixture with a new quality appears and that subjects respond to it as to a new stimulus. On the other hand, animals may show renewed interest toward the appearance of a new individual stimulus in the mixture, discriminating it from stimulus to which they have already habituated. The results of the experiments with the reverse presentation order (Fig. 6b, d) revealed a lack of dishabituation to one of the constituents of the mixture in the test. At that, clear dishabituation to control urine indicated that subjects were active and such a behavioral response was not due to low motivation to investigate the stimuli but was rather determined by the features of the stimulus. The lack of dishabituation in a test is often interpreted as an inability to discriminate between stimuli and their similarity (Johnston, 2008). In our experiments, the stimulus presented in the test (presentation 5) was part of a habituated mixture, and the lack of dishabituation to it rather indicates its discrimination and the ability of the subjects to perceive the binary mixture elementally.

These findings are in line with the results of experiment 3 where urine samples presented in habituation trials and in the tests were from one individual donor and differed only by age (shelf life in the freezer). Neither old nor fresh urine caused dishabituation, neither in an arbitrary direct or reversed order of presentations (Fig. 4a, b). Similarly, subjects habituated to the urine of castrated male donors that underwent T or E therapy showed no dishabituation when tested with the urine of the same donors collected before hormone treatment. At that, subject males intensively investigated control urine samples (Fig. 5b, d). In the first case, the samples differed only by age (but were kept in a freezer, which ensured their safety and prevented biodegradation), so it is not surprising that subject males responded to them as to similar/familiar. Since age does not affect quality of urine samples (at least it can be detected within the frames of the H-D paradigm), we assume that in the second case samples from the same castrated donors differed only by the presence/absence of T or E and their metabolites.

In fact, the urine of castrated individuals subjected to hormone therapy may represent an analog of a binary mixture. One of its constituents is a hormone-independent "basic" component of urine, and the second one appears in urine as a result of the hormone therapy, namely, hormones and their metabolites. So, when subjects habituate to "part of the mixture" in the habituation trials, they exhibit dishabituation to the "whole mixture." In reverse experiments, when habituation occurs toward the "mixture," there is no dishabituation in the test. These analogies give us even more reason to believe that individual urine samples in a binary mixture are perceived as independent and identifiable constituents, and the mixture is perceived as a sum or set of discrete units (i.e., elementally).

Our results are in line with those obtained from the use of operant conditioning and H-D procedures indicating that rats are able to distinguish one component in the mixture (Staubli et al., 1987; Linster & Smith, 1999; Wiltrout et al., 2003). In H-D experiments, rats were habituated to binary mixtures composed of dissimilar components, and then in a test they did not show dishabituation to one of the constituents of this mixture but increased their investigation behavior toward a new control chemical. These data clearly indicate that rats perceived the mixtures elementally (Wiltrout et al., 2003) and support the results of our experiment. This resemblance allows us to assert with even greater confidence that a lack of dishabituation to one of the constituents of the urine mixture is because subjects are able to discriminate it and respond to it as toward a familiar/habituated stimulus.

Thus, using the terminology adopted in research on the perception and analysis of mixtures consisting of simple substances (Linster & Smith, 1999; Wiltrout et al., 2003; Kay et al., 2005), we can say that when analyzing binary mixtures of urine from different donors, animals perceive it not configurally but rather based on elemental properties of its constituents, that is, each of the individual samples is identified and detected. This indicates that animals possess analytical abilities allowing them to perceive the blend of excretions as a sum of discrete units that can be identified and retrieved from the mixture as it was shown for blends of simple odorants (Laing & Francis, 1989; Rokni & Murthy, 2014).

One may assume that the behavioral responses of the subject hamsters described above might be because of a change in concentrations of individual samples in a mixture. However, this is hardly the case, because even a 10–150 times' dilution of the urine (Litvinova & Vasilieva, 2004a) does not influence the ability for individual discrimination in the H-D tests in Djungarian and golden hamsters (Semenova, 2024; Zakharov, 2025).

Our data and those from rats' and dogs' experiments indicate that at least the studied species are able to analyze urine mixtures and process them elementally, suggesting that the individual properties of urine are most likely determined by the unique substance(s) (Voznessenskaya et al., 1992; Krutova & Zinkevich, 2014). Indeed, if the individual properties of urine are determined by the ratio of certain substances shared by conspecifics, strains, or group members, their ratio inevitably changes when a mixture is formed, which would have to lead to the disappearance of the individual characteristics of the constituents in the mixture. Thus, we suggest that not only individual differences in the ratio of identical substances in the excreta, as was assumed for different species based on the chemical analysis of urine and secretions (Gorman, 1976; Singer et al., 1997; Brennan & Kendrick, 2006; Setchell et al., 2010;), but rather the presence of unique components encodes the individuality.

7.3 Conclusive Remarks of Experiment 4

The results obtained indicate that the analysis of behavioral response to binary urine mixtures, with all the advantages and restrictions of the H-D paradigm, may be a step in understanding how animals perceive individually distinctive information from the blends of natural excretions that presumably constitute complex scent marks. Djungarian and golden hamster males perceive binary mixture of urine elementally, that is, they are able to discriminate individual samples that constitute a mixture.

Acknowledgments. The work was performed within the framework of the State Assignment No. 075-00440-24-03 IPEE RAS, topic 1022040700480-01.6.15. "Vertebrate behavior and biocommunication."

References

Achiraman, S., Ganesh, D. S., Kannan, S., Kamalakkannan, S., Nimala, N., & Archunan, G. (2014). Response of male mice to odours of female mice in different stages of oestrous cycle: Self-grooming behaviour and the effect of castration. *Indian Journal of Experimental Biology, 52*(1), 30–35.

Ågmo, A. (2003). Unconditioned sexual incentive motivation in the male Norway rat (Rattus norvegicus). *Journal of Comparative Psychology, 117*(1), 3. https://doi.org/10.1037/0735-7036.117.1.3

Aspesi, D., & Choleris, E. (2022). Neuroendocrine underpinning of social recognition in males and females. *Journal of Neuroendocrinology, 34*(2), e13070. https://doi.org/10.1111/jne.13070

Auffarth, B. (2013). Understanding smell—The olfactory stimulus problem. *Neuroscience & Biobehavioral Reviews, 37*(8), 1667–1679. https://doi.org/10.1016/j.neubiorev.2013.06.009

Bailey, K. R., Rustay, N. R., & Crawley, J. N. (2006). Behavioral phenotyping of transgenic and knockout mice: Practical concerns and potential pitfalls. *ILAR Journal, 47*(2), 124–131. https://doi.org/10.1093/ilar.47.2.124

Ballard, C. L., & Wood, R. I. (2007). Partner preference in male hamsters: Steroids, sexual experience and chemosensory cues. *Physiology & Behavior, 91*(1), 1–8. https://doi.org/10.1016/j.physbeh.2007.01.005

Baum, M. J., & Bakker, J. (2013). Roles of sex and gonadal steroids in mammalian pheromonal communication. *Frontiers in Neuroendocrinology, 34*(4), 268–284. https://doi.org/10.1016/j.yfrne.2013.07.004

Becker, S. D., & Hurst, J. L. (2009). Female behaviour plays a critical role in controlling murine pregnancy block. *Proceedings of the Royal Society B: Biological Sciences, 276*(1662), 1723–1729. https://doi.org/10.1098/rspb.2008.1780

Berlyne, D. E. (1950). Novelty and curiosity as determinants of exploratory behaviour. *British Journal of Psychology, 41*(1), 68. https://doi.org/10.1111/j.2044-8295.1950.tb00262.x

Boyer, M. L., Jemiolo, B., Andreolini, F., Wiesler, D., & Novotny, M. (1989). Urinary volatile profiles of pine vole, Microtus pinetorum, and their endocrine dependency. *Journal of Chemical Ecology, 15*(2), 649–662. https://doi.org/10.1007/BF01014708

Brennan, P. A. (2004). The nose knows who's who: Chemosensory individuality and mate recognition in mice. *Hormones and Behavior, 46*(3), 231–240. https://doi.org/10.1016/j.yhbeh.2004.01.010

Brennan, P. A., & Kendrick, K. M. (2006). Mammalian social odours: Attraction and individual recognition. *Philosophical Transactions of the Royal Society B: Biological Sciences, 361*(1476), 2061–2078. https://doi.org/10.1098/rstb.2006.1931

Brown, R. (1985). The rodents II: Suborder Myomorpha. In *Social odours in mammals* (Vol. 1, pp. 345–457).

Brown, R. E., Singh, P. B., & Roser, B. (1987). The major histocompatibility complex and the chemosensory recognition of individuality in rats. *Physiology & Behavior, 40*(1), 65–73. https://doi.org/10.1016/0031-9384(87)90186-7

Carr, W., & Caul, W. (1962). The effect of castration in rat upon the discrimination of sex odours. *Animal Behaviour, 10*(1–2), 20–27. https://doi.org/10.1016/0003-3472(62)90126-4

Carr, W., Loeb, L., & Wylie, N. (1966). Responses to feminine odors in normal and castrated male rats. *Journal of Comparative and Physiological Psychology, 62*(2), 336–338. https://doi.org/10.1037/h0023693

Clapperton, B. K., Minot, E. O., & Crump, D. R. (1988). An olfactory recognition system in the ferret *Mustela furo* L.(Carnivora: Mustelidae). *Animal Behaviour, 36*(2), 541–553. https://doi.org/10.1016/S0003-3472(88)80025-3

Cleland,T.A.,Morse,A.,Yue,E.L.,&Linster,C.(2002).Behavioralmodelsofodorsimilarity.*Behavioral Neuroscience, 116*(2), 222–231. https://psycnet.apa.org/doi/10.1037/0735-7044.116.2.222

Coureaud, G., Thomas-Danguin, T., Sandoz, J.-C., & Wilson, D. A. (2022). Biological constraints on configural odour mixture perception. *Journal of Experimental Biology, 225*(6), jeb242274.

Ervin, K. S., Lymer, J. M., Matta, R., Clipperton-Allen, A. E., Kavaliers, M., & Choleris, E. (2015). Estrogen involvement in social behavior in rodents: Rapid and long-term actions. *Hormones and Behavior, 74*, 53–76. https://doi.org/10.1016/j.yhbeh.2015.05.023

Ferguson, J. N., Young, L. J., & Insel, T. R. (2002). The neuroendocrine basis of social recognition. *Frontiers in Neuroendocrinology, 23*(2), 200–224. https://doi.org/10.1006/frne.2002.0229

Ferkin, M. H. (1992). Time course of androgenic modulation of odor preferences and odor cues in male meadow voles, Microtus pennsylvanicus. *Hormones and Behavior, 26*(4), 512–521. https://doi.org/10.1016/0018-506X(92)90018-Q

Ferkin, M. H. (2018). Odor communication and mate choice in rodents. *Biology, 7*(1), 13. https://doi.org/10.3390/biology7010013

Ferkin, M. H., & Johnston, R. E. (1993). Roles of gonadal hormones in control of five sexually attractive odors of meadow voles (*Microtus pennsylvanicus*). *Hormones and Behavior, 27*(4), 523–538. https://doi.org/10.1006/hbeh.1993.1038

Friedrich, R. W. (2006). Mechanisms of odor discrimination: neurophysiological an behavioural approaches. *Trends in Neurosciences, 29*(1), 40–47. https://doi.org/10.1016/j.tins.2005.10.004

Gabor, C. S., Phan, A., Clipperton-Allen, A. E., Kavaliers, M., & Choleris, E. (2012). Interplay of oxytocin, vasopressin, and sex hormones in the regulation of social recognition. *Behavioral Neuroscience, 126*(1), 97. https://doi.org/10.1037/a0026464

Gheusi, G., Bluthé, R.-M., Goodall, G., & Dantzer, R. (1994). Social and individual recognition in rodents: Methodological aspects and neurobiological bases. *Behavioural Processes, 33*(1), 59–87. https://doi.org/10.1016/0376-6357(94)90060-4

Gorman, M. L. (1976). A mechanism for individual recognition by odour in *Herpestesauro punctatus* (Carnivora: Viverridae). *Animal Behaviour, 24*(1), 141–145. https://doi.org/10.1016/S0003-3472(76)80107-8

Halpin, Z. T. (1974). Individual differences in the biological odors of the mongolian gerbil (*Meriones unguiculatus*). *Behavioral Biology, 11*(2), 253–259. https://doi.org/10.1016/S0091-6773(74)90449-0

Halpin, Z. T. (1986). Individual odors among mammals: Origins and functions. In *Advances in the study of behavior* (Vol. 16, pp. 39–70). Academic Press. https://doi.org/10.1016/S0065-3454(08)60187-4

Harding, S. M., & Velotta, J. P. (2011). Comparing the relative amount of testosterone required to restore sexual arousal, motivation, and performance in male rats. *Hormones and Behavior, 59*(5), 666–673. https://doi.org/10.1016/j.yhbeh.2010.09.009

Havens, M. D., & Rose, J. D. (1992). Investigation of familiar and novel chemosensory stimuli by golden hamsters: Effects of castration and testosterone replacement. *Hormones and Behavior, 26*(4), 505–511. https://doi.org/10.1016/0018-506X(92)90017-P

Harvey, S., Jemiolo, B., & Novotny, M. (1989). Pattern of volatile compounds in dominant and subordinate male mouse urine. *Journal of Chemical Ecology, 15*(7), 2061–2072. https://doi.org/10.1007/BF01207438

Hughes, A. M., Everitt, B. J., & Herbert, J. (1990). Comparative effects of preoptic area infusions of opioid peptides, lesions and castration on sexual behaviour in male rats: Studies of instrumental behaviour, conditioned place preference and partner reference. *Psychopharmacology, 102*(2), 243–256. https://doi.org/10.1007/BF02245929

Hurst, J. L. (2009). Female recognition and assessment of males through scent. *Behavioural Brain Research, 200*(2), 295–303. https://doi.org/10.1016/j.bbr.2008.12.020

Jakupovic, J., Kang, N., & Baum, M. J. (2008). Effect of bilateral accessory olfactory bulb lesions on volatile urinary odor discrimination and investigation as well as mating behavior in male mice. *Physiology & Behavior, 93*(3), 467–473. https://doi.org/10.1016/j.physbeh.2007.10.005

Janssen, E., Göhlen, B., Behrens, D., Richter, K., & Zavazava, N. (2001). Allogeneic recombinant soluble MHC class I molecules modify urinary odor cues in rats. *Physiology & Behavior, 72*(1), 107–114. https://doi.org/10.1016/S0031-9384(00)00389-9

Jemiolo, B., Harvey, S., & Novotny, M. (1986). Promotion of the Whitten effect in female mice by synthetic analogs of male urinary constituents. *Proceedings of the National Academy of Sciences, 83*(12), 4576–4579. https://doi.org/10.1073/pnas.83.12.4576

Johnston, R. E. (1993). Memory for individual scent in hamsters (*Mesocricetus auratus*) as assessed by habituation methods. *Journal of Comparative Psychology, 107*(2), 201–207. https://doi.org/10.1037/0735-7036.107.2.201

Johnston, R. E. (2008). Individual odors and social communication: Individual recognition, kin recognition, and scent over-marking. *Advances in the Study of Behavior, 38*, 439–505.

Johnston, R. E., Chiang, G., & Tung, C. (1994). The information in scent over-marks of golden hamsters. *Animal Behaviour, 48*(2), 323–330. https://doi.org/10.1006/anbe.1994.1245.

Johnston, R. E., Derzie, A., Chiang, G., Jernigan, P., & Lee, H.-C. (1993). Individual scent signatures in golden hamsters: Evidence for specialization of function. *Animal Behaviour, 45*(6), 1061–1070. https://doi.org/10.1006/anbe.1993.1132

Karlsson, S. A., Haziri, K., Hansson, E., Kettunen, P., & Westberg, L. (2015). Effects of sex and gonadectomy on social investigation and social recognition in mice. *BMC Neuroscience, 16*(1), 83. https://doi.org/10.1186/s12868-015-0221-z

Kassambara, A. (2023). *Rstatix: Pipe-friendly framework for basic statistical tests (Version 0.7. 2)* [Computer software].

Kay, L. M., Crk, T., & Thorngate, J. (2005). A redefinition of odor mixture quality. *Behavioral Neuroscience, 119*(3), 726–733. https://doi.org/10.1037/0735-7044.119.3.726

Kelliher, K. R., & Baum, M. J. (2002). Effect of sex steroids and coital experience on ferrets' preference for the smell, sight and sound of conspecifics. *Physiology & Behavior, 76*(1), 1–7. https://doi.org/10.1016/S0031-9384(02)00691-1

Kondo, Y., & Hayashi, H. (2021). Neural and hormonal basis of opposite-sex preference by chemosensory signals. *International Journal of Molecular Sciences, 22*(15), 8311. https://doi.org/10.3390/ijms22158311

Krutova, V. I., & Zinkevich, E. P. (2003). Dogs recognize an individual odor in a mixture of odors from many individuals. *Doklady Biological Sciences, 388*(1), 56–58.

Krutova, V. I., & Zinkevich, E. P. (2014). Recognition possibilities by dogs of individual odor mixtures and their composites. *Sensornye systemy [Sensory systems], 28*(1), 84–92.

Lai, S., & Johnston, R. E. (1994). Individual odors in Djungarian hamsters (*Phodopus campbelli*). *Ethology, 96*(2), 117–126. https://doi.org/10.1111/j.1439-0310.1994.tb00887.x

Lai, S.-C., Vasilieva, N. Y., & Johnston, R. E. (1996). Odors providing sexual information in Djungarian hamsters: Evidence for an across-odor code. *Hormones and Behavior, 30*(1), 26–36. https://doi.org/10.1006/hbeh.1996.0005

Laing, D. G., & Francis, G. W. (1989). The capacity of humans to identify odors in mixtures. *Physiology & Behavior, 46*(5), 809–814. https://doi.org/10.1016/0031-9384(89)90041-3

Le Moëne, O., & Ågmo, A. (2018). The neuroendocrinology of sexual attraction. *Frontiers in Neuroendocrinology, 51*, 46–67. https://doi.org/10.1016/j.yfrne.2017.12.006

Leonard, S. T., & Ferkin, M. H. (1999). Prolactin and testosterone mediate seasonal differences in male preferences for the odors of females and the attractiveness of male odors to females. In Advances in chemical signals in vertebrates (pp. 437–443). Boston, MA: Springer US.

Linster, C., & Smith, B. H. (1999). Generalization between binary odor mixtures and their components in the rat. *Physiology & Behavior, 66*(4), 701–707. https://doi.org/10.1016/S0031-9384(99)00007-4

Litvinova, E. M. (2003). *Individual odor: Factors determining its formation and recognition in the Djungarian hamster (Phodopus sungorus)* (Ph.D. thesis). Lomonosov Moscow State University.

Litvinova, E.M., & Vasilieva, N.Y. (2001). Individual differences in the ability to recognize a single odor in a mixture of social odors in Djungarian hamsters. BXXYII International Ethological Conference, Tübingen, Germany, 22–29 August 2001. Advances in Ethology. Supplements to Ethology R. Apfelbach, M. Fendt, Kramer, B.M. Simers (36, 206).

Litvinova, E. M., & Vasilieva, N. Y. (2004a). Individual body odor in the Djungarian hamster (Phodopussungorus). 1. Sources of chemosignals. *Zoologicheskii Zhurnal, 83*(7), 876–887.

Litvinova, E. M., & Vasilieva, N. Y. (2004b). Individual odor in the Djungarian hamster Phodopussungorus. 2. The role of specific skin glands. *Zoologicheskii Zhurnal, 83*(9), 1160–1168.

Liu, Y.-J., Zhang, J.-X., Zhang, J.-H., Bao, W.-D., & Liu, D.-Z. (2010). Vomeronasal organ ablation elicits chemosensory dysfunction and abnormal behavior in mice. *Journal of Ethology, 28*(2), 263–271. https://doi.org/10.1007/s10164-009-0180-x

Luo, M., Fee, M. S., & Katz, L. C. (2003). Encoding pheromonal signals in the accessory olfactory bulb of behaving mice. *Science, 299*(5610), 1196–1201. https://doi.org/10.1126/science.1082133

Maras, P. M., & Petrulis, A. (2008). The posteromedial cortical amygdala regulates copulatory behavior, but not sexual odor preference, in the male Syrian hamster (*Mesocricetus auratus*). *Neuroscience, 156*(3), 425–435. https://doi.org/10.1016/j.neuroscience.2008.08.004

McNamara, A. M., Magidson, P. D., Linster, C., Wilson, D. A., & Cleland, T. A. (2008). Distinct neural mechanisms mediate olfactory memory formation at different timescales. *Learning & Memory, 15*(3), 117–125. https://doi.org/10.1101/lm.785608

Merkx, J. (1984). Effect of castration and subsequent substitution with testosterone, dihydrotestosterone and oestradiol on sexual preference behaviour in the male rat. *Behavioural Brain Research, 11*(1), 59–65. https://doi.org/10.1016/0166-4328(84)90008-1

Merkx, J., Slob, A. K., & Van Der Werff Ten Bosch, J. J. (1987). Attractivity of male rats induced by estradiol and progesterone. *Physiology & Behavior, 40*(6), 737–740. https://doi.org/10.1016/0031-9384(87)90276-9

Muroi, Y., Nishimura, M., & Ishii, T. (2017). The accessory olfactory system facilitates the recovery of the attraction to familiar volatile female odors in male mice. *Chemical Senses, 42*(9), 737–745. https://doi.org/10.1093/chemse/bjx045

Oliveira, R. F. (2004). Social modulation of androgens in vertebrates: Mechanisms and function. *Advances in the Study of Behavior, 34*, 165–240. https://doi.org/10.1016/S0065-3454(04)34005-2

Pankevich, D. E., Deedy, E. M., Cherry, J. A., & Baum, M. J. (2003). Interactive effects of testosterone and superior cervical ganglionectomy on attraction thresholds to volatile urinary odors in gonadectomized mice. *Behavioural Brain Research, 144*(1–2), 157–165. https://doi.org/10.1016/S0166-4328(03)00073-1

Pankevich, D. E., Baum, M. J., & Cherry, J. A. (2004). Olfactory sex discrimination persists, whereas the preference for urinary odorants from estrous females disappears in male mice after vomeronasal organ removal. *The Journal of Neuroscience, 24*(42), 9451–9457. https://doi.org/10.1523/JNEUROSCI.2376-04.2004

Pankevich, D. E., Cherry, J. A., & Baum, M. J. (2006). Effect of vomeronasal organ removal from male mice on their preference for and neural Fos responses to female urinary odors. *Behavioral Neuroscience, 120*(4), 925–936. https://doi.org/10.1037/0735-7044.120.4.925

Pannunzi, M., & Nowotny, T. (2019). Odor stimuli: Not just chemical identity. *Frontiers in Physiology, 10*, 1428. https://doi.org/10.3389/fphys.2019.01428

Papes, F., Nakahara, T. S., & Camargo, A. P. (2018). Behavioral assays in the study of olfaction: A practical guide. In B. F. M. S. De Souza & G. Antunes (Eds.), *Olfactory receptors* (pp. 289–388). Springer. https://doi.org/10.1007/978-1-4939-8609-5_21

Penn, D., & Potts, W. K. (1998). Untrained mice discriminate MHC-determined odors. *Physiology & Behavior, 64*(3), 235–243. https://doi.org/10.1016/S0031-9384(98)00052-3

Petrulis, A. (2013). Chemosignals, hormones and mammalian reproduction. *Hormones and Behavior, 63*(5), 723–741. https://doi.org/10.1016/j.yhbeh.2013.03.011

Petrulis, A., & Eichenbaum, H. (2003). The perirhinal–entorhinal cortex, but not the hippocampus, is critical for expression of individual recognition in the context of the Coolidge effect. *Neuroscience, 122*(3), 599–607. https://doi.org/10.1016/j.neuroscience.2003.08.009

Petrulis, A., & Johnston, R. E. (1999). Lesions centered on the medial amygdala impair scent-marking and sex-odor recognition but spare discrimination of individual odors in female golden hamsters. *Behavioral Neuroscience, 113*(2), 345–357. https://doi.org/10.1037/0735-7044.113.2.345

Petrulis, A., Peng, M., & Johnston, R. E. (1999). Effects of vomeronasal organ removal on individual odor discrimination, sex-odor preference, and scent marking by female hamsters. *Physiology & Behavior, 66*(1), 73–83. https://doi.org/10.1016/S0031-9384(98)00259-5

Petrulis, A., Peng, M., & Johnston, R. E. (2000). The role of the hippocampal system in social odor discrimination and scent-marking in female golden hamsters (*Mesocricetus auratus*). *Behavioral Neuroscience, 114*(1), 184–195. https://doi.org/10.1037/0735-7044.114.1.184

Powers, J. B., & Bergondy, M. L. (1983). Androgenic regulation of chemoinvestigatory behaviors in male and female hamsters. *Hormones and Behavior, 17*(1), 28–44. https://doi.org/10.1016/0018-506X(83)90013-2

R Core Team (2024). R: A Language and Environment for Statistical Computing. R Foundation for Statistical Computing, Vienna, Austria. https://www.R-project.org/.

Robarts, D. W., & Baum, M. J. (2007). Ventromedial hypothalamic nucleus lesions disrupt olfactory mate recognition and receptivity in female ferrets. *Hormones and Behavior, 51*(1), 104–113. https://doi.org/10.1016/j.yhbeh.2006.08.009

Rokni, D., & Murthy, V. N. (2014). Analysis and synthesis in olfaction. *ACS Chemical Neuroscience, 5*(10), 870–872. https://doi.org/10.1021/cn500199n

Sachs, B. D. (2007). A contextual definition of male sexual arousal. *Hormones and Behavior, 51*(5), 569–578. https://doi.org/10.1016/j.yhbeh.2007.03.011

Schellinck, H. M., Price, S. R., & Wong, M. J. (2008). Using ethologically relevant tasks to study olfactory discrimination in rodents. In J. L. Hurst, R. J. Beynon, S. C. Roberts, & T. D. Wyatt (Eds.), *Chemical signals in vertebrates 11* (pp. 71–80). Springer. https://doi.org/10.1007/978-0-387-73945-8_6

Semenova, P. D. (2024). *Analysis of excretion mixtures by golden hamster (Mesocricetus auratus) males* (M.S. thesis). Lomonosov Moscow State University.

Setchell, J. M., Vaglio, S., Moggi-Cecchi, J., Boscaro, F., Calamai, L., & Knapp, L. A. (2010). Chemical composition of scent-gland secretions in an old world monkey (*Mandrillus sphinx*): Influence of sex, male status, and individual identity. *Chemical Senses, 35*(3), 205–220. https://doi.org/10.1093/chemse/bjp105

Singer, A. G., Beauchamp, G. K., & Yamazaki, K. (1997). Volatile signals of the major histocompatibility complex in male mouse urine. *Proceedings of the National Academy of Sciences, 94*(6), 2210–2214. https://doi.org/10.1073/pnas.94.6.2210

Sokolov, V. E., Feoktistova, N. Y., & Vasilieva, N. Y. (1994). Communicative function of the secretion from additional sacculi at the opening of the cheek ouches in Siberian (*Phodopus sungorus* Pallas, 1773) and Djungarian hamsters (*Phodopus campbelli* Thomas, 1905): The role in individual discrimination. *Doklady Akademii Nauk, 335*, 118–120.

Spehr, M., Spehr, J., Ukhanov, K., Kelliher, K. R., Leinders-Zufall, T., & Zufall, F. (2006). Signaling in the chemosensory systems: Parallel processing of social signals by the mammalian main and accessory olfactory systems. *Cellular and Molecular Life Sciences CMLS, 63*(13), 1476–1484. https://doi.org/10.1007/s00018-006-6109-4

Staubli, U., Fraser, D., Faraday, R., & Lynch, G. (1987). Olfaction and the «data» memory system in rats. *Behavioral Neuroscience, 101*(6), 757–765. https://doi.org/10.1037/0735-7044.101.6.757

Stowers, L., & Logan, D. W. (2010). Sexual dimorphism in olfactory signaling. *Current Opinion in Neurobiology, 20*(6), 770–775. https://doi.org/10.1016/j.conb.2010.08.015

Stowers, L., Holy, T. E., Meister, M., Dulac, C., & Koentges, G. (2002). Loss of sex discrimination and male-male aggression in mice deficient for TRP2. *Science, 295*(5559), 1493–1500. https://doi.org/10.1126/science.1069259

Sukoff Rizzo, S. J., & Crawley, J. N. (2017). Behavioral phenotyping assays for genetic mouse models of neurodevelopmental, neurodegenerative, and psychiatric disorders. *Annual Review of Animal Biosciences, 5*(1), 371–389. https://doi.org/10.1146/annurev-animal-022516-022754

Thom, M., & Hurst, J. (2004). Individual recognition by scent. *Annales Zoologici Fennici, 41*(6), 765–787.

Thomas-Danguin, T., Sinding, C., Romagny, S., El Mountassir, F., Atanasova, B., Le Berre, E., Le Bon, A.-M., & Coureaud, G. (2014). The perception of odor objects in everyday life: A review on the processing of odor mixtures. *Frontiers in Psychology, 5*, 504. https://doi.org/10.3389/fpsyg.2014.00504

Thompson, R. F. (1986). The neurobiology of learning and memory. *Science, 233*(4767), 941–947. https://doi.org/10.1126/science.3738519

Thompson, R. F. (2009). Habituation: A history. *Neurobiology of Learning and Memory, 92*(2), 127–134. https://doi.org/10.1016/j.nlm.2008.07.011

Thor, D. H. (1980). Testosterone and persistence of social investigation in laboratory rats. *Journal of Comparative and Physiological Psychology, 94*(5), 970–976. https://doi.org/10.1037/h0077831

Thor, D. H., Wainwright, K. L., & Holloway, W. R. (1982). Persistence of attention to a novel conspecific: Some developmental variables in laboratory rats. *Developmental Psychobiology, 15*(1), 1–8. https://doi.org/10.1002/dev.420150102

Vasilieva, N. Y., & Johnston, R. E. (1996). Limits for discrimination of individual odors: Golden hamsters as a model. BAbstracts. In *ECRO XII, Zurich, August 25-31, 1996* (p. 27). Swiss Federal Institute of Technology.

Vasilieva, N. Y., Lai, S., Petrova, E. V., & Johnston, R. E. (2001). Development of species preferences in two hamsters, *Phodopus campbelli* and *Phodopus sungorus*: Effects of cross-fostering. *Ethology, 107*(3), 217–236. https://doi.org/10.1046/j.1439-0310.2001.00642.x

Voznessenskaya, V., Parfyonova, V., Zinkevich, E., & Severtzov, A. (1992). Individual odortypes. In R. L. Doty & D. Müller-Schwarze (Eds.), *Chemical signals in vertebrates 6* (pp. 503–508). Springer. https://doi.org/10.1007/978-1-4757-9655-1_77

Wickham, H., Averick, M., Bryan, J., Chang, W., McGowan, L., François, R., Grolemund, G., Hayes, A., Henry, L., Hester, J., Kuhn, M., Pedersen, T., Miller, E., Bache, S., Müller, K., Ooms, J., Robinson, D., Seidel, D., Spinu, V., Takahashi, K., Vaughan, D., Wilke, C., Woo, K., & Yutani, H. (2019). Welcome to the Tidyverse. *Journal of Open Source Software, 4*(43), 1686. https://doi.org/10.21105/joss.01686

Wilson, D. A., & Stevenson, R. J. (2003). Olfactory perceptual learning: The critical role of memory in odor discrimination. *Neuroscience & Biobehavioral Reviews, 27*(4), 307–328. https://doi.org/10.1016/S0149-7634(03)00050-2

Wiltrout, C., Dogra, S., & Linster, C. (2003). Configurational and nonconfigurational interactions between odorants in binary mixtures. *Behavioral Neuroscience, 117*(2), 236–245. https://doi.org/10.1037/0735-7044.117.2.236

Woodley, S. K. (2004). Effects of vomeronasal organ removal on olfactory sex discrimination and odor preferences of female ferrets. *Chemical Senses, 29*(8), 659–669. https://doi.org/10.1093/chemse/bjh069

Woodley, S. K., & Baum, M. J. (2003). Effects of sex hormones and gender on attraction thresholds for volatile anal scent gland odors in ferrets. *Hormones and Behavior, 44*(2), 110–118. https://doi.org/10.1016/S0018-506X(03)00126-0

Wyatt, T. D. (2014). *Pheromones and animal behavior: Chemical signals and signatures*. Cambridge University Press.

Wysocki, C., Nyby, J., Whitney, G., Beauchamp, G., & Katz, Y. (1982). The vomeronasal organ: Primary role in mouse chemosensory gender recognition. *Physiology & Behavior, 29*(2), 315–327. https://doi.org/10.1016/0031-9384(82)90021-X

Xiao, K., Kondo, Y., & Sakuma, Y. (2004). Sex-specific effects of gonadal steroids on conspecific odor preference in the rat. *Hormones and Behavior, 46*(3), 356–361. https://doi.org/10.1016/j.yhbeh.2004.05.008

Xiao, K., Chiba, A., Sakuma, Y., & Kondo, Y. (2015). Transient reversal of olfactory preference following castration in male rats: Implication for estrogen receptor involvement. *Physiology & Behavior, 152*, 161–167. https://doi.org/10.1016/j.physbeh.2015.09.016

Yang, M., & Crawley, J. N. (2009). Simple behavioral assessment of mouse olfaction. *Current Protocols in Neuroscience, 48*(1), 8–24. https://doi.org/10.1002/0471142301.ns0824s48

Young, B. D., Escalon, J. A., & Mathew, D. (2020). Odors: From chemical structures to gaseous plumes. *Neuroscience & Biobehavioral Reviews, 111*, 19–29. https://doi.org/10.1016/j.neubiorev.2020.01.009

Zakharov A.P. (2025). *Chemoanalytical abilities of male Djungarian hamsters, Phodopus sungorus (Pallas, 1773): The search for determining factors* (M.S. thesis). Lomonosov Moscow State University.

Elucidating the Role of Squalene in the Chemical Communication of Vertebrates: An *In-Silico* Study Using Various Odorant-Binding Proteins

Devaraj Sankarganesh[1(✉)], Mohanraj Gopikrishnan[1], George Priya Doss C[1], Sinsha Santhosh[1], and Gnana Thiraviam Annis Inncia[1]

[1] School of Bio Sciences and Technology, Vellore Institute of Technology (VIT), Vellore, Tamil Nadu, India
sankarganesh.d@vit.ac.in

Abstract. Pheromonal communication in vertebrates is well-established. Pheromones in vertebrates may be composed of either single molecules or mixtures of compounds, where each component, at a specific concentration, contributes to a unique odor, a phenomenon known as the "bouquet effect." In some cases, the pheromonal blend also includes a fixative compound that helps retain the pheromonal signal for a longer duration. Still, it is unknown whether such fixative compounds also contribute to the bouquet effect. Squalene has been shown to act as a sex pheromone in laboratory rats (*Rattus norvegicus*) and cows (*Bos taurus*) and as an appeasing pheromone in mammals, including pigs (*Sus scrofa*). It has also been suggested to function as a fixative in African wild dogs (*Lycaon pictus*). Given this diverse functional evidence, the precise significance of squalene in vertebrate chemical communication remains unresolved. Therefore, we assessed the binding efficacy of squalene with odorant-binding proteins (OBPs) from *R. norvegicus*, *L. pictus*, *S. scrofa*, and *B. taurus* using molecular docking as an initial *in-silico* approach to generate testable hypotheses for future biochemical and behavioral validation. We used AutoDock Vina to perform docking of squalene with these OBPs. We found the highest docking affinity between squalene and the OBP of *B. taurus*, followed by that of *L. pictus*. The docking score was comparatively lower for the OBPs of *S. scrofa* and *R. norvegicus* and did not differ significantly from each other. As expected from its chemical structure, squalene did not form hydrogen bonds with any OBP; instead, the binding interactions were dominated by hydrophobic contacts and entropic effects associated with water displacement. Taken together, we suggest that squalene is a putative signaling compound in vertebrate chemical communication and may function primarily as a fixative compound.

Keywords: Lipocalins · Olfaction · Vomeronasal organ

L. M. Schulte (ed.), *Chemical Signals in Vertebrates 16*,
https://doi.org/10.1007/978-3-032-30752-1_13

1 Introduction

Olfaction plays essential roles in social interactions among vertebrates (Taniguchi & Taniguchi, 2014). This sensory modality involves the detection of chemical signals, commonly referred to as semiochemicals, that are used during communication between different species and within a given species (Hayden & Teeling, 2014). The detection and identification of semiochemicals are therefore crucial for reproduction, survival, and early warning of potential dangers (Wang et al., 2024). The binding of odorant-binding proteins (OBPs) to the odor molecules facilitates the recognition of many odorants in an odorant environment. OBPs are found not only in the olfactory mucus but also in secretions such as urine and glandular exudates, where they can act as slow releasers or, in some cases, as semiochemicals in their own right. In the present study, however, our focus is specifically on their role as capture and transport proteins in the olfactory system (Vogt et al., 1991).

OBPs belong to the lipocalin family, which possesses a conserved topology consisting of eight antiparallel β-strands forming a β-barrel and a conserved short α-helix.. With high specificity and affinity, they bind odorant molecules, especially the small apolar molecules and lipophilic semiochemicals, therefore improving solubilization and transport in the aqueous phase of the olfactory mucus (Pelosi et al., 2014; Moitrier et al., 2022). In this process, it is believed that proteins, also known as olfactory proteins, are responsible for the initial stage of olfactory signal transduction and the protection of semiochemicals from enzymatic degradation, thereby facilitating their delivery to receptors at the dendrites of sensory neurons (Pelosi et al., 2018). The binding interaction of semiochemicals with OBPs thus ensures their presentation to receptors, ultimately evoking behavioral responses in the receiving organism (Brito et al., 2016). Over the years, researchers have investigated OBP interactions with both volatile and nonvolatile compounds (Janssenswillen et al., 2021; Pelosi & Knoll, 2022).

Squalene is a highly hydrophobic triterpene that naturally occurs in the lipids of the skin, sebaceous gland secretions, and exocrine gland secretions of mammals (Zhang et al., 2008). It is also an essential metabolic intermediate in cholesterol biosynthesis, highlighting that its occurrence is not limited to secretory pathways. Squalene has been detected in glandular secretions that correlate with reproductive and social cues in several vertebrates, particularly primates and carnivores that rely heavily on scent communication (Scordato et al., 2007). Squalene is secreted from the clitoral gland of female laboratory rats (*Rattus norvegicus*), where it varies according to the estrous cycle, which in turn modulates the behavior of the males (Achiraman et al., 2011). It has also been detected in the urine of cows (*Bos taurus*), which suggests that it may be involved in reproductive signaling (Le Danvic et al., 2015). Sebaceous secretions in cattle have also been proposed to synchronize fertile reproductive members in herds (Dutton et al., 2002). In African wild dogs (*Lycaon pictus*), high concentrations of squalene in glandular secretions have been linked to long-lasting scent marking, which helps maintain social hierarchies (Apps et al., 2012). Martín et al. (2010) also suggested that the lipoidal biphasic nature of these compounds acts as a binding substrate or matrix for more volatile odorants, thereby stabilizing scent marks and prolonging their persistence in the environment. The diversity among species has a significant impact on their odor recognition and binding efficiencies in OBPs (Gonçalves et al., 2021). Although squalene usually occurs in conjunction with other compounds, its role as a solitary active signaling compound is less investigated. Thus, it would shed light on whether squalene participates in a universal or species-restricted chemical communication if we understand its interaction with OBPs. The significant advances in structural biology and molecular dynamics are currently yielding a growing body of knowledge about how odorants interact with their receptors (Wang et al., 2025). Therefore, it needs to be demonstrated how computational work at the atomic level can contribute to understanding the molecular basis of chemical communication using these tools. As such, these advances are currently conducting a broader inquiry into the molecular dimensions of chemical communication through odorant receptor interactions (Wang et al., 2025). Accordingly, in the present study, we employed molecular docking to examine squalene-OBP interactions as a hypothesis-generating step, to complement future biochemical and behavioral experiments. We specifically investigated OBPs from *R. norvegicus*, *B. taurus*, *S. scrofa*, and *L. pictus* to assess whether squalene binding reflects broader patterns of chemical communication in vertebrates.

2 Materials and Methods

2.1 Preparation of the Receptor

Crystallographic data for *Bos taurus* OBP (PDB ID: 1GT4) (Vincent et al., 2004), *R. norvegicus* OBP (PDB ID: 3ZQ3) (Portman et al., 2014), and *S. scrofa* OBP (PDB ID: 1A3Y) were retrieved from the RCSB Protein Data Bank. At the same time, the predicted structure of *L. pictus* OBP was obtained via the AlphaFold Protein Structure Database (Jumper et al., 2021) using the UniProt ID: O18873. Following retrieval, the water molecules were removed and isolated from heteroatoms and nonstandard ligands using BIOVIA Discovery Studio v. 2020. The proteins were subsequently saved in the pdbqt format (Sharma et al., 2019).

2.2 Preparation of Ligands

From PubChem (https://pubchem.ncbi.nlm.nih.gov/), the structure of squalene (PubChem CID: 638072) was retrieved in the structured data format (SDF), and was converted into PDB format using Open Babel for further analysis (O'Boyle et al., 2011). Ligand preparation was achieved by energy minimizing, appending hydrogen atoms and charges, and finally assigning active torsions using AutoDock Tools (Morris et al., 2009). Ultimately, ligands were kept in the pdbqt format after preparation.

2.3 Active Sites Determination

The prediction of the active sites of the receptors was done using the Computed Atlas of Surface Topography of Proteins (CASTp), which may be accessed at http://sts.bioe.uic.edu/castp/index.html?2was (Tian et al., 2018).

2.4 Molecular Docking and Visualization of OBP with Squalene Interactions

AutoDock Vina and AutoDock Tools were used in molecular docking analysis. Data about squalene and OBP receptors were transferred to the Vina folder in pdbqt format for docking using AutoDock Vina. The docking results were observed using Notepad software. The ligand shape with the lowest binding free energy was identified as the optimal posture for the receptor. The data were presented in a text file format. The receptor and ligand files were accessed in the Grid menu window for docking, which was performed using AutoDock Tools; they are OBP.pdbqt and Squalene.pdbqt, respectively. The GridBox object was present in the binding site region and then saved in the gpf format. Accessing receptor and ligand retrieval from the docking menu sets the Rigid Filename for docking procedures. Subsequently, the docking parameters were saved in DPF (Docking Parameter File) format, and both the GPF and DPF files were placed in a single directory to execute AutoGrid and AutoDock using the command prompt. Using BioVIA Discovery Studio v. 2020, we visualized the position and orientations of the ligands on the active sites within the receptors. Additionally, we visualized the 2D and 3D states of the amino acid interactions within the receptor (Trott & Olson, 2010).

3 Results

Mainly, the docking protocol aimed to evaluate the protein-ligand interactions of squalene with vertebrate OBPs. For structural interpretations, all four OBPs were visualized in the same ribbon orientation, and squalene was displayed as a space-filling ligand in the binding cavity (Figs. 1 and 2). The different ligand conformations among these proteins are stressed. The hydrophobic nature of the interactions is also emphasized. Table 1 summarizes the binding free energies of squalene to the OBP receptors 1GT4, 3ZQ3, 1A3Y, and O18873. The predicted binding affinity of 1GT4 was −6.78 kcal/mol, followed by O18873 (−5.96 kcal/mol), 1A3Y (−5.02 kcal/mol), and 3ZQ3 (−5.01 kcal/mol). These values represent the docking-derived binding free energies (ΔG) predicted by the AutoDock Vina scoring function, taking into account of Van der Waals interactions, hydrophobic interactions, and torsional entropy. Consistent with the chemical structure of squalene, no hydrogen bonds were detected in any of the complexes. Instead, binding was dominated by hydrophobic contacts and entropic stabilization resulting from the displacement of water molecules.

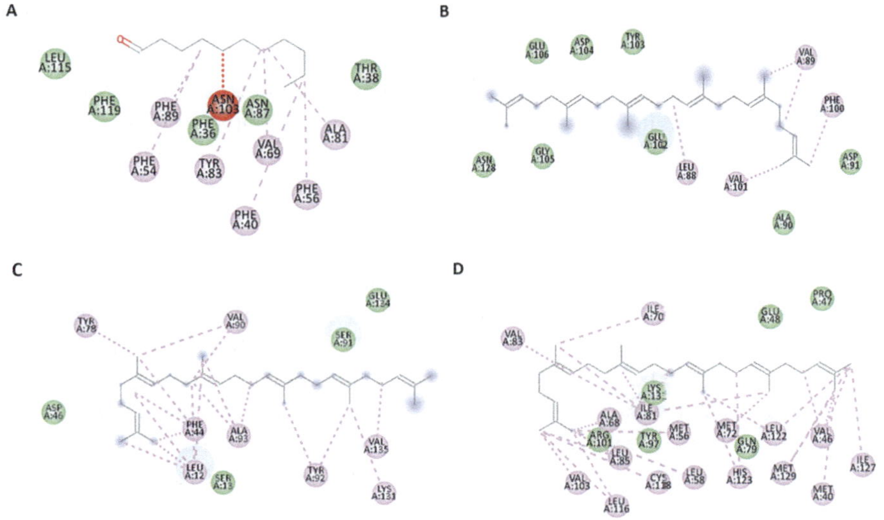

Fig. 1. Representation in 2D molecular interactions of squalene with the OBPs showing the proteins in ribbon form and squalene as a space-filling model: (**a**) 1GT4, (**b**) 3ZQ3, (**c**) 1A3Y, and (**d**) O18873. The critical hydrophobic residues relevant to ligand stabilization are highlighted.

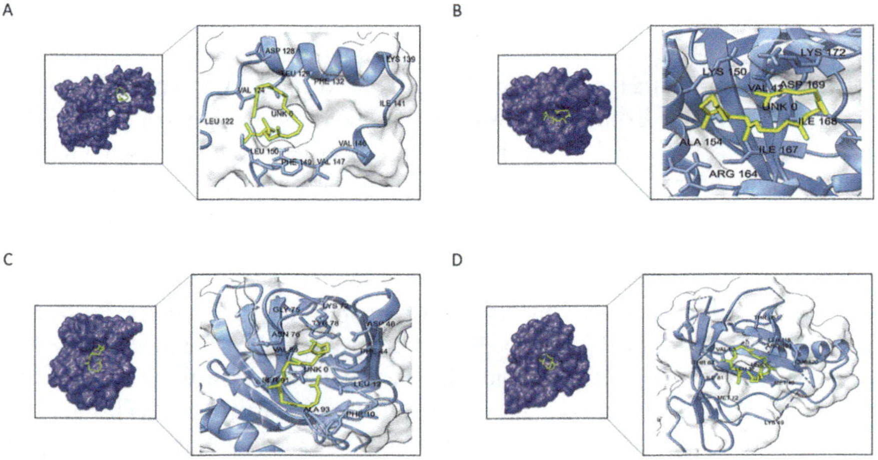

Fig. 2. 3D molecular interactions of squalene with OBPs, showing proteins in ribbon representation and squalene as a space-filling model: (**a**) 1GT4, (**b**) 3ZQ3, (**c**) 1A3Y, and (**d**) O18873.

Table 1. Docking-derived binding free energies (ΔG, kcal/mol) of squalene against four different OBPs calculated using AutoDock Vina.

Ligand	PubChem ID	1GT4	3ZQ3	1A3Y	O18873
Squalene	638072	−6.78	−5.01	−5.02	−5.96

4 Discussion

Multimodal and unimodal composite signals play crucial roles in animal communication. The complexity of these signals becomes highly accentuated as the pure scent signal itself is composed of hundreds of chemical compounds, and each chemical may have a distinct role (Greene et al., 2016). Primarily from a preliminary computational analysis perspective, the purpose of this study focused on the cross-specificity responses to composite olfactory signals of squalene and OBPs of different species, including *B. taurus*, *R. norvegicus*, *S. scrofa*, and *L. pictus*, which can be used to design future biochemical and behavioral experimental analyses. In a molecular docking analysis, squalene was reported to have the highest affinity for *B. taurus* OBP (1GT4) at −6.78 kcal/mol. However, such information does not confirm real binding and in vivo functional roles.. The strong interactions between squalene and OBP are enhanced by the presence of conserved amino acid residues in the binding site, which are critical for binding efficiency, as these residues have a notable effect on the binding of hydrophobic moieties (Rajkumar et al., 2010). These conserved residues among species are responsible for different binding interactions of OBPs with squalene, thereby affecting the overall binding strength and specificity (Campanini & de Brito, 2016). Although OBPs of *B. taurus* showed the highest binding affinity with squalene, it also interacted with the OBPs from *R. norvegicus* (3ZQ3), *S. scrofa* (1A3Y), and *L. pictus* (O18873) with comparatively lower binding energies. This suggests that the structural divergence of OBPs among species may significantly impact their interaction with squalene.

The OBPs generally possess hydrophobic binding pockets located within the lipocalin fold, which interact with hydrophobic ligands through entropic contributions, thereby promoting the expulsion of water molecules from the binding pocket (Pelosi et al., 2014; Hong et al., 2022). This is likely due to variations in ecological niches and social structure, where the role of chemical signaling can vary considerably among the studied species (Monaco & Zanotti, 1992; Yang et al., 2022). For instance, the very conservative Tyr82 in the pig OBP is thought to be helpful in binding several ligands. The substitution of another residue in the protein-binding pocket with similar physicochemical properties, such as phenylalanine, resulted in changes in the cavity structure, which is consistent with the assigned function (Nagnan-Le Meillour et al., 2009). Interestingly, Tyr83 residues in bovine OBPs are thought to constitute the gate of the binding pocket and are involved in the release process (Gonçalves et al., 2021). In other words, ligand perception owes its utmost importance to the hydrophobic binding pockets of OBPs, while the non-covalent intermolecular forces characterize the binding affinity toward the ligand. Under no circumstances in our study where OBPs found to form hydrogen bonds with any of the entities in the test. Squalene is a hydrocarbon and, thus, does not provide any functional groups necessary to form hydrogen bonds. The main interactions of OBPs with their ligands include hydrogen bonding, hydrophobic interactions, and van der Waals interactions. Consequently, squalene was able to interact with OBPs in a hydrophobic manner, which indicates that hydrogen bonding does not contribute much to the binding affinity of squalene due to the lack of polar groups (hydroxyl or amine) that could offer hydrogen bonding (Hauss et al., 2002). This observation is also supported by the docking results, which suggest that binding occurs predominantly through hydrophobic complementarity and van der Waals interactions. At the same time, hydrogen bonding is found to play a minor role, consistent with the chemical behavior of squalene. The hydrophobic nature of squalene aligns well with the hydrophobic residues within the OBP binding pockets, favoring a conformational adaptation that enhances the interaction (Schiefner et al., 2015). According to Qu et al. (2022), α-ionone and dihydro-β-ionone (β-ionone analogs) exhibit significant affinity for HaxyOBP15. Such high binding affinities are attributed mainly to their hydrophobic nature, as these compounds lack any polar side chains in their structures and exhibit a marked hydrophobic effect. The existence of such highly hydrophobic interactions is said to play a crucial role in binding specific odorant molecules to OBPs, and this may, in turn, synergize to affect the general olfactory mechanism. Together, these results provide evidence of the binding competencies of squalene via hydrophobic complementarity with the OBP binding pockets.

Recently, squalene has been proposed to act as a pheromone-like molecule, conveying airborne cues over prolonged time, and is presumably involved in social interactions and reproductive behavior between animals (Bala & Babbar, 2024). Squalene is recognized as a sex pheromone in carnivores and a scent-marking material in primates, thus demonstrating its broad range of chemical signaling across species (Scordato et al., 2007; Zhang et al., 2008). Aside from its pheromone signaling, the proposed function of squalene, due to its hydrophobic nature, high molecular weight, and low vapor pressure, might be to act as a fixative for small molecules (Apps et al., 2012). Alternatively, this property stabilizes and prolongs the presence of volatile compounds, thereby enhancing chemical communication by ensuring that signaling molecules remain in an active form for an extended period (Greene et al., 2016). Squalene may help retain odorant molecules on the surface of the nasal cavity, thereby enhancing the duration and intensity of the scent, as it is a well-known fixer and lipophilic antioxidant. Furthermore, squalene may stabilize other lipid fractions by inhibiting oxidation and enhancing the stability of various compounds in a mixture (Zhang et al., 2008; Martín et al., 2010). Squalene was present in the clitoral glands of laboratory rats and its concentration was varying across the estrous cycle. It acts as a signaling molecule influencing male behaviors through the olfactory-vomeronasal system (Achiraman et al., 2011). Computer analysis suggests that there may be a difference in the binding specificity of squalene to OBPs across species, which requires biochemical and behavioral validation to confirm functional involvement. The fixative nature of squalene further enhances its potential as a signaling molecule, likely regulating reproductive behaviors across diverse species. In molecular docking, the 1GT4 was found to exhibit a better prediction of binding specificity for squalene compared to 3ZQ3, 1A3Y, and O18873. In this study, only docking experiments were performed, which may not accurately represent the molecule's real-life environment and dynamic interactions within the organism. Therefore, further experimental exploration of how squalene interactions influence physiological and behavioral effects on animal communication is necessary.

References

Achiraman, S., Archunan, G., Abirami, B., Kokilavani, P., Suriyakalaa, U., SankarGanesh, D., Kamalakkannan, S., Kannan, S., Habara, Y., & Sankar, R. (2011). Increased squalene concentrations in the clitoral gland during the estrous cycle in rats: An estrus-indicating scent mark? *Theriogenology, 76*(9), 1676–1683. https://doi.org/10.1016/j.theriogenology.2011.06.033

Apps, P., Mmualefe, L., & McNutt, J. W. (2012). Identification of volatiles from the secretions and excretions of African wild dogs (Lycaon pictus). *Journal of Chemical Ecology, 38*(11), 1450–1461. https://doi.org/10.1007/s10886-012-0206-7

Bala, B., & Babbar, B. K. (2024). Behavioural responses and characterization of preputial gland extract of black rat, Rattus rattus. *Biologia, 79*(7), 2079–2093. https://doi.org/10.1007/s11756-024-01685-7

Campanini, E. B., & de Brito, R. A. (2016). Molecular evolution of odorant-binding proteins gene family in two closely related Anastrepha fruit flies. *BMC Evolutionary Biology, 16*(1), 198. https://doi.org/10.1186/s12862-016-0775-0

Dutton, A., Mattiacci, L., Amadò, R., & Dorn, S. (2002). A novel function of the triterpene squalene in a tritrophic system. *Journal of Chemical Ecology, 28*(1), 103–116. https://doi.org/10.1023/a:1013514903036

Gonçalves, F., Ribeiro, A., Silva, C., & Cavaco-Paulo, A. (2021). Biotechnological applications of mammalian odorant-binding proteins. *Critical Reviews in Biotechnology, 41*(3), 441–455. https://doi.org/10.1080/07388551.2020.1853672

Greene, L. K., Grogan, K. E., Smyth, K. N., Adams, C. A., Klager, S. A., & Drea, C. M. (2016). Mix it and fix it: Functions of composite olfactory signals in ring-tailed lemurs. *Royal Society Open Science, 3*(4), 160076. https://doi.org/10.1098/rsos.160076

Hauss, T., Dante, S., Dencher, N. A., & Haines, T. H. (2002). Squalane is in the midplane of the lipid bilayer: Implications for its function as a proton permeability barrier. *Biochimica et Biophysica Acta, 1556*(2–3), 149–154. https://doi.org/10.1016/s0005-2728(02)00346-8

Hayden, S., & Teeling, E. C. (2014). The molecular biology of vertebrate olfaction. *The Anatomical Record, 297*(11), 2216–2226. https://doi.org/10.1002/ar.23031

Hong, B., Chang, Q., Zhai, Y., Ren, B., & Zhang, F. (2022). Functional characterization of odorant binding protein PyasOBP2 from the jujube bud weevil, Pachyrhinus yasumatsui (Coleoptera: Curculionidae). *Frontiers in Physiology, 13*, 900752. https://doi.org/10.3389/fphys.2022.900752

Janssenswillen, S., Roelants, K., Carpentier, S., de Rooster, H., Metzemaekers, M., Vanschoenwinkel, B., Proost, P., & Bossuyt, F. (2021). Odorant-binding proteins in canine anal sac glands indicate an evolutionarily conserved role in mammalian chemical communication. *BMC Ecology and Evolution, 21*(1), 182. https://doi.org/10.1186/s12862-021-01910-w

Jumper, J., Evans, R., Pritzel, A., Green, T., Figurnov, M., Ronneberger, O., Tunyasuvunakool, K., Bates, R., Žídek, A., Potapenko, A., Bridgland, A., Meyer, C., Kohl, S. A. A., Ballard, A. J., Cowie, A., Romera-Paredes, B., Nikolov, S., Jain, R., Adler, J., Back, T., Petersen, S., Reiman, D., Clancy, E., Zielinski, M., Steinegger, M., Pacholska, M., Berghammer, T., Bodenstein, S., Silver, D., Vinyals, O., Senior, A. W., Kavukcuoglu, K., Kohli, P., & Hassabis, D. (2021). Highly accurate protein structure prediction with AlphaFold. *Nature, 596*(7873), 583–589. https://doi.org/10.1038/s41586-021-03819-2

Le Danvic, C., Gérard, O., Sellem, E., Ponsart, C., Chemineau, P., Humblot, P., & Nagnan-Le Meillour, P. (2015). Enhancing bull sexual behavior using estrus-specific molecules identified in cow urine. *Theriogenology, 83*(9), 1381–1388. https://doi.org/10.1016/j.theriogenology.2015.02.004

Martín, J., Barja, I., & López, P. (2010). Chemical scent constituents in feces of wild Iberian wolves (Canis lupus signatus). *Biochemical Systematics and Ecology, 38*(6), 1096–1102. https://doi.org/10.1016/j.bse.2010.10.014

Moitrier, L., Belloir, C., Lalis, M., Hou, Y., Topin, J., & Briand, L. (2022). Ligand binding properties of odorant-binding protein OBP5 from Mus musculus. *Biology, 12*(1), 2. https://doi.org/10.3390/biology12010002

Monaco, H. L., & Zanotti, G. (1992). Three-dimensional structure and active site of three hydrophobic molecule-binding proteins with significant amino acid sequence similarity. *Biopolymers, 32*(4), 457–465. https://doi.org/10.1002/bip.360320425

Morris, G. M., Huey, R., Lindstrom, W., Sanner, M. F., Belew, R. K., Goodsell, D. S., & Olson, A. J. (2009). AutoDock4 and AutoDockTools4: Automated docking with selective receptor flexibility. *Journal of Computational Chemistry, 30*(16), 2785–2791. https://doi.org/10.1002/jcc.21256

Nagnan-Le Meillour, P., Lagant, P., Cornard, J. P., Brimau, F., Le Danvic, C., Vergoten, G., & Michalski, J. C. (2009). Phenylalanine 35 and tyrosine 82 are involved in the uptake and release of ligand by porcine odorant-binding protein. *Biochimica et Biophysica Acta, Proteins and Proteomics, 1794*(8), 1142–1150. https://doi.org/10.1016/j.bbapap.2009.04.012

O'Boyle, N. M., Banck, M., James, C. A., Morley, C., Vandermeersch, T., & Hutchison, G. R. (2011). Open babel:An open chemical toolbox. *Journal of Cheminformatics, 3* , https://doi.org/10.1186/1758-2946-3-33

Pelosi, P., & Knoll, W. (2022). Odorant-binding proteins of mammals. *Biological Reviews Cambridge Philosophical Society, 97*(1), 20–44. https://doi.org/10.1111/brv.12787

Pelosi, P., Iovinella, I., Zhu, J., Wang, G., & Dani, F. R. (2018). Beyond chemoreception: diverse tasks of soluble olfactory proteins in insects. *Biological Reviews Cambridge Philosophical Society*, *93*(1), 184-200. https://doi.org/10.1111/brv.12339

Pelosi, P., Iovinella, I., Felicioli, A., & Dani, F. R. (2014). Soluble proteins of chemical communication: An overview across arthropods. *Frontiers in Physiology, 5*, 320. https://doi.org/10.3389/fphys.2014.00320

Portman, K. L., Long, J., Carr, S., Briand, L., Winzor, D. J., Searle, M. S., & Scott, D. J. (2014). Enthalpy/entropy compensation effects from cavity desolvation underpin broad ligand binding selectivity for rat odorant binding protein 3. *Biochemistry, 53*(14), 2371–2379. https://doi.org/10.1021/bi5002344

Qu, C., Yang, Z. K., Wang, S., Zhao, H. P., Li, F. Q., Yang, X. L., & Luo, C. (2022). Binding affinity characterization of four antennae-enriched odorant-binding proteins from Harmonia axyridis (Coleoptera: Coccinellidae). *Frontiers in Physiology, 13*, 829766. https://doi.org/10.3389/fphys.2022.829766

Rajkumar, R., Karthikeyan, K., Archunan, G., Huang, P. H., Chen, Y. W., Ng, W. V., & Liao, C. C. (2010). Using mass spectrometry to detect buffalo salivary odorant-binding protein and its post-translational modifications. *Rapid Communications in Mass Spectrometry, 24*(22), 3248–3254. https://doi.org/10.1002/rcm.4766

Schiefner, A., Freier, R., Eichinger, A., & Skerra, A. (2015). Crystal structure of the human odorant binding protein, OBPIIa. *Proteins, 83*(6), 1180–1184. https://doi.org/10.1002/prot.24797

Scordato, E. S., Dubay, G., & Drea, C. M. (2007). Chemical composition of scent marks varies with the testosterone levels of ring-tailed lemurs (Lemur catta). *Behavioral Ecology and Sociobiology, 61*(4), 775–790. https://doi.org/10.1093/chemse/bjm018

Sharma, S., Kumar, P., & Chandra, R. (2019). Molecular dynamics simulation of nanocomposites using BIOVIA materials studio. In S. Sharma (Ed.), *Lammps and Gromacs* (pp. 329–341). Elsevier.

Taniguchi, K., & Taniguchi, K. (2014). Phylogenic studies on the olfactory system in vertebrates. *Journal of Veterinary Medical Science, 76*(6), 781–788. https://doi.org/10.1292/jvms.13-0650

Tian, W., Chen, C., Lei, X., Zhao, J., & Liang, J. (2018). CASTp 3.0: Computed atlas of surface topography of proteins. *Nucleic Acids Research, 46*(W1), W363–W367. https://doi.org/10.1093/nar/gky473

Trott, O., & Olson, A. J. (2010). AutoDock Vina: Improving the speed and accuracy of docking with a new scoring function, efficient optimization, and multithreading. *Journal of Computational Chemistry, 31*(2), 455–461. https://doi.org/10.1002/jcc.21334

Vincent, F., Ramoni, R., Spinelli, S., Grolli, S., Tegoni, M., & Cambillau, C. (2004). Crystal structures of bovine odorant-binding protein in complex with odorant molecules. *European Journal of Biochemistry, 271*(19), 3832–3842. https://doi.org/10.1111/j.1432-1033.2004.04315.x

Vogt, R. G., Prestwich, G. D., & Lerner, M. R. (1991). Odorant-binding-protein subfamilies associate with distinct classes of olfactory receptor neurons in insects. *Journal of Neurobiology, 22*(1), 74–84. https://doi.org/10.1002/neu.480220108

Wang, Y., Qin, C., Wang, X., Wang, T., Dong, J., & Lu, Y. (2024). Odorant-binding proteins as recognition elements for smell exploration. *Biochemical Engineering Journal, 205*, 109284. https://doi.org/10.1016/j.bej.2024.109284

Wang, J., Zhang, Q., Fan, W., Shi, Q., Mao, J., Xie, J., Chai, G., & Zhang, C. (2025). Deciphering olfactory receptor binding mechanisms: A structural and dynamic perspective on olfactory receptors. *Frontiers in Molecular Biosciences, 11*, 1498796. https://doi.org/10.3389/fmolb.2024.1498796

Yang, R., Li, D., Yi, S., & Wang, M. (2022). Evolutionarily conserved odorant-binding proteins participate in establishing tritrophic interactions. *iScience, 25*(7), 104664. https://doi.org/10.1016/j.isci.2022.104664

Zhang, J. X., Liu, D., Sun, L., Wei, R., Zhang, G., Wu, H., Zhang, H., & Zhao, C. (2008). Potential chemosignals in the anogenital gland secretion of giant pandas, Ailuropoda melanoleuca, associated with sex and individual identity. *Journal of Chemical Ecology, 34*(3), 398–407. https://doi.org/10.1007/s10886-008-9441-3

Gas Chromatography-Olfactometry in the Analysis of Axillary Odors: Insights and Outlook

Helene M. Loos[1,2,3] (✉), Laleh Kiavar[3], and Sümeyye Carkit-Yilmaz[3]

[1] Institute of Nutritional and Food Sciences, Rheinische Friedrich-Wilhelms-Universität Bonn, Bonn, Germany
[2] Sensory Analytics and Technologies, Fraunhofer Institute for Process Engineering and Packaging IVV, Freising, Germany
[3] Chair of Aroma and Smell Research, Friedrich-Alexander-Universität Erlangen-Nürnberg, Erlangen, Germany
helene.loos@fau.de

Abstract. Axillary odor is certainly one of the most frequently encountered odors of the human body. Its production is known to depend on the axillary microbiome (determining release of odorants from odorant precursors), sex (influencing quantities of odorant precursors), and an individual's genotype regarding the ABCC11 gene, encoding an ATP-binding cassette protein responsible for membrane transport of odorant precursors in apocrine sweat glands. Evidence from behavioral studies suggests that it also varies with other factors, for instance, the emotional or health state of the sender, which might be related to different metabolic processes. In this chapter, we provide an overview of research conducted so far to elucidate the chemical composition of axillary odor with a focus on studies applying gas chromatography-olfactometry for targeting volatile compounds bearing an odor. Knowledge on the chemical nature of compounds contributing to axillary odor has substantially increased from the 1960s to the 2000s. Nonetheless, still little is known on the variability of body odors both within an individual and between individuals and the temporal fluctuations in emission of axillary odorants.

Keywords: Apocrine sweat · Armpit · Odor analysis · Steroids · Carboxylic acids · Sulfanylalkanols

1 Introduction

Chemical senses allow for the detection of chemical stimuli in the environment. Due to the high diversity of chemical stimuli, and several methodological and technical difficulties regarding the recording and standardized presentation of these stimuli, the state of knowledge in the field of chemosensation is still inferior compared to other senses like vision or hearing. Nonetheless, substantial progress has been made in the past few decades, leading to an increased understanding of the chemical stimuli underlying different taste and smell percepts. This chapter focuses on the sense of smell in humans,

L. M. Schulte (ed.), *Chemical Signals in Vertebrates 16*,
https://doi.org/10.1007/978-3-032-30752-1_14

and especially on one of its functions, namely, its involvement in human chemical communication. (Odor-active) volatile organic substances emitted from the human body can convey information about the sender's traits and states such as individuality, sex, developmental stage, emotional and health state (Doty, 1981; Loos et al., 2025 2023). A special emphasis in this field of research has been placed on information transmitted by the axillary odor. In this chapter, we use the term axillary odor with two different meanings: (i) the odor (percept) generated through odor-active substances emitted from the human axilla and (ii) volatiles and odor-active substances emitted from the human axilla. Volatile compounds in general are all compounds that are in gas phase at ambient temperature, independent of their capacity to interact with olfactory receptors. Volatility of a given substance is generally considered to be a prerequisite for airborne chemical communication, notably long-range communication, since these compounds can be easily carried through the air to the receiver's nasal cavity. Odor-active substances are a subset of volatile compounds that can be detected by the human olfactory system through interaction with receptors. These interactions take place in the olfactory epithelium, located at the top of the nasal cavity, and are the prerequisite for odor perception at above-threshold concentrations and chemical communication processes mediated by the human olfactory system. Currently, technical systems able to comprehensively analyze odor-active substances are not yet available. Therefore, a technique combining instrumental analysis with human sensory evaluation by trained panelists, called gas chromatography-olfactometry, is currently still used to study those compounds.

Repeatedly gained evidence suggests that axillary odor can transmit information about emotion (Chen & Haviland-Jones, 2000; de Groot et al., 2020), sickness (Olsson et al., 2014; Tognetti et al., 2023), and age (Mitro et al., 2012; Owsienko et al., 2024b), among others, and thereby influence the receivers' behavior and physiology (see Loos et al., 2025 for a comprehensive review). Additionally, axillary odor is influenced by diet (Havlicek & Lenochova, 2006). As in many other examples of chemical communication, the elucidation of the chemical nature of semiochemicals involved in human chemical communication is a complex and necessarily multidisciplinary endeavor (Wyatt, 2015). In this chapter, we will focus on the identification of odor-active compounds, which is one of four methodological approaches needed to find out more about human body odors and about the potential involvement of body odors in human chemical communication. As depicted in Fig. 1, we need to (i) determine the behavioral and (psycho)physiological effects of axillary odors in receivers to evaluate their role in chemical communication; (ii) characterize (identify, quantify, and determine temporal release patterns) volatile substances in axillary odor, as we can expect that volatile substances can be transferred from the sender to the receiver and therefore contribute to airborne chemical communication; (iii) characterize odor-active substances in axillary odor, as these can be expected to be involved in communication processes mediated by the sense of smell; and (iv) evaluate axillary odor samples by a trained panel to achieve a standardized description of their smell characteristics or by untrained panelists to assess their olfactory perception. Of course, the final step will be the comparison of the isolated chemicals' effects on human behavior and physiology with those of the original substrates, in their naturally occurring concentrations (cf. Wyatt, 2015).

Within the current chapter, we discuss the application of a specific technique called gas chromatography-olfactometry (GC-O) on axillary odors. Therefore, we first intro-

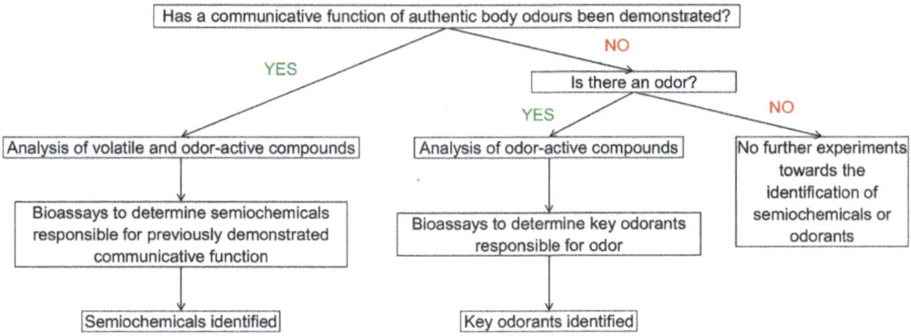

Fig. 1. Steps to achieve an understanding of the chemical identity of body odorants and semiochemicals.

duce the current state of knowledge regarding the chemistry of axillary odors, then summarize findings obtained with GC-O, and finally provide an outlook regarding future optimization of sampling techniques.

2 Chemistry of Axillary Odors

Whereas the relationship between bacterial composition and axillary odor types has been recognized as early as in the 1950s (Shelley et al., 1953), and acids and steroids have been described as contributors to axillary odor from the 1960s/1980s on (Smith et al., 1969; Nixon et al., 1988), the major insights into the biochemistry beyond axillary odor formation have been obtained in the early 2000s. The axilla harbors mainly eccrine sweat glands, apocrine sweat glands, and sebaceous glands. Eccrine sweat glands are active from birth and are involved in thermoregulatory and to some extent also emotional sweating (Wilke et al., 2007), but no odor is released from watery eccrine sweat even after 96 h (Shelley et al., 1953). Apocrine sweat glands become active at the onset of puberty, are a primary cause of axillary odor, and secrete a viscous fluid, among others, as response to emotional stimuli (Wilke et al., 2007). In contrast to eccrine sweat glands, apocrine sweat glands are associated with hair follicles, as are the sebaceous glands which secrete sebum, a lipid-rich secretion helping to maintain the skin lipid barrier (Nicolaides, 1974). The secretion of apocrine sweat glands contains odorless precursors which are cleaved by bacterial enzymes, leading to the release of odor-active substances. The secretion of axillary precursors depends on the genotype (Martin et al., 2010), with different prevalence in different geographical regions (Yoshiura et al., 2006). The ratio of precursor substances differs between men and women (Troccaz et al., 2009). It is now widely accepted that the main substance classes contributing to axillary odors are thiols, acids, and steroids (Natsch & Emter, 2020). These emerge from bacterial cleavage of precursors (Fig. 2). In addition, the ketone 2,3-butanedione has been proposed to contribute to axillary odor (Hara et al., 2015).

In sum, knowledge regarding the biochemical basis of axillary odors has substantially increased in the past decades. However, the variability of axillary odors is still

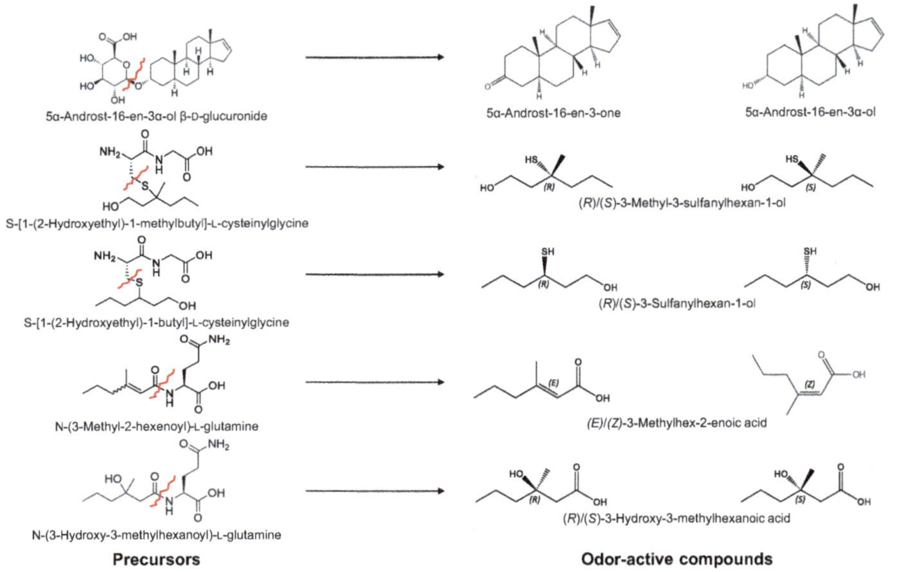

Fig. 2. Principle of odor generation through bacterial action on exemplary precursors from initially odorless apocrine sweat. Bacteria reported to be involved in cleavage of the precursors are *Corynebacteria* and *Anaerococcus* in case of glutamine conjugates, *Corynebacteria* and *Staphylococci* in case of cysteinylglycine conjugates, and *Staphylococci* and *Propionibacteria* in case of the glucuronide. (Natsch & Emter, 2020; Starkenmann et al., 2013; Starkenmann, 2017; Natsch, 2017).

under-researched. Axillary odors can be subject to short-term (e.g., during a day) and long-term (e.g., during aging) changes, and the odors emanating from the left and right axilla may not be the same at a given point in time. It can therefore be assumed that the composition of axillary odor changes, with some odorants being more prevalent and hence dominating the overall percept at one point in time and other odorants being more prevalent at another point in time. Since the axillary skin harbors eccrine, apocrine, and sebaceous glands, it can further be hypothesized that axillary odor, next to the substances being released by bacterial enzymes from apocrine sweat, also contains some by-products of human metabolism, which could be brought to the skin's surface by sebaceous glands, for instance. More recently, sebum has been proposed as a biofluid for gaining insights into lipid metabolism, and its composition has been shown to be altered in certain disease conditions like Parkinson's disease (Sinclair et al., 2021). Many researchers interested in the role of axillary odors in human chemical communication have used textile materials to sample the axillary odors and present them to receivers in behavioral studies. Since this is a direct contact sampling method, it can be expected that sebum is sampled together with apocrine and eccrine sweat. Therefore, compounds originating from sebum could also contribute to information transfer in these experimental studies. In the following, we will summarize insights into the odorant composition of axillary odor samples, as determined through GC-O.

3 Gas Chromatography-Olfactometry in the Analysis of Axillary Odors

GC-O is a technique that allows for the continuous monitoring of the odor quality of a GC eluate by a person trained in describing smell percepts with a given flavor language. The basis of such a flavor language is a consensus obtained within the group of trained panelists, e.g., via defining reference compounds, materials, or concepts for a given odor descriptor (e.g., the smell of 1-octen-3-one for "mushroomlike"). The first application of GC-O with a professional perfumer has been described in the 1960s (Fuller et al., 1964). Since then, GC-O was mainly applied in research on food aroma, leading to the elucidation of several odor-active substances in food. Among these are so-called generalists and specialists, the former occurring in many types of food, the latter being more specific to certain raw materials or processing steps (Dunkel et al., 2014). The application of GC-O is one part of a more comprehensive approach to determine the odorants being relevant for a given olfactory percept. First, GC-O is used to rank the odorants according to their potential relevance in the sample of interest. For this, different approaches can be used, one of which is a dilution analysis method evaluating which odorants are still perceived in high dilutions (reviewed by van Ruth, 2001). Then, the identity and concentration of suggested relevant odorants in the original sample (generally in mass/mass or mass/volume) is determined. Certain criteria such as retention indices, odor quality, and mass spectrum are applied for identification (Molyneux & Schieberle, 2007). Quantification is, for instance, achieved by establishing a calibration curve with an internal standard. For identification and quantification, mass spectrometric or other instrumental detection techniques are needed in addition to GC-O. When the concentration of a target odorant in a sample is known, together with the odor threshold of the odorant in the sample matrix, the so-called odor activity value (OAV) can be calculated as the ratio of "odorant concentration in the sample" to "odor threshold of the odorant in the sample matrix." The theoretical minimum value of OAV is 0 (when the odorant is not present in the sample), and the theoretical upper limit is determined by the respective odor threshold. Substances with OAV <1 are generally assumed to contribute little or indirectly to a sample's odor, while substances with OAV >1 are generally considered in recombination and omission experiments, needed to evaluate whether a synthetic mixture resembles the original odor (Grosch, 2001). This approach was successfully applied to different foods and later also to materials of daily use (Wiedmer et al., 2017; Strangl et al., 2018). It has also been used to elucidate odorants in human milk (Buettner, 2007; Loos et al., 2019), follicular fluid and vaginal odors (Hartmann et al., 2013), and urine (Troccaz et al., 2013; Wagenstaller & Buettner, 2013). The application of GC-O has also been important in the elucidation of odorants responsible for axillary odor.

In the early 1990s, Preti and colleagues published a series of manuscripts on the chemical composition of axillary odor samples. Analyzing substances extracted from cotton pads worn by male volunteers, including an organoleptic evaluation of the GC eluate, they suggested that (*E*)/(*Z*)-3-methyl-2-hexenoic acid has a major impact on axillary odor, together with some further odor-active substances: n-hexanoic acid, 3-methylhexanoic acid, 4-ethylpentanoic acid, (*Z*)-3-methyl-2-hexenoic acid, n-heptanoic acid, 2-methylheptanoic acid, (*E*)-3-methyl-2-hexenoic acid, n-octanoic acid, 4-ethylheptanoic acid, 7-octenoic acid, 4-ethyloctanoic acid, an unknown unsaturated C9 and C10 acid, 4-ethylnonanoic acid, 9-decenoic acid, and phenylacetic acid (Zeng et al., 1991). In a subsequent study on female axillary odor samples, they additionally propose the compounds 2-methylhexanoic acid, (*E*)-3-methyl-2-pentenoic acid, 2-piperidone, (*E*)-3-methyl-2-octenoic acid, and 10-undecenoic acid as contributors to axillary odor—"from data reported in male samples" (Zeng et al., 1996).

GC-O analyses of cotton pad extracts were also the basis for the identification of 3-hydroxy-3-methylhexanoic acid, being present only in samples with a spicy axillary odor, as compared to samples with a sour, acidic odor (Hasegawa et al., 2004). One year before this observation, this typically sweaty smelling compound had been identified by Natsch and colleagues, based on the GC-O analysis of axillary secretions sampled with 40% ethanol (Natsch et al., 2003). Natsch and colleagues also used GC-O to determine the contribution of a large variety of different acids to axillary odor (see Table 1 in Natsch et al., 2006), based on the analysis of extracts of cotton pads treated with N_{α}-acyl-glutamine aminoacylase (Natsch et al., 2006; Kuhn & Natsch, 2008). GC-O also led to the identification of sulfanylalcohols, based on the analysis of hydrolyzed axillary secretions (Natsch et al., 2004), sweat sampled during sauna (Hasegawa et al., 2004), and incubated sweat (Troccaz et al., 2004).

A slightly different approach, inspired by the comprehensive approach described above, including the ranking of perceived odorants in a sample by a dilution analysis method, has been applied on laundry samples (Munk et al., 2000; Munk et al., 2001; Takeuchi et al., 2012) and cotton pads obtained from infants and postpubertals (Owsienko et al., 2024a). In these studies, next to acids and steroids, also several aldehydes and ketones have been identified, possibly originating from degradation of lipids contained in sebum. In addition, several odor-active substances potentially originating from personal care products were detected. Headspace analysis of cotton pads after fenugreek ingestion additionally demonstrated an altered odorant composition of axillary samples dependent on the diet (Mebazaa et al., 2011). These results indicate that cotton pads used for studies on human chemical communication can be considered to contain a mixture of substances originating from axillary sweat and sebum, and additionally to be influenced by the donor's environment, if sampled at home (e.g., personal care products, furniture, etc.). Cotton pads exposed to the donor's room air also exhibited several of the detected substances (Fig. 3).

4 Conclusion and Outlook

GC-O is a technique that proved to be crucial for elucidating odor-active compounds in axillary odor. In the 1990s and 2000s, it has been applied to identify a large number of acids and several sulfanylalkanols that majorly contribute to the axillary odor. In general, hydrolysis or incubation with bacterial strains or enzymes has been applied to liberate odor-active compounds from their odorless precursors, and acidification was used to protonate acids and thiols for optimal extraction.

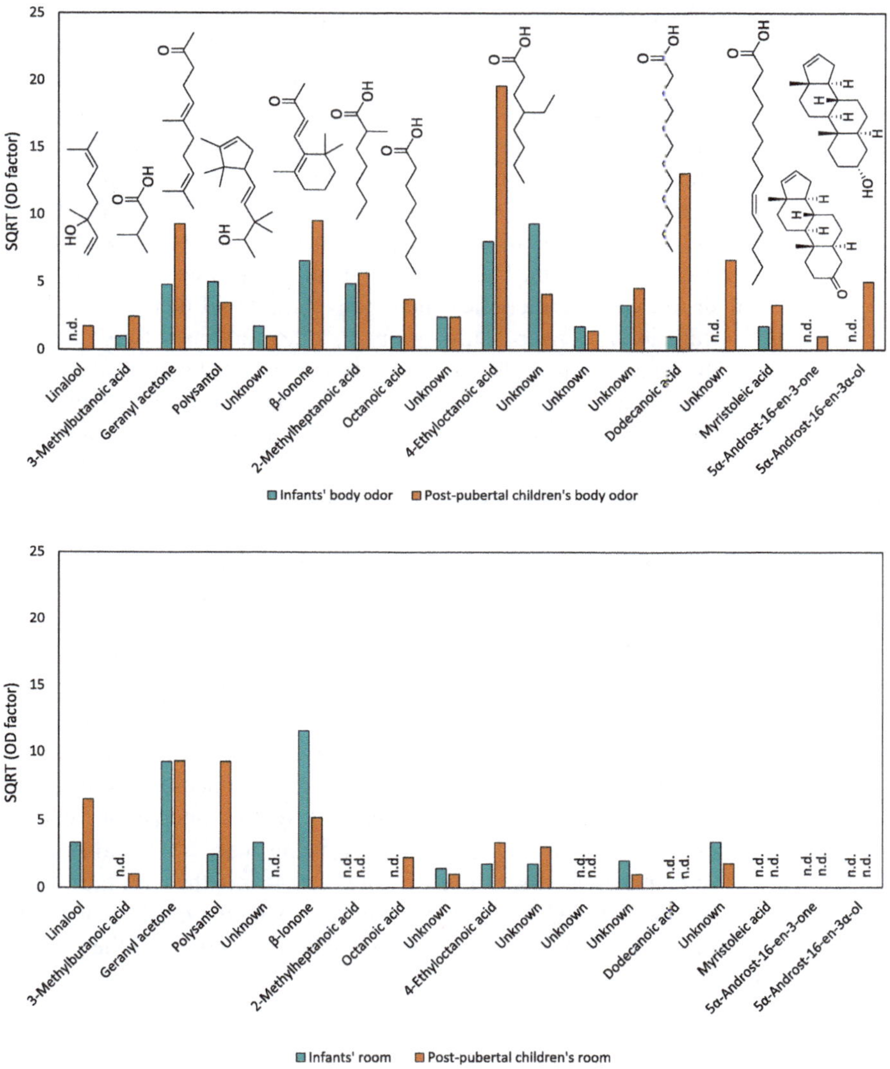

Fig. 3. Odor dilution (OD) factors (SQRT: square root) of compounds detected in children body odor samples (top) and samples from the children's rooms (bottom; Owsienko, Goppelt, et al., 2024a). *n.d.* not detected.

A limited number of publications exist reporting a less targeted approach to investigate axillary odors through GC-O. The available studies suggest that cotton pads worn under the axilla contain, in addition to the known axillary odor compounds, a range of aldehydes, ketones, heterocycles, and other substances. Future studies will shed light on the variability of axillary odors—either odors emanating from textile materials, as these might be of relevance in daily interactions in today's societies, or odors emanating from the axilla itself, as these might represent more "natural" conditions. For the latter, direct sampling methods should be applied, for instance, extraction of odor-active compounds from the gas phase above the skin (von Hartungen et al., 2004). In addition, sampling of liquid sweat followed by incubation or hydrolysis to release odorants from their precursors might lead to further insights (Troccaz et al., 2004), as the recovery of especially sulfur substances from cotton pads is relatively low (Starkenmann, 2017).

Future studies might also consider temporal release patterns of odorants by the human body (Roberts et al., 2020). Release can thereby be influenced by the duration and timing of the release itself, but also by the physicochemical properties of the substances and the release mechanism. The involved chemicals' physicochemical characteristics affect their release characteristics and reactivity and will therefore determine whether they can function as longer-lasting or quick, temporary cues. Substances persisting longer on the human skin might serve as enduring cues that convey stable information about an individual, such as genetic background or health status. Such information might therefore be conveyed by substances with lower volatility and higher molecular weight or by substances that are continuously produced by the human body. Another mechanism allowing for longer-lasting release might be the sustained degradation of abundant less volatile compounds present on the skin, resulting in the gradual production of more volatile compounds. This also effectively lengthens the time frame during which the information is available to the receivers. Low molecular weight and highly volatile compounds tend to evaporate rapidly, serving as instantaneous signals that transmit fleeting information. Such signals could allow for quick social interactions and adaptive reactions. Release of odorants from the human body can be further assumed to depend on body-related parameters (e.g., skin pH, skin temperature, quantity of sebum, skin microbiome), as well as environmental conditions such as humidity, airflow, temperature, and ozone levels (Wang et al., 2022). The release patterns of volatile substances from the human body is also influenced by their interactions with surfaces, including skin, clothes, and ambient substrates (Wang et al., 2022). Adsorption to these surfaces can either decrease the availability of substances in the airborne compose or increase their presence through a slower release. Clarifying the mechanisms behind chemical communication in humans also requires an understanding of the timing of odorant release.

References

Buettner, A. (2007). A selective and sensitive approach to characterize odour-active and volatile constituents in small-scale human milk samples. *Flavour and Fragrance Journal, 22*(6), 465–473. https://doi.org/10.1002/ffj.1822

Chen, D., & Haviland-Jones, J. (2000). Human olfactory communication of emotion. *Perceptual and Motor Skills, 91*(3), 771–781. https://doi.org/10.2466/pms.2000.91.3.771

de Groot, J. H. B., Kirk, P. A., & Gottfried, J. A. (2020). Encoding fear intensity in human sweat. *Philosophical Transactions of the Royal Society B: Biological Sciences, 375*(1800), 20190271. https://doi.org/10.1098/rstb.2019.0271

Doty, R. L. (1981). Olfactory communication in humans. *Chemical Senses, 6*(4), 351–376. https://doi.org/10.1093/chemse/6.4.351

Dunkel, A., Steinhaus, M., Kotthoff, M., Nowak, B., Krautwurst, D., Schieberle, P., & Homann, T. (2014). Nature's chemical signatures in human olfaction: A foodborne perspective for future biotechnology. *Angewandte Chemie International Edition, 53*(28), 7124–7143. https://doi.org/10.1002/anie.201309508

Fuller, G. H., Steltenkamp, R., & Tisserand, G. A. (1964). The gas chromatograph with human sensor: Perfumer model. *Annals of the New York Academy of Sciences, 116*(2), 711–724. https://doi.org/10.1111/j.1749-6632.1964.tb45106.x

Grosch, W. (2001). Evaluation of the key odorants of foods by dilution experiments, aroma models and omission. *Chemical Senses, 26*(5), 533–545. https://doi.org/10.1093/chemse/26.5.533

Hara, T., Kyuka, A., & Shimizu, H. (2015). Butane-2,3-dione: The key contributor to axillary and foot odor associated with an acidic note. *Chemistry & Biodiversity, 12*(2), 248–258. https://doi.org/10.1002/cbdv.201400272

Hartmann, C., Triller, A., Spehr, M., Dittrich, R., Hatt, H., & Buettner, A. (2013). Sperm-activating odorous substances in human follicular fluid and vaginal secretion: Identification by gas chromatography-olfactometry and Ca2+ imaging. *ChemPlusChem, 78*(7), 695–702. https://doi.org/10.1002/cplu.201300008

Hasegawa, Y., Yabuki, M., & Matsukane, M. (2004). Identification of new odoriferous compounds in human axillary sweat. *Chemistry & Biodiversity, 1*(12), 2042–2050. https://doi.org/10.1002/cbdv.200490157

Havlicek, J., & Lenochova, P. (2006). The effect of meat consumption on body odor attractiveness. *Chemical Senses, 31*(8), 747–752. https://doi.org/10.1093/chemse/bjl017

Kuhn, F., & Natsch, A. (2008). Body odour of monozygotic human twins: A common pattern of odorant carboxylic acids released by a bacterial aminoacylase from axilla secretions contributing to an inherited body odour type. *Journal of the Royal Society Interface, 6*(33), 377–392. https://doi.org/10.1098/rsif.2008.0223

Loos, H. M., Reger, D., & Schaal, B. (2019). The odour of human milk: Its chemical variability and detection by newborns. *Physiology & Behavior, 199*, 88–99. https://doi.org/10.1016/j.physbeh.2018.11.008

Loos, H. M., Schaal, B., Pause, B. M., Smeets, M. A. M., Ferdenzi, C., Roberts, S. C., de Groot, J., Lübke, K. T., Croy, I., Freiherr, J., Bensafi, M., Hummel, T., & Havlíček, J. (2023). Past, present, and future of human chemical communication research. *Perspectives on Psychological Science, 20*(1), 20–44. https://doi.org/10.1177/17456916231188147

Loos, H. M., Schaal, B., Pause, B. M., Smeets, M. A. M., Ferdenzi, C., Roberts, S. C., de Groot, J., Lübke, K. T., Croy, I., Freiherr, J., Bensafi, M., Hummel, T., & Havlíček, J. (2025). Past, present, and future of human chemical communication research. Perspectives on Psychological Science, 20(1), 20–44. https://doi.org/10.1177/17456916231188147

Martin, A., Saathoff, M., Kuhn, F., Max, H., Terstegen, L., & Natsch, A. (2010). A functional ABCC11 allele is essential in the biochemical formation of human axillary odor. *Journal of Investigative Dermatology, 130*(2), 529–540. https://doi.org/10.1038/jid.2009.254

Mebazaa, R., Rega, B., & Camel, V. (2011). Analysis of human male armpit sweat after fenugreek ingestion: Characterisation of odour active compounds by gas chromatography coupled to mass spectrometry and olfactometry. *Food Chemistry, 128*(1), 227–235. https://doi.org/10.1016/j.foodchem.2011.02.063

Mitro, S., Gordon, A. R., Olsson, M. J., & Lundström, J. N. (2012). The smell of age: Perception and discrimination of body odors of different ages. *PLoS One, 7*(5), e38110. https://doi.org/10.1371/journal.pone.0038110

Molyneux, R. J., & Schieberle, P. (2007). Compound identification: A journal of agricultural and food chemistry perspective. *Journal of Agricultural and Food Chemistry, 55*(12), 4625–4629. https://doi.org/10.1021/jf070242j

Munk, S., Münch, P., Stahnke, L., Adler-Nissen, J., & Schieberle, P. (2000). Primary odorants of laundry soiled with sweat/sebum: Influence of lipase on the odor profile. *Journal of Surfactants and Detergents, 3*(4), 505–515. https://doi.org/10.1007/s11743-000-0150-z

Munk, S., Johansen, C., Stahnke, L. H., & Adler-Nissen, J. (2001). Microbial survival and odor in laundry. *Journal of Surfactants and Detergents, 4*(4), 385–394. https://doi.org/10.1007/s11743-001-0192-2

Natsch, A. (2017). Biochemistry and genetics of human axilla odor. In A. Buettner (Ed.), *Springer Handbook of Odor* (pp. 123–124). Springer International Publishing.

Natsch, A., & Emter, R. (2020). The specific biochemistry of human axilla odour formation viewed in an evolutionary context. *Philosophical Transactions of the Royal Society B: Biological Sciences, 375*(1800), 20190269. https://doi.org/10.1098/rstb.2019.0269

Natsch, A., Gfeller, H., Gygax, P., Schmid, J., & Acuna, G. (2003). A specific bacterial aminoacylase cleaves odorant precursors secreted in the human axilla*. *Journal of Biological Chemistry, 278*(8), 5718–5727. https://doi.org/10.1074/jbc.M210142200

Natsch, A., Schmid, J., & Flachsmann, F. (2004). Identification of odoriferous sulfanylalkanols in human axilla secretions and their formation through cleavage of cysteine precursors by a C-S lyase isolated from axilla bacteria. *Chemistry & Biodiversity, 1*(7), 1058–1072. https://doi.org/10.1002/cbdv.200490079

Natsch, A., Derrer, S., Flachsmann, F., & Schmid, J. (2006). A broad diversity of volatile carboxylic acids, released by a bacterial aminoacylase from axilla secretions, as candidate molecules for the determination of human-body odor type. *Chemistry & Biodiversity, 3*(1), 1–20. https://doi.org/10.1002/cbdv.200690015

Nicolaides, N. (1974). Skin lipids: Their biochemical uniqueness. *Science, 186*(4158), 19–26. https://doi.org/10.1126/science.186.4158.19

Nixon, A., Mallet, A. I., & Gower, D. B. (1988). Simultaneous quantification of five odorous steroids (16-androstenes) in the axillary hair of men. *Journal of Steroid Biochemistry, 29*(5), 505–510. https://doi.org/10.1016/0022-4731(88)90185-9

Olsson, M. J., Lundström, J. N., Kimball, B. A., Gordon, A. R., Karshikoff, B., Hosseini, N., Sorjonen, K., Olgart Höglund, C., Solares, C., Soop, A., Axelsson, J., & Lekander, M. (2014). The scent of disease: Human body odor contains an early chemosensory cue of sickness. *Psychological Science, 25*(3), 817–823. https://doi.org/10.1177/0956797613515681

Owsienko, D., Goppelt, L., Hierl, K., Schäfer, L., Croy, I., & Loos, H. M. (2024a). Body odor samples from infants and post-pubertal children differ in their volatile profiles. *Communications Chemistry, 7*(1), 53. https://doi.org/10.1038/s42004-024-01131-4

Owsienko, D., Schwinn, L., Eskofier, B. M., Kiesswetter, E., & Loos, H. M. (2024b). Sensory evaluation of axillary odour samples of younger and older adults by a trained panel. *Flavour and Fragrance Journal, 39*(1), 3–9. https://doi.org/10.1002/ffj.3762

Roberts, S. C., Misztal, P. K., & Langford, B. (2020). Decoding the social volatilome by tracking rapid context-dependent odour change. *Philosophical Transactions of the Royal Society B: Biological Sciences, 375*(1800), 20190259. https://doi.org/10.1098/rstb.2019.0259

Shelley, W. B., Hurley, H. J., & Nichols, A. C. (1953). Axillary odor: Experimental study of the role of bacteria, apocrine sweat, and deodorants. *A.M.A Archives of Dermatology and Syphilology, 68*(4), 430–446. https://doi.org/10.1001/archderm.1953.01540100070012

Sinclair, E., Trivedi, D. K., Sarkar, D., Walton-Doyle, C., Milne, J., Kunath, T., Rijs, A. M., de Bie, R. M. A., Goodacre, R., Silverdale, M., & Barran, P. (2021). Metabolomics of sebum reveals lipid dysregulation in Parkinson's disease. *Nature Communications, 12*(1), 1592. https://doi.org/10.1038/s41467-021-21669-4

Smith, K., Thompson, G. F., & Koster, H. D. (1969). Sweat in schizophrenic patients: Identification of the odorous substance. *Science, 166*(3903), 398–399. https://doi.org/10.1126/science.166.3903.398

Starkenmann, C. (2017). Analysis and chemistry of human odors. In A. Buettner (Ed.), *Springer Handbook of Odor* (pp. 121–122). Springer International Publishing.

Starkenmann, C., Mayenzet, F., Brauchli, R., & Troccaz, M. (2013). 5α-Androst-16-en-3α-ol β-D-glucuronide, precursor of 5α-Androst-16-en-3α-ol in human sweat. *Chemistry & Biodiversity, 10*(12), 2197–2208. https://doi.org/10.1002/cbdv.201300286

Strangl, M., Schlummer, M., Maeurer, A., & Buettner, A. (2018). Comparison of the odorant composition of post-consumer high-density polyethylene waste with corresponding recycled and virgin pellets by combined instrumental and sensory analysis. *Journal of Cleaner Production, 181*, 599–607. https://doi.org/10.1016/j.jclepro.2018.01.137

Takeuchi, K., Hasegawa, Y., Ishida, H., & Kashiwagi, M. (2012). Identification of novel malodour compounds in laundry. *Flavour and Fragrance Journal, 27*(1), 89–94. https://doi.org/10.1002/ffj.2088

Tognetti, A., Williams, M. N., Lybert, N., Lekander, M., Axelsson, J., & Olsson, M. J. (2023). Humans can detect axillary odor cues of an acute respiratory infection in others. *Evolution, Medicine, and Public Health, 11*(1), 219–228. https://doi.org/10.1093/emph/eoad016

Troccaz, M., Starkenmann, C., Niclass, Y., van de Waal, M., & Clark, A. J. (2004). 3-Methyl-3-sulfanylhexan-1-ol as a major descriptor for the human axilla-sweat odour profile. *Chemistry & Biodiversity, 1*(7), 1022–1035. https://doi.org/10.1002/cbdv.200490077

Troccaz, M., Borchard, G., Vuilleumier, C., Raviot-Derrien, S., Niclass, Y., Beccucci, S., & Starkenmann, C. (2009). Gender-specific differences between the concentrations of nonvolatile (R)/(S)-3-Methyl-3-Sulfanylhexan-1-ol and (R)/(S)-3-Hydroxy-3-methyl-Hexanoic acid odor precursors in axillary secretions. *Chemical Senses, 34*(3), 203–210. https://doi.org/10.1093/chemse/bjn076

Troccaz, M., Niclass, Y., Anziani, P., & Starkenmann, C. (2013). The influence of thermal reaction and microbial transformation on the odour of human urine. *Flavour and Fragrance Journal, 28*(4), 200–211. https://doi.org/10.1002/ffj.3143

van Ruth, S. M. (2001). Methods for gas chromatography-olfactometry: A review. *Biomolecular Engineering, 17*(4), 121–128. https://doi.org/10.1016/S1389-0344(01)00070-3

von Hartungen, E., Wisthaler, A., Mikoviny, T., Jaksch, D., Boscaini, E., Dunphy, P. J., & Märk, T. D. (2004). Proton-transfer-reaction mass spectrometry (PTR-MS) of carboxylic acids: Determination of Henry's law constants and axillary odour investigations. *International Journal of Mass Spectrometry, 239*(2), 243–248. https://doi.org/10.1016/j.ijms.2004.09.009

Wagenstaller, M., & Buettner, A. (2013). Characterization of odorants in human urine using a combined chemo-analytical and human-sensory approach: A potential diagnostic strategy. *Metabolomics, 9*(1), 9–20. https://doi.org/10.1007/s11306-012-0425-5

Wang, N., Ernle, L., Bekö, G., Wargocki, P., & Williams, J. (2022). Emission rates of volatile organic compounds from humans. *Environmental Science & Technology, 56*(8), 4838–4848. https://doi.org/10.1021/acs.est.1c08764

Wiedmer, C., Velasco-Schön, C., & Buettner, A. (2017). Characterization of off-odours and potentially harmful substances in a fancy dress accessory handbag for children. *Scientific Reports, 7*(1), 1807. https://doi.org/10.1038/s41598-017-01720-5

Wilke, K., Martin, A., Terstegen, L., & Biel, S. S. (2007). A short history of sweat gland biology. *International Journal of Cosmetic Science, 29*(3), 169–179. https://doi.org/10.1111/j.1467-2494.2007.00387.x

Wyatt, T. D. (2015). The search for human pheromones: The lost decades and the necessity of returning to first principles. *Proceedings of the Royal Society B: Biological Sciences, 282*(1804), 20142994. https://doi.org/10.1098/rspb.2014.2994

Yoshiura, K.-I., Kinoshita, A., Ishida, T., Ninokata, A., Ishikawa, T., Kaname, T., Bannai, M., Tokunaga, K., Sonoda, S., Komaki, R., Ihara, M., Saenko, V. A., Alipov, G. K., Sekine, I., Komatsu, K., Takahashi, H., Nakashima, M., Sosonkina, N., Mapendano, C. K., Ghadami,

M., Nomura, M., Liang, D.-S., Miwa, N., Kim, D.-K., Garidkhuu, A., Natsume, N., Ohta, T., Tomita, H., Kaneko, A., Kikuchi, M., Russomando, G., Hirayama, K., Ishibashi, M., Takahashi, A., Saitou, N., Murray, J. C., Saito, S., Nakamura, Y., & Niikawa, N. (2006). A SNP in the ABCC11 gene is the determinant of human earwax type. *Nature Genetics, 38*(3), 324–330. https://doi.org/10.1038/ng1733

Zeng, X. N., Leyden, J. J., Lawley, H. J., Sawano, K., Nohara, I., & Preti, G. (1991). Analysis of characteristic odors from human male axillae. *Journal of Chemical Ecology, 17*(7), 1469–1492. https://doi.org/10.1007/BF00983777

Zeng, X.-N., Leyden, J. J., Spielman, A. I., & Preti, G. (1996). Analysis of characteristic human female axillary odors: Qualitative comparison to males. *Journal of Chemical Ecology, 22*(2), 237–257. https://doi.org/10.1007/BF02055096

Finding the "Needle in the Haystack": A Comparison of Statistical Tools for Untargeted Chemical Analysis

Brigitte M. Weiß[1,2,3] (✉), Erik Kusch[1,4,5], Marlen Kücklich[1,2,3], and Anja Widdig[1,2,3]

[1] Research Group Behavioural Ecology, Institute of Biology, University of Leipzig, Leipzig, Germany

[2] Research Group Primate Behavioural Ecology, Department of Human Behaviour, Ecology and Culture, Max-Planck-Institute for Evolutionary Anthropology, Leipzig, Germany

[3] German Center for Integrative Biodiversity Research (iDiv), Leipzig, Germany

brigitte.schloegl@uni-leipzig.de

[4] Natural History Museum, University of Oslo, Oslo, Norway

[5] Climate and Nature Risk Research Group, CICERO Centre for International Climate Research, Oslo, Norway

Abstract. Deciphering chemical cues and signals frequently resembles the search for a "needle in a haystack" of chemical compounds. Particularly in non-model vertebrate species, the complexity of chemical profiles combined with little prior knowledge about potential semiochemicals leads to various challenges, including the statistical analysis. Statistical tools need to be able to pinpoint accurately if and which compounds out of hundreds of candidates are associated with certain biological traits. We therefore investigated eight commonly used statistical methods (analysis of similarity, Mantel test, permutational MANOVA, Procrustes, GLMM, random forest classification, sparse partial-least-squares discriminant analysis, and similarity percentage analysis) for how they perform under the challenges of untargeted chemical analysis. We created more than 2000 simulated data sets that systematically varied in sample size, the number and intensity of compounds in a chemical profile, the number and intensity of compounds associated with a trait, and the presence of noise. Four discrete and four continuous traits were included, half of which contained simulated associations with certain compounds while the other half did not. For each data set, we determined which compounds were found to be associated with these traits and compared the test results to the actually simulated associations. False positives were close to the nominal level for all tests in the control condition, but power to detect simulated associations varied. GLMMs with vectorized data matrix showed high power even when few or relatively small compounds were associated with a trait. Along with random forest classification, GLMMs also performed best in pinpointing which of the compounds were associated with the respective trait. Finally, we applied all statistical tests to real mammalian chemical profiles. Based on these analysis, we suggest that combining two to three statistical tests (e.g., GLMM, analysis of similarity, random forest analysis) covers different analytical aspects best and allows for the most robust conclusions.

Authors Marlen Kücklich and Anja Widdig have equally contributed to this chapter.

L. M. Schulte (ed.), *Chemical Signals in Vertebrates 16*,
https://doi.org/10.1007/978-3-032-30752-1_15

Keywords: Chemical profiles · Untargeted analysis · Biomarker identification · Dissimilarity · Multidimensional analysis

1 Introduction

The constant improvement and increasing availability of analytical tools such as gas chromatography-mass spectrometry (GC-MS) have greatly facilitated studies in chemical ecology, expanding the range of study species, contexts, and questions both in the lab and field. The resulting data sets are typically characterized by complex multidimensional chemical profiles with dozens to hundreds of compounds, low abundances, and lots of noise (e.g., Nair et al., 2018; Weiß et al., 2018a). Particularly in non-model species and novel contexts, many of these compounds will be structurally and/or functionally unknown: they may not have been chemically characterized before and thus cannot be identified by comparison with available reference libraries (Charpentier et al., 2012; Nair et al., 2018), and there will be little to no prior knowledge about which compounds are of interest, i.e., on which particular compounds one should focus the (statistical) analysis. As a result, chemical ecologists will frequently need to adopt an untargeted analytical approach in which dozens to hundreds of candidate compounds are considered for statistical analysis.

In statistical analysis of chemical profiles, researchers will usually want to know at least one of the following: if, how, and how strongly traits relate to chemical profiles or parts thereof. This requires tools that can be used for hypothesis testing by, e.g., computing *p*-values or other measures of statistical inference, identifying (groups of) biomarkers (i.e., those compounds out of a chemical profile that are associated with a given trait), and/or computing some measure for the strength of association between traits and chemical profiles (see Hervé et al., 2018).

A range of statistical tools that employ different computational approaches is available for these tasks. An informed choice of appropriate statistical tools and adequate interpretation of their results thus represents one of the challenges faced by chemical ecologists, particularly because systematic overviews of tools for metabolomic data analysis have become available only quite recently. For instance, Hervé et al. (2018) illustrate the application of several well-established as well as novel multivariate ordination methods in practical examples covering a range of questions of varying complexity. Similarly, Brückner and Heethoff (2017) use a published data set to guide the reader through several common and less common ordination methods suitable for the analysis of compositional (or relational) data. Both guides cover the pretreatment of data, summarize the main attributes and assumptions of the selected tools, and illustrate which tools are suitable for which kinds of research questions, thereby providing valuable guidance in the choice and application of tools. When choosing among those tools that are theoretically suitable for one's question, however, hardly any guidance is available. In such cases, the choice of tools may be based on hearsay or what was used in previous studies rather than on knowledge about the strengths and weaknesses of the respective tools. Unfortunately, systematic evaluations and comparisons of tools for metabolomics data are still very scarce and limited to only few aspects and tools. In the most comprehensive study available to date, Trainor et al. (2017) evaluated the performance of seven classification techniques by comparing misclassification rates in simulated and real data sets, whereby aspects such as variance, error structure, or missing values affected which of the classification tools performed best. On a much smaller scale, Nair et al. (2018) compared classification accuracy and root-mean-square error obtained with two clustering, two classification, and two regression tools for one real data set. Other evaluations focused on very specific aspects such as how to impute missing values in liquid chromatography (LC)-MS data (Kokla et al., 2019) or comparing different data normalization techniques for the pretreatment of data (Noonan et al., 2018), but without comparing statistical tools for the subsequent analyses.

Hence, more research is needed on how statistical tools perform with metabolomics data across different kinds of data and effect structures to facilitate choosing among them and to be able to better interpret results. We systematically compared the performance of eight statistical tools that compute *p*-values and identify biomarkers and/or some kinds of effect size measures. We applied them to data sets with different data structures and kinds of effects to assess how these tools deal with potential issues such as small or few affected compounds, low effect sizes, noise, variation in sample size, or the number of compounds considered for the analysis. Our selection of tools covered classification as well as regression tools. We did not evaluate clustering tools, which may be more appropriate when groupings (i.e., clusters) are not known prior to analysis. We also focused on tools with the possibility to account for control variables. With these aspects in mind, we finally selected tools we repeatedly came across in the mammalian chemical ecology literature. In particular, we evaluated analysis of similarity (AnoSim), Mantel tests, a permutational MANOVA (perMANOVA), Procrustes analysis, a linear mixed model approach (GLMM), random forest (RF) classification, sparse partial-least-squares discriminant analysis (sPLS-DA), and similarity percentage analysis (SimPer).

AnoSim, Mantel tests, perMANOVA, and Procrustes all employ a dissimilarity (or distance) approach (see Table 1) in which dissimilarity scores, such as Bray-Curtis indices, are computed between pairs of samples. These dissimilarity scores are then related to the predictor of interest. They use permutations to compute *p*-values and, as such, are

Table 1. Overview of evaluated tools. "Approach" summarizes the computational approach, while "n pred." and "pred. scale" describe the number and scale of predictors the respective tool is suitable for. "*P*-value" and "biomarker" indicate whether the tools can be used for hypothesis testing or biomarker identification. "Measure" describes the effect size or accuracy measures provided by the respective tools.

Tool	Approach	n pred.	pred. scale	*P*-value	Biomarker	Measure
AnoSim	Dissimilarity	Single	Discrete	Yes	No	r
Mantel	Dissimilarity	Single	Continuous	Yes	No	r
perMANOVA	Dissimilarity	Multiple	All	Yes	No	r2, sum square
Procrustes	Dissimilarity	Single	All	Yes	No	r, sum square
GLMM	Multidimensional	Multiple	All	Yes	Yes	Model est.
RF classification	Multidimensional	Single	Discrete	No	Yes	Error rate
sPLS-DA	Multidimensional	Single	Discrete	No	Yes	Error rate
SimPer	Multidimensional	Single	All	No	Yes	–

suitable for hypothesis testing but are unsuitable for pinpointing which of the compounds in a chemical profile are associated with a given predictor (i.e., identifying biomarkers) because of the performed dimension reduction. Accordingly, these tools are suitable when the question focuses on the overall similarity between samples, rather than on specific compounds (see also Hervé et al., 2018). AnoSim, Mantel tests, and Procrustes provide a measure for the strength of association between sample similarity and the trait of interest called "r" that can be interpreted as an effect size measure similar to a correlation coefficient, while the perMANOVA estimates r^2, which represents the partial coefficient of determination if estimated after accounting for other terms in the model. r^2 also represents an effect size measure but is framed as the proportion of variance explained by a predictor or set of predictors. AnoSim can only be performed for discrete and Mantel tests for continuous predictors, while perMANOVA and Procrustes analysis can handle predictors of both scales. Among these tools, only the perMANOVA can handle multiple predictors within the same analysis, a requirement for many complex data sets (see Table 1).

We also evaluated a vectorized linear mixed model approach adopted from community ecology (see Jamil et al., 2013), because of its flexibility in incorporating multiple test and control predictors into the same analysis, thereby accommodating a large range of study designs. It retains the multidimensionality of the chemical profile data by vectorizing the data matrix and fitting the sample and compound identity as random effects. The effects of interest in this approach are not the fixed effects of the predictors but their interplay with the compound and thus the random interactions/slopes of the predictors within compound (see Weiß et al., 2018a). Unlike the other tools, this GLMM approach allows for hypothesis testing and biomarker identification within a single analysis (Weiß et al., 2018a).

RF classification, sPLS-DA, and SimPer analysis can be used to make predictions which of the compounds in a chemical profile are most strongly associated with a given trait (i.e., biomarker identification). RF is a machine learning algorithm that builds decision trees from samples and computes variable importance scores by permuting these decision trees (see Nair et al. 2018 and references therein). These scores can then be used to rank variables (i.e., compounds) by their predicted importance. RF classification also computes error rates that describe how accurately samples can be assigned to the respective predictor levels based on the compounds entered into the analysis. RF algorithms can be used for classification or regression, but we only evaluated RF classification in this study. sPLS-DA is a modification of the ubiquitous PLS-DA classification technique (see Trainor et al., 2017 and references therein). It employs variable selection, i.e., the removal of uninformative compounds, and classification within one step (Lê Cao et al., 2011), thereby performing some (but no complete) dimensionality reduction. sPLS-DA computes the loadings of the variables (i.e., compounds) that were retained into one or more components. Within each component, the variables can be ranked by importance based on their loadings. Similar to RF, sPLS-DA also computes classification accuracy expressed as error rates. Finally, SimPer analysis allows ranking compounds by order of importance by determining how much each compound contributes to the average Bray-Curtis dissimilarities in groups of sampling units (Clarke, 1993). SimPer analysis does not provide error rates or other measures of classification accuracy.

We simulated over 2000 data sets with known properties and effects. These data sets varied in data set attributes such as sample size, number of compounds, or the size distribution of compounds across the data set, as well as effect attributes such as the number, size, and prevalence of effects and the relative size of the compounds with implemented effects or the presence of noise. To not further strain the complexity of our evaluation, we do not cover preprocessing of chemical data (which is covered elsewhere, e.g., Brückner & Heethoff, 2017; Hervé et al., 2018; Noonan et al., 2018) and focus on analysis of the commonly used compositional data rather than presence/absence data or concentrations. Specific characteristics and analytical requirements of presence/absence and concentration-like data are described in Hervé et al. (2018). In sum, we compared the performance of statistical tools for the analysis of compositional data with respect to false positives, power, and biomarker identification under different data and effect structures. We further compared effect size measures and error rates within tools across different data and effect structures to facilitate interpretation of results. We also used all tools on two real data sets to apply the lessons from the simulated data sets to real-life scenarios, for which the "real" number, size, and prevalence of effects are unknown but which will likely be more complex than the simulated data sets. With these evaluations, we hope to support chemical ecologists in finding appropriate, powerful statistical tools for addressing their research questions.

2 Methods

2.1 Data Simulation

We simulated a total of 2080 data sets varying in data and effect structure using a self-written *R* script.

2.1.1 Data Structure. We varied data structure with respect to sample size (100 or 150), number of compounds (120 or 180), and the size distribution of the compounds. Based upon typical peak areas in real data sets measured within previous studies of our group (Weiß et al., 2018a, b; Jänig et al., 2019; Kücklich et al., 2019), we defined six size classes for the peak areas of simulated compounds as well as their distributions across the data set (see Table 2). We specified two versions of size distribution, with version 1 biased toward compounds of size class 3 ("small") and version 2 toward compounds of size class 2 ("minute").

Peak areas for each simulated compound were created by randomly sampling a value from the peak area range of the respective size class using the *R* function "sample." This value served as the mean and half the mean as the standard deviation for creating n values (where n equals the specified sample size, i.e., 100 or 150) using the *R* function "rnorm." A pre-specified proportion of values was then replaced by zeros (see Table 2) to mimic measures below the detection limit. We again used real data sets to define the proportion of zero values in the smallest two size classes and used these values to extrapolate the proportion of zeros in the other size classes using the *R* function "predict.lm." The entire process was repeated 120 or 180 times, respectively, to

Table 2. Definition and distribution of size classes for simulated compounds. “Min” and “max” define the minimum and maximum peak area divided by 1000 and “0-rate” the proportion of zeros implemented. “V1” and “V2” denote the two versions of size distribution, given either in % or in absolute numbers (n) per data set, with values in parenthesis indicating data sets with 120 or 180 compounds, respectively.

Size class	Min/1000	Max/1000	0-rate	% V1	% V2	n(120) V1	n(120) V2	n(180) V1	n(180) V2
Tiny	2	25	0.35	12.5	18	15	21	23	32
Minute	20	100	0.25	21	30	25	36	37	54
Small	80	300	0.15	33	23	40	28	60	41
Medium	200	1′000	0.05	17	15	20	18	30	27
Large	800	12′000	0	12.5	10	15	12	22	18
Largest	10′000	500′000	0	4	4	5	5	8	8

obtain one data set. Finally, we reshuffled the rows of the data set to avoid blocks of same size compounds or zeros. In this manner, we created 20 data sets for each of the eight possible combinations of sample size, n compounds, and size distribution, resulting in 160 “base” data sets containing randomly created peak areas without implemented effects.

2.1.2 Predictor Variables. For data sets with 100 samples, we assigned the samples to 10, and for data sets with 150 samples to 15 different individuals, thus resembling 10 measures per subject per data set. We then created one predictor variable for each combination of scale (discrete or continuous), design (within- or between-subjects), and type (test or control), resulting in a total of eight different predictor variables. As exemplary labels of the test predictors, we used sex, diet, age, and sample number; control predictors were labeled dummy 1–4. “Sex” represented a discrete between-subjects test predictor with two levels (five males, five females at sample size 100; eight males, seven females at sample size 150). “Diet” represented a discrete within-subjects test predictor with two levels, with levels distributed randomly within individuals. “Age” represented a continuous between-subjects test predictor, with values ranging from one to nine across individuals and constant values within individuals. “Sample number” represented a continuous within-subjects test predictor, ranging from one to ten within each individual. Individual identities were only used for assigning predictor variable values to samples, but not for the actual implementation of effects (see below). Control variables (i.e., dummy 1–4) were created accordingly and so as to not be confounded with their corresponding test predictors (i.e., levels/values distributed differently within/across individuals than in the test predictors). The same set of eight predictor variables was used for all subsequent analyses.

2.1.3 Effect Structure. We implemented effects for each of the four test predictors but not for the control predictors. We varied the structure of the implemented effects with respect to the number of compounds affected per predictor, the effect size (i.e., magnitude of change), the size class of affected compounds, the prevalence (or consistency) of effects across the data set, and the presence of noise (see Table 3). We combined the different parameter values in a semi-balanced manner, resulting in 13 different effect structures consisting of 12 varying effect compositions and one control condition without implemented effects.

Table 3. Implemented effect structures. Values in italics highlight the characteristic attributes of the respective effect condition.

nr.	Short description	n effects	Effect size	Size class	Prevalence	n noise
0	No effects	*0*	–	–	–	–
1	Weak effects	6	*Weak*	All	95	0
2	Intermediate effects	6	Intermediate	All	95	0
3	Strong effects	6	*Strong*	All	95	0
4	Smallest size classes	6	Intermediate	*Tiny, minute*	95	0
5	Medium size classes	6	Intermediate	*Small, medium*	95	0
6	Largest size classes	3	Intermediate	*Large, largest*	95	0
7	Many effects	*11*	Intermediate	All	95	0
8	Few effects	*2*	Intermediate	Small, medium	95	0
9	75% prevalence	6	Intermediate	All	*75*	0
10	50% prevalence	6	Intermediate	All	*50*	0
11	Low noise	6	Intermediate	All	95	*3*
12	High noise	6	Intermediate	All	95	*10*

We implemented the 12 effect conditions onto each of the 160 control data sets to be able to systematically compare the role of different effects while being able to account for the underlying random data, resulting in 13*160 = 2080 different data sets. Hence, condition 0 (no effects) represented the respective unaltered control data sets created as described above. In all other conditions, we created effects by supplying different means to the condition-specific number of compounds randomly selected from the condition-specific size classes. In particular, we simulated weak effects by providing a mean from either the bottom or top 10 percentile of the peak area range for the respective compound size class. Intermediate and strong effects were simulated by randomly selecting the mean from the peak area range of one (intermediate) or two (strong) size classes higher (or lower) than the actual compound size class. Standard deviations were specified to be mean/2. The effects created in this manner were aligned with the values of the respective test predictors and were implemented for the specified proportion of samples. For conditions 1–3 (weak/intermediate/strong effects), we implemented one effect per predictor in each compound size class (i.e., six affected compounds per predictor) in 95% of the samples. Conditions 4–6 (smallest/medium/largest size classes) resembled condition 2 (intermediate effects, prevalence 95%), but effects were implemented in the two smallest, two intermediate, or two largest size classes, respectively. For conditions 4–5, we implemented three effects per predictor and size class (i.e., six per predictor), but because of the lower number of "large" and "largest" size compounds (see Table 2), we only implemented two effects per predictor in the "large" and one in the "largest" size class (i.e., three per predictor) in condition 6. Condition 7 (many effects) resembled condition 2 (intermediate effects in all size classes, prevalence 95%), but with twice as many effects in all but the "largest" size class (i.e., 11 effects per predictor). Condition 8 (few effects) resembled condition 5 (intermediate effects in the two intermediate size classes, prevalence 95%) but contained only one effect per size class (i.e., two per predictor). Conditions 9 and 10 (prevalence 75% and 50%) again resembled condition 2 (one intermediate effect per predictor in each compound size class), but effects were implemented in only 75% (condition 9) or 50% (condition 10) of the samples. Finally, conditions 11–12 (low and high noise) were implemented on top of condition 2 of the respective base data set and contained either three (condition 11) or ten (condition 12) compounds from size class "small" for which we implemented a strong increase in 30% of samples independent of predictor variables to mimic noise created by, e.g., contaminants in the data set.

2.2 Real Data Sets

In addition to the simulated data sets with known effects but comparably simple data structure, we also used two real data sets. Data set 1 was obtained from two captive populations of meerkats (*Suricata suricatta*) within the frame of a methodological study on sampling methods for mammalian body odor (Weiß et al., 2018a). In brief, a total of 101 body odor samples were collected with thermal desorption tubes from the anogenital region of 24 individual meerkats. After semiautomated profiling and subtraction of compounds consistently found in blank samples, 111 compounds were used for statistical analysis to assess how variation in sampling parameters such as distance, volume, or flow rate as well as population differences affect the chemical composition of body odor.

Data set 2 was obtained from a captive population of common marmosets (*Callithrix jacchus*) in a study on olfactory fertility cues (Kücklich et al., 2019). In brief, 212 genital samples were collected from 12 female common marmosets during their follicular, peri-ovulatory, and luteal menstrual cycle phase. A total of 160 compounds were used for statistical analysis after chemical profiling and blank subtraction. Menstrual cycle stage, age of the females (1–11 years), and whether they already gave birth to infants (yes or no) were the fertility predictors of interest.

2.3 Computations

All simulated and real data sets were analyzed in the *R* environment (versions 3.4.1–3.4.3, R Core Team 2017) using the packages "vegan" (Oksanen et al., 2016), "lme4" (Bates et al., 2015), "randomForest" (Liaw & Wiener, 2002), and "mixOmics" (Rohart et al., 2017). Unless noted otherwise, all analyses were conducted with relative peak areas (i.e., % area out of the sum of all peak areas in a sample) that were log(x + 1)-transformed prior to analysis (Stoffel et al., 2015) to reduce the impact of large compounds on the results. Whenever possible, we accounted for individual identity (ID) in the analyses by fitting ID as stratum or random effect.

2.3.1 Analysis of Similarity. AnoSims were conducted separately for each discrete predictor variable, with dissimilarities between samples computed as pairwise Bray-Curtis indices. Bray-Curtis indices are a widely adopted similarity measure that computes a form of standardized absolute deviation between samples based on presence/absence as well as abundance of each compound without making distributional assumptions (Clarke 1999). We computed AnoSims with 999 permutations and specified individual ID as strata. If strata are specified, observations are permuted only within individuals and thus allow accounting for repeated measures. We used the function "anosim" provided in the package "vegan" for analyses without strata but used a customized script (written by Lars Kulik) when using individual ID as strata. This customized script was based on the AnoSim implemented in "vegan" but used a slightly different permutation regime to better account for repeated measures and accommodate the strata option for within- and between-subjects predictors alike. It produced the same r-values but different (usually more conservative) *p*-values than the function implemented in "vegan." For all analyses, we extracted r-values as a measure of effect size and *p*-values for inference.

2.3.2 Mantel Test. Mantel tests were performed separately for each continuous predictor using the function "mantel" in the package "vegan." Bray-Curtis dissimilarities were computed using the function "vegdist"; tests were performed with the correlation method "pearson" and 999 permutations. We ran tests with specifying individual ID as strata for within- but not for between-subjects predictors because stratified permutations are restricted to occur only within individuals. Given that predictors like sex do not vary within an individual, no valid permutations can be generated and the test would yield a trivial *p*-value of 1. For all tests, we extracted *r* and *p*-values.

2.3.3 Permutational MANOVA. PerMANOVAs were conducted for all predictors based on Bray-Curtis dissimilarities using the function "adonis2" in the package "vegan." The adonis2 function allows fitting all predictors within the same analysis and to determine significance for all terms together or for single terms. We thus computed the overall *p*-value of the full model (using the argument by = NULL) as well as the *p*-values of the single terms while all other predictors remained in the model (by = margin). We performed 999 permutations and ran all tests with defining individuals as permutation blocks. For all analyses, we extracted r^2 and *p*-values.

2.3.4 Procrustes. Procrustes analyses were performed separately for each discrete and continuous predictor using the function "protest" in the package "vegan." For this purpose, discrete predictors were dummy-coded numerically (i.e., as 0–1). We performed 999 permutations and ran all tests of within-subject predictors with, and of between-subjects predictors without, defining individuals as permutation blocks, as the permutation regime again would have led to *p*-values of between-subjects predictors to be estimated as 1. For all analyses, we extracted *r* and *p*-values.

2.3.5 General Linear Mixed Model (GLMM). GLMMs were conducted for all predictors based on a vectorized data matrix following the approach of Jamil et al. (2013). Models were fit with a Gaussian error structure using the function "lmer" in the package "lme4." To improve model fit, standardized peak areas were additionally arcsine-transformed before log-transformation (overall transformation: log((asin(sqrt(x/100))) + 0.01) to reduce the boundary effects common in ratios and percentages (Bate & Clark, 2014; Laurencelle & Cousineau, 2023). The rows and columns (i.e., the samples and compounds) of the underlying matrix were fitted as random intercepts to overcome the pseudo-replication and heteroscedastic variance introduced by vectorizing the matrix (Jamil et al., 2013). To account for repeated measures per individual, we further fitted individual ID as random intercept. The eight predictor variables were fitted as fixed effects, and their random interactions (for discrete predictors) or random slopes (for continuous predictors) fitted within the random intercept of compound, the latter constituting the effects of interest. We used likelihood ratio tests to compute the overall *p*-value of the full model (containing all predictors) in comparison with a null model (excluding the random intercepts and slopes of the predictors), which can reduce multiple testing issues associated with testing multiple predictors individually (Forstmeier & Schielzeth, 2011). We further computed the *p*-values of the single random interactions and slopes. To do so, we removed one random interaction/slope term at a time while all other predictors remained in the model and compared the reduced models to the full model with a likelihood ratio test. For all analyses, we extracted *p*-values for inference and computed unadjusted and adjusted false positive rates. For the latter, we scored all results of individual predictors as nonsignificant if the associated full-null model comparison was statistically nonsignificant. Furthermore, we obtained slope estimates per compound for each test predictor. We then determined the order of importance by ranking the absolute slope estimates of all compounds from largest to smallest and extracting the ranks of those compounds containing a simulated effect.

2.3.6 Random Forest Classification. Random forest classifications were performed separately for each discrete predictor using the function "tuneRF" in the package "randomForest." We specified astarting value of 8 for the number of variables tried at each split that was inflated (or deflated) by 1 at each iteration.

We further specified 20′000 trees for tuning and a relative improvement in OOB (Out-of-Bagerror estimate) of 0.0001 for the search to continue. The optimal values determined during tuning were used to run a "forest" by setting the argument doBest to true. We ran all tests with individual ID as strata and extracted the OOB as a measure of classification accuracy and importance scores to determine the order of importance for each compound containing a simulated effect.

2.3.7 Sparse Partial-Least-Squares Discriminant Analysis. sPLS-DAs were conducted separately for each discrete predictor using the functions "tune.splsda" and "splsda" in the package "mixOmics." The function "tune.splsda" was run with five components to be included in the model, fivefold cross-validation repeated 50 times, and the Mahalanobis distance metric. The number of variables (i.e., compounds) to include in the tuning model ranged from 1 to 10 in steps of 1 and from 20 to 120 in steps of 10. The tuned parameters (numbers of components and variables to select) were subsequently used to run the "splsda" function. Similar to Mantel tests and Procrustes analysis, accounting for repeated measures of individuals (using the argument "multilevel") was only possible for the within-subject predictors "diet" and "dummy2." We extracted error rates as a measure of classification accuracy and loadings per component to determine the order of importance for the compounds containing a simulated effect. Due to the very long computation times of the sPLS-DA, we ran the analyses on only half of the data sets, i.e., on 10 (instead of 20) of the repetitions for each data-structural and effect condition (10 repetitions * 8 data structures * 14 effect conditions = 1120 data sets).

2.3.8 Similarity Percentage Analysis. SimPer analyses were conducted separately for each test predictor using the function "simper" in the package "vegan." All tests were run with the default of 999 permutations, and we extracted the average contribution of each compound to determine the order of importance for each compound containing a simulated effect. Results were identical whether or not individuals were defined as permutation blocks (as determined for all eight predictors in 140 data sets), which is why we subsequently computed all results without defining permutation blocks because of the much lower computation time.

2.4 Evaluation of Performance

To compare performance across statistical tools in the light of variation in data and effect structure, we investigated false positives (type I error) and statistical power for all tools that compute p-values, i.e., AnoSim, PerMANOVA, Mantel tests, Procrustes analyses, and GLMMs. To assess false positives, we determined whether or not results were statistically significant ($p < 0.05$) in the "no effects" control condition for all predictors and, where applicable, for full-null model comparisons. We computed type I error rates per statistical tool, predictor, data structure, and effect structure as the number of statistically significant results out of all tests for which no effects were simulated. To assess statistical power, we determined whether or not results of test predictors were statistically significant ($p < 0.05$) in those effect conditions in which effects were implemented. Accordingly, we computed rates of power per statistical tool, predictor, data structure, and effect structure as the number of statistically significant results out of all tests for which effects were simulated.

To assess how well the multidimensional tools, i.e., GLMMs, random forest classification, sPLS-DA, and SimPer, were able to pinpoint which compounds were affected by a predictor, we further determined the order of importance of the simulated biomarkers as computed by the respective tools. To make the orders of importance of simulated biomarkers comparable across data sets with different numbers of implemented effects, we first determined the mean order biomarkers would take in the respective effect condition if all biomarkers were scored top (e.g., 3.5 for biomarkers ranked 1–6 in a condition with six implemented effects). We then computed a relative order of importance for each biomarker as the computed order of importance divided by the respective mean order of importance and computed mean relative orders of importance per statistical tool, data structure, and effect structure.

In addition, we evaluated effect sizes and measures such as classification accuracy in the light of variation in data and effect structure *within* each tool as these measures were not readily comparable across tools.

3 Results

3.1 False Positives

In the control condition (i.e., without implemented effects), the false positive rate was close to the nominal level of 0.05 for all tools, data structures, and predictors (Fig. 1). When scoring all results for individual predictors as nonsignificant if the full-null model

	AnoSim	Mantel	PerMAN	PerMAN full-null	Procrustes	GLMM	GLMM full-null
full-null	na	na	na	0.056	na	Na	0.069
Sex	0	na	0.063	0.006	0.075	0.063	0.025
Diet	0.02	na	0.069	0.019	0.075	0.069	0.019
Age	na	0	0.038	0	0.069	0.013	0
Sample_Nr	na	0.025	0.025	0.013	0.044	0.038	0
Dummy1	0	na	0.031	0	0.081	0.081	0.019
Dummy2	0.05	na	0.044	0.013	0.050	0.081	0.019
Dummy3	na	0.044	0.044	0.013	0.069	0.069	0.025
Dummy4	na	0.038	0.031	0.006	0.025	0.025	0.013

< 0.05 | < 0.06 | < 0.07 | < 0.08 | < 0.09

Fig. 1. Heat map representing the false positive rates per statistical tool for each predictor variable in a control condition (i.e., without implemented effects). Predictors not testable with a given tool are denoted with "na." Darker blue colors indicate a higher rate of significant results (i.e., more false positives), whereas lighter shades indicate lower false positive rates. Full-null refers to the results of the full-null model comparisons, only available for PerMANOVA (PerMAN) and GLMM. The respective second columns of PerMANOVA and GLMM show the results for the individual predictors after conducting a full-null model comparison first. Predictors "sex," "diet," "dummy1," and "dummy2" represent discrete predictors.

comparison (conducted for PerMANOVA and GLMM) was nonsignificant, false positive rates of individual predictors were even lower (all < 0.025) than without a preceding full-null model comparison (see Fig. 1).

3.2 Power

Statistical power varied considerably between statistical tools. Mantel tests generally performed poorly, but also Procrustes analyses and PerMANOVAs showed power rates below the recommended rate of 0.8 across all data structures. AnoSim showed an acceptable power rate around 0.8 for most data structures but was outperformed by the GLMM approach, which had an excellent power > 0.9 in almost all types of data sets. Overall, power appeared to be only little affected by sample size, the number of considered compounds, and the size distribution of the compounds (Fig. 2).

In addition to the statistical tool, the effect structure strongly influenced power (Fig. 3). The most noticeable influence was a dramatic decrease in power in all tools using dissimilarity scores (i.e., reducing the dimension of the data to 1) when effects were implemented only in compounds of the low size class (i.e., tiny and minute compounds, condition 4). Although this also led to a small decrease in power in the GLMM, this was the only tool that retained good power in this condition. When effects were implemented in compounds of the medium size class only (i.e., small and medium compounds, conditions 5 and 8), performance clearly increased for all tools compared to condition 4 but still remained poor in all four dissimilarity tools (Fig. 3).

sample size	compounds	size version	AnoSim	Mantel	PerMAN	Procrustes	GLMM
100	120	medium	0.806	0.583	0.761	0.688	0.909
100	120	small	0.815	0.515	0.754	0.677	0.927
100	180	medium	0.783	0.506	0.706	0.661	0.903
100	180	small	0.777	0.463	0.696	0.646	0.888
150	120	medium	0.813	0.552	0.732	0.660	0.969
150	120	small	0.831	0.606	0.780	0.684	0.980
150	180	medium	0.790	0.573	0.698	0.649	0.948
150	180	small	0.825	0.542	0.715	0.659	0.968

> 0.9
> 0.8
> 0.7
> 0.6
> 0.5
< 0.5

Fig. 2. Heat map representing average statistical power across the four test predictors for each statistical tool and data set version. Darker blue colors indicate a lower rate of significant results (i.e., lower power), whereas lighter shades indicate greater power.

nr	effect size	prev.	n	size class	noise	AnoSim	Mantel	PerMAN	Procr.	GLMM
1	weak	95	6	all	0	0.916	0.628	0.838	0.752	0.873
2	intermediate	95	6	all	0	0.991	0.744	0.933	0.870	0.998
3	strong	95	6	all	0	1.000	0.728	0.869	0.889	0.984
4	intermediate	95	6	low	0	0.028	0.038	0.116	0.064	0.867
5	intermediate	95	6	medium	0	0.547	0.322	0.588	0.398	0.933
6	intermediate	95	3	large	0	0.994	0.734	0.883	0.866	0.981
7	intermediate	95	11	all	0	0.988	0.853	0.963	0.891	1.000
8	intermediate	95	2	medium	0	0.328	0.097	0.330	0.170	0.794
9	intermediate	75	6	all	0	0.978	0.581	0.800	0.772	0.975
10	intermediate	50	6	all	0	0.906	0.334	0.584	0.564	0.834
11	intermediate	95	6	all	3	0.991	0.744	0.931	0.873	0.998
12	intermediate	95	6	all	10	0.994	0.706	0.931	0.878	0.998

power	> 0.9	> 0.8	> 0.7	> 0.6	> 0.5	< 0.5

Fig. 3. Heat map representing average statistical power across the four test predictors for each statistical tool and effect conditions. Darker blue colors indicate a lower rate of significant results (i.e., lower power), whereas lighter shades indicate greater power. nr = number of the respective effect condition, prev. = prevalence (%), Procr. = Procrustes.

When fewer effects were implemented (condition 5 vs. 8; Fig. 3), power decreased noticeably for all statistical tools, whereby this decrease was most pronounced in Mantel tests and least in GLMMs. Again, only GLMMs retained decent power values in this condition. In all tools, power increased with effect size, whereby this was primarily the case when increasing effect size from weak to medium (condition 1 vs. 2). Similarly, a lower prevalence of effects across samples led to decreased power, which was most apparent when decreasing prevalence from 75% (condition 9) to 50% (condition 10). Again, prevalence affected power in all tools, but more in Mantel tests, PerMANOVA, and Procrustes analysis than in AnoSim or GLMMs. Finally, the inclusion of additional noise compounds did not seem to have a pronounced effect on power (Fig. 3).

3.3 Biomarker Identification

The order of importance assigned to compounds with implemented effects varied considerably between statistical tools and, to a lesser extent, with data and effect structure. Overall, RF classification outperformed all other investigated tools and showed excellent detection of simulated biomarkers, as indicated by lower scores (Fig. 4). GLMMs generally performed well too, while sPLS-DA detected biomarkers only moderately well and SimPer analysis poorly.

All tools except SimPer showed a slight improvement of performance (i.e., lower scores for order of importance) when sample size increased from 100 to 150 (Fig. 4). When the number of compounds included in the analyses increased from 120 to 180, GLMMs and sPLS-DA showed a slight and SimPer a clear reduction in performance, while the number of compounds did not seem to affect the accuracy of RF classifications (Fig. 4).

In all tools, performance was generally better (i.e., order of importance lower) when effect size was higher (see conditions 1–3 in Fig. 5). Performance particularly dropped in GLMMs and sPLS-DA, where weak effects represented the effect structure with the worst detection of biomarkers for these two tools. To a much lesser extent, the performance of GLMMs and sPLS-DA also decreased with decreasing prevalence of effects from 95% (condition 2) to 75% (condition 9) to 50% (condition 10). All tools further performed worse when only few effects were implemented (condition 8), whereby the number of effects particularly affected the performance of SimPer analysis. Biomarker detection with SimPer analysis further appeared to be strongly affected by the size class of the compounds into which effects were implemented, with SimPer performing moderately well for large size classes (condition 6) and very poorly for the lower size classes (condition 4).

The size-related performance of SimPer analyses was also corroborated when relating performance to the specific compound size of each implemented effect (Fig. 6) rather than the general effect condition (as shown in Fig. 5). SimPer analyses performed well when effects were implemented in the largest compounds but rapidly deteriorated

sample size	compounds	size	GLMM	RF	sPLS-DA	SimPer
100	120	medium	3.3	2.0	7.1	13.6
100	120	small	3.4	1.9	6.8	11.8
100	180	medium	4.4	1.8	9.1	20.1
100	180	small	4.6	2.0	11.3	17.5
150	120	medium	2.5	1.3	5.6	13.9
150	120	small	2.6	1.4	5.7	12.0
150	180	medium	3.3	1.3	7.1	20.2
150	180	small	3.1	1.5	8.7	17.3

< 3
< 5
< 10
< 15
< 20
> 20

Fig. 4. Heat map representing average order of importance designated to the compounds with implemented effects by four statistical tools for eight different data structures. A value of 1 would imply that all compounds with implemented effects were ranked top by the respective tool. Darker blue colors indicate higher values (i.e., lower importance), whereas lighter shades indicate a high importance. *RF* random forest classification.

nr	effect size	prev.	n	size class	noise	GLMM	RF	sPLS-DA	SimPer
1	weak	95	6	all	0	10.1	1.9	19.1	19.7
2	medium	95	6	all	0	2.4	1.7	6.8	15.4
3	strong	95	6	all	0	1.3	1.1	3.5	11.9
4	medium	95	6	low	0	2.8	1.4	3.4	30.2
5	medium	95	6	medium	0	3.0	1.9	6.1	14.0
6	medium	95	3	large	0	1.6	1.6	6.7	5.8
7	medium	95	11	all	0	2.1	1.8	5.6	9.8
8	medium	95	2	medium	0	5.4	1.9	14.3	30.4
9	medium	75	6	all	0	3.2	1.6	8.9	14.9
10	medium	50	6	all	0	5.5	1.7	9.8	15.0
11	medium	95	6	all	3	2.4	1.7	6.9	15.7
12	medium	95	6	all	10	2.5	1.7	6.8	16.3

< 3
< 5
< 10
< 15
< 20
> 20

Fig. 5. Heat map representing average order of importance designated to the compounds with implemented effects by four statistical tools in 12 different effect conditions. A value of 1 would imply that all compounds with implemented effects were ranked top by the respective tool. Darker blue colors indicate higher values (i.e., lower importance), whereas lighter shades indicate a high importance. nr = number of the respective effect condition, prev. = prevalence (%), RF = random forest classification. The number of effects and size class are only provided for description of the respective effect condition but were not considered for further analysis.

compound size	GLMM	RF	sPLS-DA	SimPer
tiny	3.2	1.2	5.0	28.6
minute	5.6	2.0	10.5	25.4
small	4.8	2.4	11.7	19.2
medium	2.9	2.0	7.3	10.3
large	1.7	1.2	5.5	4.6
largest	1.4	0.8	4.9	2.4

< 3
< 5
< 10
< 15
< 20
> 20

Fig. 6. Heat map representing average order of importance designated to the compounds with implemented effects by four statistical tools for six different compound sizes (ranked from smallest to largest). A value of 1 would imply that all compounds with implemented effects were ranked top by the respective tool. Darker blue colors indicate higher values (i.e., lower importance), whereas lighter shades indicate a high importance. RF = random forest classification.

with decreasing size of the compounds (Fig. 6). Although accuracy was slightly better in the larger size classes in the other statistical tools too, none of them showed a linear change in the order of importance related to compound size.

3.4 Effect Sizes and Error Rates

All evaluated dissimilarity tools provide estimates that are informative about the magnitude of association between chemical profiles and a given trait. These estimates are either derived differently (e.g., r vs. r^2 values) or computed for different types of predictor variables (categorical, continuous, or both), which does not make a comparison *across* tools particularly meaningful. However, a comparison across conditions *within* a given tool can help with interpretation of the results and identifying further strengths and weaknesses, which is why we determined average r and r^2 values for the four dissimilarity tools for each effect condition as well as the control condition without implemented effects.

In the control condition, AnoSim and Mantel tests produced estimates very close to 0, PerMANOVA values around 0.01, and Procrustes analysis 0.1, i.e., one to two orders of magnitude higher than the other tools. These tool-specific "baselines" need to be taken into account when assessing estimates in the effect conditions. In fact, for all four tools, average effect size estimates were higher in the effect conditions than in the control condition, with the notable exception of condition 4, in which effects were only implemented in the smallest compound size classes (Fig. 7). Estimates were also small (albeit higher than in the control condition) when effects were implemented in medium-sized compounds only (conditions 5 and 8), particularly if effects were few (condition 8; see Fig. 7). Compared to other conditions, estimates were also rather small when prevalence was lower (conditions 9 and 10).

Among the multivariate tools, sPLS-DA and RF provide error measures (sPLS-DA as rate and RF as percentage) that depict how well trait categories can be distinguished based on chemical profiles. As expected, both tools showed high error rates of around 50% in the control condition without implemented effects (Fig. 8). Notably, trait levels were discriminated much better in all effect conditions than in the control condition. In particular, unlike for power and effect size, smaller size of affected compounds (condition 4) did not appear to have noteworthy effects on the discriminability between trait levels. Classification errors were only slightly elevated in conditions 8 (few effects in medium-sized compounds) and 10 (only 50% prevalence), but even in these scenarios both tools generally performed well (Fig. 8).

nr	effect size	prev.	n	size class	noise	AnoSim r	Mantel r	PerMAN r2	Procr. r
0	no	0	0	na	0	0.000	-0.002	0.009	0.097
1	weak	95	6	all	0	0.283	0.116	0.068	0.377
2	intermed	95	6	all	0	0.481	0.125	0.098	0.468
3	strong	95	6	all	0	0.528	0.118	0.108	0.468
4	intermed	95	6	low	0	0.008	0.004	0.010	0.098
5	intermed	95	6	medium	0	0.111	0.060	0.030	0.220
6	intermed	95	3	large	0	0.499	0.128	0.111	0.462
7	intermed	95	11	all	0	0.515	0.144	0.117	0.474
8	intermed	95	2	medium	0	0.045	0.022	0.017	0.140
9	intermed	75	6	all	0	0.292	0.068	0.062	0.337
10	intermed	50	6	all	0	0.173	0.035	0.039	0.242
11	intermed	95	6	all	3	0.478	0.124	0.096	0.467
12	intermed	95	6	all	10	0.472	0.121	0.090	0.465

AnoSim Procr.	Mantel PerMAN
> 0.4	> 0.20
< 0.4	< 0.20
< 0.3	< 0.15
< 0.2	< 0.10
< 0.1	< 0.05
< 0.01	< 0.01

Fig. 7. Heat map representing average effect size, measured as r or r^2, for four statistical tools in 12 different effect conditions and a control condition (nr. 0) without implemented effects. Darker blue colors indicate lower values (i.e., lower effect sizes), whereas lighter shades indicate higher effect sizes. Note that a comparison of values between statistical tools is not particularly meaningful due to differences in the types of predictors and, in part, different effect size measures.

nr	effect size	prev.	n	size class	noise	sPLS-DA error	RF error (%)
0	no	0	0	na	0	0.468	50.8
1	weak	95	6	all	0	0.033	4.2
2	intermed	95	6	all	0	0.023	2.0
3	strong	95	6	all	0	0.011	0.6
4	intermed	95	6	low	0	0.033	2.8
5	intermed	95	6	medium	0	0.048	4.5
6	intermed	95	3	large	0	0.032	4.2
7	intermed	95	11	all	0	0.008	0.7
8	intermed	95	2	medium	0	0.103	13.2
9	intermed	75	6	all	0	0.062	5.5
10	intermed	50	6	all	0	0.117	10.9
11	intermed	95	6	all	3	0.024	2.0
12	intermed	95	6	all	10	0.022	2.0

sPLS-DA	RF
< 0.05	< 5
< 0.1	< 10
< 0.2	< 20
< 0.3	< 30
< 0.4	< 40
> 0.4	> 40

Fig. 8. Heat map representing average error rate (sPLS-DA) or % of errors (RF) in 12 different effect conditions and a control condition without implemented effects. Darker blue colors indicate higher values (i.e., higher errors), whereas lighter shades indicate low error (i.e., high accuracy). *RF* random forest classification.

In the GLMMs, the estimates of the random slope terms for compounds quantify the variability in the slopes of a given predictor across compounds, with large estimates indicating that the relative abundance of some compounds is affected strongly and that of others only weakly by a given predictor. While random slopes represent a measure that describes the heterogeneity in the data associated with a given predictor, it is not standardized and thus difficult to compare across different models. Accordingly, we did not compare slope estimates between effect conditions for GLMMs. SimPer analysis provides no measure that is indicative of effect size or accuracy of discrimination between trait values.

3.5 Empirical Data

We compared the results of all statistical tools for two empirical data sets on mammalian body odor profiles. Both data sets were obtained by sampling body odor with thermal desorption tubes and subsequent GC-MS analysis.

3.5.1 Meerkat Data. The first data set (Weiß et al., 2018c) was used to evaluate the effects of different sampling parameters on noninvasively obtained body odor samples of meerkats (*Suricata suricatta*). As described in detail in Weiß et al. (2018a), a total of 101 samples from 24 individuals in two captive populations of meerkats were used for statistical analysis. Samples were collected from the anogenital region of the animals with different sampling parameters. After a comparison with blank samples, the log-transformed, relative areas of 111 compounds were entered into statistical analysis. In particular, Weiß et al. (2018a) applied AnoSim, Mantel tests, and GLMMs to assess effects of sampling volume (discrete predictor with two levels), flow rate, the number of subsamples, and sampling distance (all

Table 4. *P*-values obtained for the predictors in the meerkat data set with five statistical tools.

Term	AnoSim	Mantel	PerMAN	Procrustes	GLMM
Full-null	Na	Na	< 0.001	Na	< 0.001
Volume	0.672	Na	0.317	0.437	0.165
Flow rate	Na	0.010	0.001	0.020	< 0.001
n subsamples	Na	0.065	0.030	0.005	< 0.001
Distance	Na	0.431	0.005	0.025	< 0.001
Population	0.001	Na	0.895	0.004	< 0.001

continuous predictors) on the composition of chemical profiles. Furthermore, population differences (discrete predictor with two levels) were tested to demonstrate the applicability of the sampling method to biological questions.

We used the results of the AnoSim and GLMM as published in Weiß et al. (2018a) and recomputed the data with all other statistical tools evaluated here, including Mantel tests because different *R* functions were used in the original study and the simulation study. Inference based on *p*-values computed for a full model (PerMANOVA and GLMM only) and the respective predictors was similar across tools in most cases (Table 4). All tools suggested no significant effect of sampling volume on odor profiles and significant effects (or at least a trend) of flow rate and the number of subsamples. Effects of sampling distance and population were detected as statistically significant by all tools except one, with the Mantel test being clearly nonsignificant for sampling distance and PerMANOVA for population differences.

We further compared the top 10% of compounds (i.e., the first 11 of the 111 compounds) identified by the respective tools as being associated with the statistically significant discrete predictor "population." For the sPLS-DA analysis, we considered the 11 top compounds in each of the three components (or all, if fewer than 11) identified in the analysis. GLMM, RF, and sPLS-DA component 1 showed considerable overlap in their top-ranked compounds, while there was less overlap with SimPer or the other two sPLS-DA components (Fig. 9). GLMM, RF, and sPLS-DA particularly agreed on the highest-ranked compounds, with all three tools placing RT19 first. Also RT33 appeared in the top list of all four tools, albeit only at the end of the SimPer top list. Each tool listed some compounds that did not appear in any of the other top lists (GLMM, 3; RF, 4; sPLS-DA comp1, 3; SimPer, 5).

3.5.2 Marmoset Data. The second data set was used to investigate olfactory fertility cues in common marmosets (*Callithrix jacchus*). As described in detail in Kücklich et al. (2019), a total of 212 samples from 12 female marmosets collected from the anogenital region in different cycle stages were used for statistical analysis. After a comparison with blank samples, the log-transformed, relative areas of 160 compounds were entered into statistical analysis. In particular, Kücklich et al. (2019) applied AnoSim and GLMM to assess effects of cycle stage (discrete predictor with three levels), parity (discrete predictor with two levels), and female age (modeled as age class in AnoSim and continuously in GLMM). Again, we used the results of the AnoSim and GLMM as published in Kücklich et al. (2019) and recomputed the data with all other statistical tools evaluated here.

Nr	RT	GLMM	RF	sPLS-DA	SimPer	Nr	RT	GLMM	RF	sPLS-DA	SimPer
5	2.3				10	204	21.3		13	5 [2]	
11	2.7	11				227	25.3			2 [3]	
13	3.0	15		6 [2]		230	25.6		7		
14	3.1			2 [2]		243	27.1		10		
19	3.7	1	1	1 [1]	8	263	30.0			7 [2]	
21	4.4			11 [2]		276	32.6			9 [2]	
30	5.6	4	8	3 [1]	12	282	33.3	16	4	16 [1]	
33	6.2	2	5	2 [1]	11	283	33.7		11		
36	6.8	5	3	14 [2]		284	33.8	14	2	9 [1]	
56	10.4		9	4 [1]		292	38.7	13		10 [1]	1
57	10.4		14	8 [1]		315	43.3				3
71	12.0			4 [2]		316	43.4			4 [3]	
78	12.6				5	317	43.4	9	12	1 [2]	
81	12.8				4	319	43.5	6			
84	13.0			11 [1]	9	324	44.4	7			
108	14.3				6	328	45.5	10		3 [2]	2
135	15.4			5 [1]		332	46.2	3	6	7 [1]	
150	16.4			10 [2]		333	46.4	8		8 [2]	7
195	20.1			6 [1]	14						

1-3
4-6
7-11
>11

Fig. 9. Order of importance for all compounds listed among the top 10% of compounds associated with the predictor "population" by at least one of four statistical tools. Darker green shades indicate lower values (i.e., better ranking/higher importance), whereas lighter shades indicate a lower importance. Nr describes the numeric label of the respective compound, using the same labels as in Table 3 in Weiß et al. (2018a), where putative identifications for the majority of compounds are provided. RT describes the mean retention time (in minutes) of the respective compound. *RF* random forest classification. Compounds loaded onto three different components in sPLS-DA as indicated by superscripts. Gray color fonts denote compounds outside the top 10% but within the top 15% of compounds.

Table 5. *P*-values obtained for the predictors in the marmoset data set with five statistical tools.

Tool	AnoSim	Mantel	PerMANOVA	Procrustes	GLMM
Full-null	Na	Na	0.723	Na	< 0.001
Cycle stage	0.081	Na	0.707	0.920	0.999
Parity	0.435	Na	0.484	0.002	< 0.001
Age	0.028	0.003	0.363	0.001	< 0.001

For the marmoset data set, inference based on *p*-values varied considerably more between tools than in the meerkat data. While results of Procrustes analyses and the GLMM led to the same conclusions, results differed for two out of three AnoSim results and three out of four PerMANOVA results (Table 5). Most notably, PerMANOVA did not detect statistically significant effects in the full model or any of the single predictor variables, while Procrustes and GLMM results suggested highly significant effects of parity and female age on chemical profiles.

Nr	RT	GLMM	RF	sPLSDA	SimPer	Nr	RT	GLMM	RF	sPLSDA	SimPer
6	2.4		1		14	113	16.3		12		
8	2.8		9		2	118	16.9		4		15
10	3.1				7	133	18.6			2	
38	7.8		8		10	172	24.3		10		
39	7.8	3				177	25.8	13			
42	8.1	4			4	179	25.9	9			
47	9.2	6			8	188	27.8	7			
54	9.9	16				199	28.9	2	3		6
55	10.1		6			210	30.0				16
64	11.1		14			219	31.4		16		
65	11.2	15	11			229	32.7				11
69	11.5		13		13	234	33.2	10			
80	12.3	8				268	40.1		5		
82	12.5		2	1	9	269	40.7	12			
84	12.8				1	275	42.5	14			12
89	13.5		15		5	278	43.0	11			
106	15.3		7		3	284	46.6	5			
109	15.8	1									

1-4
5-8
9-16
>16

Fig. 10. Order of importance for all compounds listed among the top 10% of compounds associated with the predictor "parity" by at least one of four statistical tools. Darker green shades indicate lower values (i.e., higher importance), whereas lighter shades indicate a lower importance. Nr describes the numeric label of the respective compound and RT its mean retention time in minutes. Fourteen of these compounds have been putatively identified in Table 3 of Kücklich et al. (2019). Most of these can be matched by their unique mean RTs except RT nr 39 (mean RT 7.8), which was identified as 1,2-dimethyl-benzene. *RF* random forest classification.

Inference remained variable when comparing the top 10% of compounds (i.e., the first 16 of the 160 compounds) identified by the respective tools as being associated with the discrete predictor "parity," which was statistically significant when analyzed with Procrustes or GLMM and a weak trend when analyzed with AnoSim. None of the compounds appeared in all four top lists, and there was no clear agreement on the most important compound, although two compounds, RT82 and RT199, were ranked close to the top by two and appeared in the list of a third tool (Fig. 10). No other compounds appeared in more than two of the four top lists. Interestingly, the tool performing best in the simulated data (RF) had the largest overlap with the tool performing worst in the simulation (SimPer), with more than half of the top-ranked compounds shared between lists.

4 Discussion

Our data simulations showed that statistical tools as well as structural properties of data sets or types of effects may affect how well relationships between traits and chemical profiles can be detected.

4.1 False Positives

Importantly, all tools performed satisfactorily in the absence of any implemented effects. Nonetheless, following a full-null model approach for PerMANOVAs and GLMMs further reduced false positive rates, corroborating other studies advocating a full-null model approach when multiple predictors are evaluated to reduce the rate of false positives arising from cryptic multiple testing (e.g., Forstmeier & Schielzeth, 2011).

4.2 Power

Of the five tools that compute *p*-values for the association between traits and chemical profiles, GLMMs were the only tool with good to excellent power across different data and effect structures. The multivariate tool thus was superior to all four investigated dissimilarity tools (i.e., AnoSim, Mantel test, PerMANOVA, and Procrustes), among which AnoSim generally performed better and Mantel tests worse than the other dissimilarity tools. All dissimilarity tools performed particularly bad when few and/or smaller compounds were affected by a trait, suggesting that such effects are easily masked by other, unaffected compounds during the process of dimension reduction. Accordingly, dissimilarity-based approaches are probably inappropriate for detecting associations between traits and chemical profiles resulting from only a small suite of affected compounds, particularly if these are of tiny to small size. These tools might be more suitable for detecting associations with traits that are likely to affect chemical mixtures rather than specific compounds (such as individual identity or kinship; Wyatt, 2010), a scenario not represented in our simulated data sets.

4.3 Biomarker Identification

The correct identification of biomarkers differed considerably between statistical tools. In particular, RF classification was excellent across different data and effect structures. RF was also among the tools with the lowest classification error under the different scenarios investigated by Trainor et al. (2017) and was suggested to be the best tool for imputation (i.e., replacing missing values; Kokla et al., 2019). RF thus appears to be the tool of choice across a wide range of scenarios, although one potential drawback is its limited possibility to take into account additional predictors and control variables. If this is required, GLMMs appear to be a good alternative or addition, as they also performed well unless all implemented effects were weak. Even in this scenario though, GLMMs outperformed the remaining investigated tools, i.e., sPLS-DA and SimPer. SimPer analysis performed considerably worse than all other tools and only detected biomarkers well when large or very large compounds were affected, which is not often to be expected in biological data sets. The poor performance of SimPer is in line with a study by Warton et al. (2012), who describe that the SimPer algorithm does not account correctly for mean-variance relationships and therefore confounds between group differences and within group variation.

For most tools, data and effect structures only slightly affected biomarker identification. The most noticeable effect was that of effect size, with biomarkers being identified less accurately when effect size was lower. We therefore suggest to adjust confidence in the order of importance of putative biomarkers depending on effect size or comparable measures for a given trait-profile relationship, as inference regarding putative biomarkers should be more robust when effect sizes are rather high (or error rates low). Results also suggest that biomarkers were detected slightly more accurately when sample sizes were larger. So, as with any other statistical results, also sample size should be considered for one's confidence in the biomarker estimates.

One difficulty in using the order of importance (or any similar measures) for identifying biomarkers is to decide how many of the top-ranked compounds to consider further, because in empirical data sets the number of compounds robustly associated with a given trait is unknown. Frequently used approaches include determining the "elbow" in a scree plot or using standard deviations, numbers or percentages as cut-off values, all of which are rather subjective or arbitrary and carry the risk of selecting false positives (i.e., unaffected compounds) or missing relevant biomarkers. GLMMs using functions that compute slope estimates along with confidence or credible intervals (such as brm from the *R* package brms) provide an alternative to using the order of importance or similar measures for pinpointing biomarkers by assessing for each trait and compound whether the CIs overlap 0 (or overlap between trait levels).

4.4 Effect Sizes and Error Rates

As would be expected from effect size measures, the r and r^2 values computed by the dissimilarity tools generally increased with increasing simulated effect size, the number of affected compounds, and the proportion of samples for which effects were simulated. However, effect sizes were also clearly related to the size of the compounds into which effects were implemented, although the simulated effect size (i.e., the relative change in compound abundance) was the same across the different compound size classes (see conditions 4–6 in Fig. 7). This implies that these tools primarily seem to pick up upon absolute rather than relative changes in compound abundance, which can lead to the masking (and thus underestimation) of effects if primarily smaller compounds are affected.

Furthermore, effect size measures need to be interpreted in the light of the tool-specific baselines in the control condition. In the absence of simulated effects, AnoSim and Mantel tests routinely estimated r values very close to 0, implying that also estimates only slightly larger than 0 may be indicative of some effect (either a weak one or one that is primarily mediated by small compounds). In contrast, r values of around 0.1 were not uncommon in Procrustes analysis even in the absence of simulated effects, suggesting that any estimates below 0.1 have a high chance of being just spurious associations but no genuine effects. Effect size estimates of the PerMANOVA averaged 0.01 in the control condition. For the interpretation of the r^2 values provided by the adonis2 function, it further needs to be kept in mind that they represent the proportion of variance explained by a given predictor (with the effects of all predictor variables and the residual variance adding up to 1) and thus cannot be directly compared to r values.

In contrast to the effect size measures provided by the dissimilarity tools, the evaluated multivariate tools seemed to be less affected by data and effect structural properties than dissimilarity tools and thus may be the better choice for assessing the strength of association between a trait and chemical profiles.

4.5 Empirical Data

The application of all tools on two empirical data sets highlighted that the degree of accordance between tools may depend greatly on the specific data set. In the meerkat data set, the main conclusions would have been similar irrespective of the tools used, with the exception of SimPer, which—in line with the simulation results—differed most from the other tools. In contrast, the results were much more variable for the marmoset data. Inference derived from Procrustes analysis was well in accordance with that from GLMMs but less with the other dissimilarity tools, which is somewhat curious given that the dissimilarity tools often provided similar results for the simulated data. There was also less agreement about putative biomarkers in the marmoset data than in the meerkat data. Based on the results of all four multivariate tools, one would probably only consider three or four of the compounds to be more or less robustly associated with the predictor parity. It is particularly curious that the presumably best (RF) and presumably worst (SimPer) tool showed the highest agreement. However, the classification error in the RF analysis was considerably lower in the meerkat data set (3%) than in the marmoset data set (35.9%). Along with the greater variation in inference in the marmoset compared to the meerkat data set, this could be indicative of more pronounced and/or robust effects in the latter case, potentially owing to the greater focus on methodological compared to biological aspects in the meerkat study than in the marmoset study. Either way, these results point toward the value of using multiple tools for improving inference (see also Nair et al., 2018).

Importantly, inference based on multiple tools must not be misused for fishing for results. When used responsibly and reported transparently, it can be highly valuable for judging the robustness of results in a field where a multitude of methodological decisions can have (pronounced) effects on results (Gould et al., 2025).

4.6 Limitations and Outlook

Conclusions about the suitability of statistical tools for different kinds of questions and data sets should be made with various aspects in mind. First of all, we strictly dichotomized p-values into statistically significant/nonsignificant, thereby ignoring whether results were only marginally or clearly above/below the alpha threshold of 0.05. We chose to do so to be able to provide a comprehensible overview about false positives and power with different tools under different conditions but would like to point out that p-values are continuous measures and should usually be interpreted accordingly, particularly if values are close to the alpha level of 0.05. After all, the message derived from a p-value of 0.051 is almost identical to that of a p-value of 0.049.

Furthermore, we simulated multiple test and dummy predictors but only had one replicate per type of predictor (continuous/categorical and within–/between-subjects variation) to not add another layer of complexity to the study design. Results regarding power, detection of biomarkers, and effect sizes may thus be somewhat conditional on the specific expressions of our simulated predictor variables. Simulating more than just one variable for each type of predictor would represent a valuable follow-up study to assess the generalizability of our results and if the type of predictor affects the patterns presented here. Within this context, it would also be important to investigate dummy predictors in detail to assess whether effects simulated for one variable may "bleed through" to non-affected variables and cause elevated false positive rates in unaffected compounds. In the present study, we focused on false positive rates in the absence of any implemented effects and only roughly screened the results for dummy predictors when effects were simulated for test predictors. This preliminary investigation suggested the expected low type I error rates in many, but not all of our simulated scenarios. Particularly one continuous dummy predictor appeared to be prone to high rates of false positives when analyzed with highly sensitive tools such as GLMMs. However, assessing whether this was a chance result or if certain types of variables or tools are generally more susceptible to spurious associations in unaffected predictors would require additional simulations with more diverse test and dummy predictors. This clearly exceeds the scope and feasibility of the present simulation study but would be an important aspect to address in follow-up studies to be able to assess whether the conclusions drawn here across predictors are the same across different kinds and numbers of predictor variables.

Other aspects that we did not take into account, because they would have added even more dimensions of complexity to an already complex design, include repeated measures study designs and the parameter space of the employed statistical tools. For instance, although we theoretically created data sets with repeated measures per individual, we did not consider ID for the actual simulation of the effects, so that the simulated data essentially represent a design without repeated measures. Given the relevance of repeated measures for many questions, it would thus be interesting in future work to assess how tools that can account for repeated measures compare to tools that cannot or can only do so in a limited way. With respect to the investigated parameter space, particularly tools such as RF or sPLS-DA can be applied with a range of different settings (e.g., number of splits, trees, cross-validations, etc.) that may further affect performance (see also Trainor et al., 2017). We kept the settings constant within each applied tool, but follow-up studies could focus on the most promising techniques and most critical conditions (i.e., effect conditions that were associated with a lot of variation in performance) to address how different tool-specific settings do—or do not—modulate the patterns described here. We also did not assess in detail whether assumptions of the respective tests were fulfilled for each of the simulated data sets, a step that should certainly be performed when running tests on one's empirical data. The permutational tools we applied should be robust to non-normality, but in, e.g., perMANOVAs, p-values are sensitive to heterogeneous dispersion between groups. Our use of the Bray-Curtis distance measure on log-transformed proportions makes it unlikely that we have introduced strong differences in dispersion, but a valuable aspect to address in the future would be a detailed analysis of performance that takes into account the adherence to test assumption for each single analysis that was performed.

Finally, we have only covered tools we have frequently come across or that seemed particularly suitable to us, but there are certainly other options available. Statistical procedures and computational possibilities also continuously develop further. Along with the rapid development of AI tools, this will require a constant monitoring and reassessment of what may be the optimal approaches for one's question. For instance, the proper way of testing for nonzero variance in the random part of GLMMs is still debated (e.g., Bolker et al., 2009; Pick et al., 2023). We used an approach well-established for testing the significance of fixed effects, i.e., LRTs. While our results suggest that LRTs were also rather suitable for assessing the significance of random slope terms, other options may be identified in the future that are more suitable. Similarly, we ran the GLMMs with a Gaussian error distribution and arcsine transformation to reduce the bottom-ceiling effects typically associated with the analysis of ratios or percentages (such as relative areas). However, a centered log-ratio (clr) transformation might be more appropriate given that it should account not only for the lower and upper bounds in the data but also for the fact that relative areas sum to 1 in each sample (Van den Boogaart & Tolosana-Delgado, 2013). Again, our results suggest that our approach did not noticeably compromise model performance, but it should be investigated in the future if model fit and performance can indeed be improved further by employing a clr transformation. With increasing computational capacities, mixed model approaches such as multi-membership models may further become a powerful alternative to AnoSims and PerMANOVAs for assessing which predictors affect dissimilarities between pairs of samples, as multi-membership models provide the full flexibility of mixed models and thus can be applied to a wider range of study designs (e.g., Weiß et al., 2025).

4.7 Implications for Tool Selection and Conclusions

Conditional on the data and effect structures simulated in this study, some tools are clearly recommendable for non-targeted analysis of chemical profiles and others not. The GLMM approach is a powerful all-rounder that allows addressing many aspects within a single analytical approach and is applicable to a wide range of study designs. Although the possibilities to deduce the strength of effects are more limited than for other tools (at least for comparisons across studies), GLMMs performed very well and better than the dissimilarity tools in terms of power. It is the only tool that can be used for pinpointing biomarkers within the same analytical approach and, although outperformed by RF classification, also performed well for biomarker identification in most simulated scenarios. It remains to be investigated if its high sensitivity also makes it more prone to false positives in certain types of unaffected variables when associations between chemical profiles and other traits exist. It is a continuously developing approach, and developments such as a wider range of applicable error distributions or possibilities to compute compound-specific CIs for each trait should be considered to further improve inference. Along with sPLS-DA, however, GLMMs were the most computationally intense method investigated in this study. Models with similar numbers of samples and compounds as investigated here will typically run for several hours or even days.

If the primary aim is to compute whether or not chemical profiles as a whole are statistically associated with a particular trait, AnoSim also showed good power in many scenarios and provides an easy to interpret effect size measure (r). AnoSim requires very little computational power but can only be computed for single categorical predictors at a time, limiting its suitability for certain questions and study designs. The PerMANOVA is more flexible in this regard, as it allows fitting multiple predictors that can be categorical or continuous, but was slightly less powerful than AnoSim. Procrustes analysis and Mantel tests were (much) less powerful and/or flexible than their alternatives. It should be kept in mind that all dissimilarity tools performed very poorly when only the smallest compound classes were affected by a given trait.

If the primary aim is biomarker identification and the study design allows its application, RF classification can clearly be recommended because it performed excellent across different scenarios and was barely affected by data or effect structures. It can only be applied to single categorical predictors though and thus is not suitable for more complex designs. RF regression also exists, but it remains to be investigated if this performs similarly well as RF classification. sPLS-DA was the computationally most intense tool and was outperformed by RF classification and GLMMs. SimPer results did not reflect the simulated biomarkers at all in the vast majority of scenarios, which is why we do not recommend its use.

Ultimately, the selection of statistical tools will certainly depend on the study design and which particular aspects (*p*-values or similar measures of inference, biomarker identification, effect sizes) need to be covered. For many questions, it may be helpful to base inference on a combination of tools, so that the tools can serve to verify the respective other (e.g., GLMMs combined with RF; see also Nair et al., 2018) or to cover slightly different aspects such as dissimilarities between profiles vs. compound-specific effects. Furthermore, our results indicate that also information about aspects such as the relative size of affected compounds may help in assessing the robustness of one's results.

In conclusion, our evaluation shows the impact the choice of statistical tools—even if theoretically appropriate for one's questions—may have on the analysis of chemical profiles and that this impact may depend on data sets and types of effects. It is therefore pivotal to report and use all available information (e.g., *p*-values *and* effect sizes, results from different analyses) for the interpretation of results and for evaluating the robustness of one's inferences. Our study thus highlights the importance of carefully choosing one's statistical approach following the principles of good scientific practice, critically considering methodological limitations, and remaining open to new (statistical) developments.

Acknowledgments. We thank Lars Kulik for providing an adapted *R* script for ANOSIM. Claudia Wiesner and her lab at the Institute of Analytical Chemistry, Leipzig University, provided valuable chemical expertise for the chemical studies that inspired this statistical evaluation. Marta Manser at the University of Zurich and Ruben Holland at Zoo Leipzig supported the meerkat study, and Almuth Einspanier at the Faculty of Veterinary Medicine, Leipzig University, the marmoset study that served as empirical data examples. BMW was funded by the German Research Foundation, DFG (grant no. SCHL 2011/2-1 awarded to B. M. Weiß). The GC-MS instrument used to analyze the empirical data was funded by the European Fund for Regional Structure Development, EFRE ("Europe funds Saxony," grant #100195810 to A. Widdig). Leipzig University provided further logistic and financial support.

References

Bate, S. T., & Clark, R. A. (2014). *The design and statistical analysis of animal experiments.* Cambridge University Press.

Bates, D., Mächler, M., Bolker, B., & Walker, S. (2015). Fitting linear mixed-effects models using lme4. *Journal of Statistical Software, 67*, 1–48.

Bolker, B. M., Brooks, M. E., Clark, C. J., Geange, S. W., Poulsen, J. R., Stevens, M. H. H., & White, J. S. S. (2009). Generalized linear mixed models: A practical guide for ecology and evolution. *Trends in Ecology and Evolution, 24*, 127–135.

Brückner, A., & Heethoff, M. (2017). A chemo-ecologists' practical guide to compositional data analysis. *Chemoecology, 27*, 33–46.

Charpentier, M. J. E., Barthes, N., Proffit, M., Bessière, J. M., & Grison, C. (2012). Critical thinking in the chemical ecology of mammalian communication: Roadmap for future studies. *Functional Ecology, 26*, 769–774.

Clarke, K. R. (1993). Non-parametric multivariate analyses of changes in community structure. *Australian Journal of Ecology, 18*(1), 117–143.

Clarke, K. R. (1999). Nonmetric multivariate analysis in community-level ecotoxicology. *Environmental Toxicology and Chemistry, 18*(2), 118–127.

Forstmeier, W., & Schielzeth, H. (2011). Cryptic multiple hypotheses testing in linear models: Overestimated effect sizes and the winner's curse. *Behavioral Ecology and Sociobiology, 65*, 47–55.

Gould, E., Fraser, H. S., Parker, T. H., et al. (2025). Same data, different analysts: Variation in effect sizes due to analytical decisions in ecology and evolutionary biology. *BMC Biology, 23*, 1–36.

Hervé, M. R., Nicolè, F., & Lê Cao, K.-A. (2018). Multivariate analysis of multiple datasets: A practical guide for chemical ecology. *Journal of Chemical Ecology, 44*, 215–234.

Jamil, T., Ozinga, W. A., Kleyer, M., & ter Braak, C. J. F. (2013). Selecting traits that explain species–environment relationships: A generalized linear mixed model approach. *Journal of Vegetation Science, 24*(6), 988–1000.

Jänig, S., Weiß, B. M., Birkemeyer, C., & Widdig, A. (2019). Comparative chemical analysis of body odor in great apes. *American Journal of Primatology, 81*(6), e22976.

Kokla, M., Virtanen, J., Kolehmainen, M., Paananen, J., & Hanhineva, K. (2019). Random forest-based imputation outperforms other methods for imputing LC-MS metabolomics data: A comparative study. *BMC Bioinformatics, 20*, 492.

Kücklich, M., Weiß, B. M., Birkemeyer, C., Einspanier, A., & Widdig, A. (2019). Chemical cues of female fertility states in a non-human primate. *Scientific Reports, 9*(1), 13716.

Laurencelle, D., & Cousineau, D. (2023). Analysis of proportions using arcsine transform with any experimental design. *Frontiers in Psychology, 13*, 1045436.

Lê Cao, K.-A., Boitard, S., & Besse, P. (2011). Sparse PLS discriminant analysis: Biologically relevant feature selection and graphical displays for multiclass problems. *BMC Bioinformatics, 12*, 253.

Liaw, A., & Wiener, M. (2002). Classification and regression by randomForest. *R News, 2*(3), 18–22.

Nair, J. V., Shanmugam, P. V., Karpe, S. D., Ramakrishnan, U., & Olsson, S. (2018). An optimized protocol for large-scale in situ sampling and analysis of volatile organic compounds. *Ecology and Evolution, 8*(11), 5924–5936.

Noonan, M. J., Tinnesand, H. V., & Buesching, C. D. (2018). Normalizing gas-chromatography–mass spectrometry data: Method choice can alter biological inference. *BioEssays, 40*, 1700210.

Oksanen, J., Blanchet, F. G., Friendly, M., et al (2016). *Vegan: Community Ecology Package*. https://CRAN.R-project.org/package=vegan.

Pick, J. L., Kasper, C., Allegue, H., Dingemanse, N. J., Laskowski, K. L., Lima, M. R., Schielzeth, H., Westneat, D. F., Wright, J., & Araya-Ajoy, Y. G. (2023). Describing posterior distributions of variance components: Problems and the use of null distributions to aid interpretation. *Methods in Ecology and Evolution, 14*(10), 2557–2574.

R Core Team. (2017). *A language and environment for statistical computing*. R Foundation for Statistical Computing. https://www.R-project.org/

Rohart, F., Gautier, B., Singh, A., & Lê Cao, K.-A. (2017). mixOmics: An R package for 'omics feature selection and multiple data integration. *PLoS Computational Biology, 13*, e1005752.

Stoffel, M. A., Caspers, B. A., Forcada, J., Giannakara, A., Baier, M., Eberhart-Phillips, L., Müller, C., & Hoffman, J. I. (2015). Chemical fingerprints encode mother–offspring similarity, colony membership, relatedness, and genetic quality in fur seals. *Proceedings of the National Academy of Sciences of the United States of America, 112*(36), E5005–E5012.

Trainor, P. J., DeFilippis, A. P., & Rai, S. N. (2017). Evaluation of classifier performance for multiclass phenotype discrimination in untargeted metabolomics. *Metabolites, 7*, 30.

Van den Boogaart, K. G., & Tolosana-Delgado, R. (2013). *Analyzing Compositional Data with R, volume 122* (pp. 1–200). Springer.

Warton, D. I., Wright, S. T., & Wang, Y. (2012). Distance-based multivariate analyses confound location and dispersion effects. *Methods in Ecology and Evolution, 3*(1), 89–101.

Weiß, B. M., Marcillo, A., Manser, M., Holland, R., Birkemeyer, C., & Widdig, A. (2018a). A non-invasive method for sampling the body odour of mammals. *Methods in Ecology and Evolution, 9*(2), 420–429.

Weiß, B. M., Kücklich, M., Thomsen, R., Henkel, S., Jänig, S., Kulik, L., Birkemeyer, C., & Widdig, A. (2018b). Chemical composition of axillary odorants reflects social and individual attributes in rhesus macaques. *Behavioral Ecology and Sociobiology, 72*(4), 65.

Weiß, B. M., Marcillo, A., Manser, M., Holland, R., Birkemeyer, C., & Widdig, A. (2018c). Data from: A non-invasive method for sampling the body odour of mammals. *Methods in Ecology and Evolution, 9*(2), 420–429.

Weiß, B. M., Birkemeyer, C., Kücklich, M., & Widdig, A. (2025). Chemical signatures of social information in barbary macaques. *Scientific Reports, 15*, 1902.

Wyatt, T. D. (2010). Pheromones and signature mixtures: Defining species-wide signals and variable cues for identity in both invertebrates and vertebrates. *Journal of Comparative Physiology A, 196*(10), 685–700.

Part IV

Environmental and Applied Chemical Ecology

Odors in the City: Intraspecific Chemical Communication in Urban Environments

Javier delBarco-Trillo[1](✉)

[1] School of Biological Sciences, Illinois State University, Normal, IL, USA
jdelbar@ilstu.edu

Abstract. Under increasing rates of urbanization worldwide, many vertebrate species are transitioning into these expanding urban environments. Animals entering urban environments are exposed and need to respond to several stressors and ecological challenges. Many of these stressors can have a direct impact on the production and functionality of chemical signals. Here I comprehensively reviewed the literature on how urbanization can affect any aspect of chemical communication, including signal production and transmission and receiver responses. The only urban studies assessing the composition of chemical signals have been conducted in two species of birds, providing conflicting results. Most of the available research has concentrated on scent marking in urban mammals. Overall, urbanization leads to higher rates of scent marking, and the spatial deposition of scent marks mostly depends on the level of human disturbance and species-specific social structure. As for the responses of receivers to the odors of conspecifics, there are results supporting both impaired responses and odors of urban senders being highly attractive to opposite-sex members from both urban and nonurban populations. We clearly need more studies on chemical communication in urban populations to better understand how the signaling behavior of an increasing number of species is affected across evolving urban environments.

Keywords: Chemical communication · Human disturbance · Latrines · Odors · Scent glands · Scent marking · Territorial marking · Urban environments · Urbanization

1 Introduction

Over 50% of the world's human population already lives in cities, and it is estimated that by 2030 that number will approach 70%. The exponential rise in the human population, coupled with an increasing proportion of people living in urban areas, is driving a rapid expansion in both the number and size of cities worldwide. This will lead to many native species of animals being forced to cope with these expanding urban environments. Animals colonizing urban environments are exposed to a series of stressors and

L. M. Schulte (ed.), *Chemical Signals in Vertebrates 16*,
https://doi.org/10.1007/978-3-032-30752-1_16

ecological challenges, the combination of which can result in phenotypic changes in these urban individuals. Changes in behavior and in animal communication are particularly susceptible to the effects of urbanization (Sepp et al., 2020).

Indeed, most of the research conducted on phenotypic traits in urban populations has focused on animal communication (Sepp et al., 2020), mainly singing in birds and to a lesser degree visual communication (Yeh, 2004; Derryberry & Luther, 2021; Cronin et al., 2022; Gomes et al., 2022). The fact that most communication research on urban animals has involved the investigation of acoustic and visual communication (Heinen-Kay et al., 2021) does not necessarily reflect an obvious importance of these signaling modalities across all species found in urban environments, but more the fact that urban research has been (and still is) dominated by avian biologists. Not surprisingly then, there is a large body of literature on how noise pollution in cities leads to changes in birdsong parameters and how such changes may affect mate choice and male-male competition (Slabbekoorn & Peet, 2003; Proppe et al., 2011; Schmidt et al., 2014; Derryberry & Luther, 2021). The overall finding in these studies is that bird species change their vocalizations to optimize signal transmission, mainly to overcome noise pollution. Generally, noise pollution in urban environments forces many species of birds to sing louder, which facilitates the transmission of the signal. However, such changes can affect intraspecific communication and thus sexual selection processes (Winandy et al., 2021). Some attributes of these songs, like their structure, can also change in response to population density, which is normally higher in urban environments (Narango & Rodewald, 2016). The strong focus on birdsong in urban birds has also spearheaded studies on vocal communication in other taxa. For example, urban túngara frogs (*Physalaemus pustulosus*) change their calls and attract more females (Halfwerk et al., 2019). When translocated to forests, urban males may adjust their calls to avoid predation and parasitism (which can be relatively low in cities), but forest frogs translocated to urban sites cannot change their calls to better attract females (Halfwerk et al., 2019). Changes in vocal communication driven by urbanization, however, are taxa dependent (Gomes et al., 2022). Whereas birds can change many song parameters (e.g., frequency, amplitude, and note duration), insects do not tend to change their vocal signals (Gomes et al., 2022).

Even though olfactory communication has received little attention in the context of urbanization (Birnie-Gauvin et al., 2016), there are many reasons to expect olfactory communication to be greatly affected in urban populations compared to nonurban populations. Animals living in cities are exposed to many biotic and abiotic factors that can affect olfactory communication. They are exposed to high chemical, noise, and light pollution (Birnie-Gauvin et al., 2016; Heinen-Kay et al., 2021; Cronin et al., 2022); can uptake high levels of endocrine disruptors (e.g., from plastics) and pharmaceuticals for humans and pets, e.g., contraceptives (Birnie-Gauvin et al., 2016); tend to live in high population densities, which may require a higher and more efficient transfer of information (Cronin et al., 2022); tend to experience high levels of stress, associated, for example, to noise pollution and to human activity (Birnie-Gauvin et al., 2016); are exposed to dramatic changes in diet, including high-fat high-sugar human food (Birnie-Gauvin et al., 2016; Cronin et al., 2022); can be exposed to different types and intensities of diseases and parasites (Cronin et al., 2022); tend to experience higher average and extreme temperatures (Cronin et al., 2022); have smaller home ranges, which can lead to changes in scent marking behavior (O'Donnell & delBarco-Trillo, 2020); tend to be more nocturnal than their nonurban counterparts, which may select for higher reliance on olfactory communication (Gaynor et al., 2018); and experience high subpopulation structuring and isolation, which reduces gene flow and results in local episodes of inbreeding (Johnson & Munshi-South, 2017). Any of these parameters by themselves can have significant effects on the composition of chemical signals, scent marking behaviors, or the processing of chemical signal by the receivers (Crawford et al., 2011; Rosenthal & Stuart-Fox, 2012; Blocker & Ophir, 2013; Van Geffen et al., 2015a, b; Henneken & Jones, 2017). If we consider that the effects from all these different parameters are likely to be compounded, we should expect significant and profound changes in olfactory communication in urban animals.

Pollutants, endocrine disruptors, pharmaceuticals, and other anthropogenic stressors can affect chemical communication at three different levels: signal production, signal transmission, and signal reception and processing (Vom Saal et al., 1995; Park et al., 2001; Crawford et al., 2011; Rosenthal & Stuart-Fox, 2012; Henneken & Jones, 2017). First, anthropogenic contaminants and other anthropogenic factors can disrupt the production and composition of chemical signals (Blocker & Ophir, 2013; Henneken & Jones, 2017), but how this occurs in urban populations remains to be investigated. Nonchemical signal quality tends to decrease in urban environments (Hutton & McGraw, 2016), e.g., urban birds normally have duller coloration and reduced conspicuousness of sexually selected plumage traits (Leveau, 2021), but it is unclear whether a similar trend can be expected in the case of chemical signals. Second, the transmission of chemical signals can be affected through behavioral changes in scent marking, and pollutants can also bind to compounds from the chemical signal, reducing the available quantity or salience of these compounds (Rosenthal & Stuart-Fox, 2012). Third, brain function can be affected by urbanization, and pollutants can also bind to brain receptors involved in the detection of chemical signals, minimizing or disrupting their detection and processing (Rosenthal & Stuart-Fox, 2012).

Here, I performed a comprehensive review about intraspecific olfactory communication in urban environments across taxa, to determine what we know about this topic and what remains to be investigated.

2 Methods

I performed a dual search approach, using both a standardized Web of Science (WoS) search and searches in a previously curated (but unpublished) EndNote library with a combined focus on urban ecology and chemical communication. This approach was used to maximize the number of studies covered in this review, by enriching the standardized search with relevant results known to me that might not have been otherwise found through a standard search in WoS.

I started the literature search on May 14, 2024, using a previously compiled EndNote library of references on urban ecology and chemical communication, selecting 1116 references that contained a combination of two terms anywhere in the text ("urban" and then one of the following terms: "communication," "chemical," "scent," "odor," or "odour"). From these, I retained 33 references that considered urbanization and any aspect of chemical communication. Most discarded references included the term "disturbance" instead of "urban" or were focused on other sensory modalities than olfaction (mainly vocal communication).

That same day I performed a search in WoS Core Collection (Science Citation Index Expanded; accessed through the Milner Library, Illinois State University), entering the term "urban*" and one of the following terms: "olfaction," "olfactory communication," "chemical compound," "chemical ecology," "chemical signal," "semiochemical," "pheromon*," "scent*," "odor*," "odour*," "smell," "volatile," "GC-MS," "glandular secretion," "urine," or "urinar*." Each pair of terms were entered under the field "Topic." Search words with an asterisk allow for different related words to appear in the search results (e.g., the term "urban*" searched publications containing the words urban, urbanised, urbanized, urbanisation, urbanization, etc.). For each search, I only retained results under pertinent Web of Science categories, mainly "Ecology," "Zoology," "Biology," "Entomology," "Evolutionary Biology," "Ornithology," "Reproductive biology," "Physiology," "Anatomy & Morphology," "Biodiversity Conservation," "Endocrinology & Metabolism," and "Psychology Biological." This initial filtering led to the following results for each search: 39 results using "olfaction"; 14 results using "olfactory communication"; 691 results using "chemical compound"; 121 results using "chemical ecology"; 67 results using "chemical signal"; 10 results using "semiochemical"; 123 results using "pheromon*"; 82 results using "scent*"; 144 results using "odor*"; 44 results using "odour*"; 82 results using "smell"; 532 results using "volatile"; 286 results using "GC-MS"; 1 result using "glandular secretion"; 326 results using "urine"; and 152 results using "urinar*." After removing duplicate results, I had 1615 results to consider. From these, I first manually disregarded most results because (a) they were nonanimal studies or (b) because they clearly did not address any aspect of communication in the context of urbanization or anthropogenic environments, only retaining 41 results. Of these 41 results, 8 results were already in the initial search described above and were thus also discarded. The remaining 33 references from this second search were pooled with the 33 references from the first search. In the process of reading these 66 references and their cited literature, I identified 6 further references that appeared relevant. This resulted in a first batch of 72 references that I analyzed in detail.

Of these 72 a priori appropriate references, I discarded 48 that turned out not to be relevant (they did not address any aspect of olfactory communication in the context of urbanization). In addition, 13 references were only partly relevant, including (a) reviews on urban ecology only briefly mentioning the need to study olfactory communication in urban areas (Birnie-Gauvin et al., 2016; Hutton & McGraw, 2016; Henneken & Jones, 2017; Cronin et al., 2022); (b) papers simply trying to determine whether urban animals accumulate higher levels of pollutants (Perugini et al., 2009); (c) studies on the effects of contaminants on chemical communication, but in a broad context, and not specifically urbanization (Blocker & Ophir, 2013); and (d) studies on chemical ecology but not on intraspecific communication, e.g., urbanization affecting anti-predatory chemical defenses in toads (Bokony et al., 2019); avoidance of snake scents by geckos on different urban substrates (Cornelis et al., 2019); neophobia to neutral, unfamiliar odors in lizards (Walsh et al., 2018); responses in great tits to plant volatiles used as foraging cues (Rubene et al., 2022; Rubene et al., 2023); and plant-insect interactions (Masui et al., 2023). This further process of exclusion resulted in only 13 relevant references.

Next, I performed another set of searches in WoS on June 25, 2024. In this case, I compiled all the references citing any of the 13 relevant studies identified above as well as any of 4 reviews briefly mentioning the importance of studying chemical signals in urban contexts (Birnie-Gauvin et al., 2016; Hutton & McGraw, 2016; Henneken & Jones, 2017; Cronin et al., 2022). These searches produced 510 results (473 after removing duplicates). Disregarding irrelevant studies (see above) resulted in 22 a priori relevant studies, 7 of which I had already identified in previous searches. After reviewing the resulting 15 publications in detail, I identified only 2 that addressed communication in the context of urbanization (Whittaker et al., 2011; Alexander & Nichols, 2020). On June 28, 2024, I considered the 75 publications citing these 2 new relevant references, but none of those 75 publications turned out to be relevant.

In summary, I only found 15 relevant publications describing original research on intraspecific chemical communication in the context of urbanization. These are discussed below.

3 Results and Discussion

Despite several authors mentioning the importance of assessing chemical communication within urban environments (Birnie-Gauvin et al., 2016; Hutton & McGraw, 2016; Henneken & Jones, 2017; Cronin et al., 2022), very little research has been conducted on this topic. It is surprising how some studies addressing multimodal signaling ignore chemical communication altogether (Partan et al., 2010). And this is especially surprising because chemical communication is the main mode of communication for many species found in urban landscapes, such as mammals. Also, the effects of urbanization on some other sensory modalities may increase the reliance of some species on olfactory communication. For example, urban animals tend to have duller colorations and be more nocturnal (Gaynor et al., 2018; Leveau, 2021), which may decrease the effectiveness of visual communication, consequently increasing the use of chemical signals (Secondi et al., 2009). One potential explanation for the scarcity of studies on chemical communication in urban areas is that most researchers with the expertise to study chemical communication have so far not been captivated by the research questions posed by urban populations and their adaptations and responses to urbanization. Below I present and discuss all the available studies I found, and what main types of questions remain to be investigated, with the expectation of capturing the interest of experts on chemical signals toward such research questions.

Anthropogenic disturbances can impact the production of chemical signals, their transmission, and their processing by receivers (Rosenthal & Stuart-Fox, 2012). Overall, the few available studies have investigated the production and the transmission of chemical signals, as well as the responses to odors from conspecifics, but not the processing of chemical signals within the brain of signal receivers.

Urbanization may exert a strong effect on the production of chemical signals. This effect may be more disruptive than adaptive or functional. Research on acoustic communication, the sensory modality best studied in urban environments, clearly shows that several traits of acoustic signals (e.g., frequency, amplitude, and timing of the calls) can change to minimize the masking effects of anthropogenic noise (Birnie-Gauvin et al., 2016). Chemical signals can also change in an adaptive manner in urban environments, being shaped, for example, by urban-associated factors like predation pressure and resource availability. Also, in urban populations experiencing higher population densities than in nonurban populations, chemical signals can be predicted to become more complex in order to transfer more information to successfully handle more interconnected social networks (Donihue et al., 2020). But generally, it is expected that anthropogenic factors, such as endocrine disruptors, toxins, and substantial diet changes, will lead to disruptive, maladaptive changes in the composition of chemical signals.

The only studies analyzing the composition of chemical signals in urban populations involve the uropygial or preen gland secretions of birds. Uropygial gland secretions have a more constant composition compared, for example, to rich mammalian secretions (Díez-Fernández et al., 2021), so any differences in relation to urbanization would indicate a strong effect of urban stressors on the production of such secretions. In European blackbirds (*Turdus merula*), no significant differences were found between the secretions of urban and forest animals, although the number of compounds tended to be lower in the urban animals (Díez-Fernández et al., 2021). These secretions seem to be strongly driven by diet, and the lack of significant differences in European blackbirds may have been due to the studied urban and forest populations having similar diets in both types of habitats (Díez-Fernández et al., 2021). However, in dark-eyed juncos (*Junco hyemalis*), volatile profiles differed between the preen gland secretions of urban and montane populations, even when individuals from these two populations were raised under the same conditions and diet, indicating a possible genetic basis for the preen oil chemistry in this system (Whittaker et al., 2010). The intraindividual variation in the composition of volatiles at different time points was also lower in urban juncos than in montane ones, which can be related to the fact that urban juncos have reduced genetic diversity compared to montane juncos (Whittaker et al., 2010).

More dramatic than changes in the chemical composition of odors are possible differences driven by urbanization involving the glands producing those odors. Both male and female Indian gerbils (*Tatera indica indica*) can have a midventral scent gland, although the presence of this gland differs by sex and by habitat (Prakash & Idris, 1982). This gland is reported to be present in about 35% of females living in grasslands (nonurban habitats), but only in about 3% of urban females (Prakash & Idris, 1982). And whereas almost all sampled grassland males expressed the midventral scent gland, only 78.3% of urban males did (Prakash & Idris, 1982). The main explanation proposed by the authors for these dramatic changes relates to population density. Urban gerbils are more gregarious and live in much higher population densities, compared to singly or pair living in nonurban habitats (Idris & Prakash, 1986, 1987). If the main function of the midventral secretions is to advertise the presence of individuals or their sexual receptivity, there may be little benefit to developing an energetically costly gland in urban populations, where interactions with potential mates are frequent (Prakash & Idris, 1982).

Most of the available studies on urban chemical communication have to do with the transmission of chemical signals, mainly scent marking. In pied tamarins (*Saguinus bicolor*) inhabiting several urban forests in the city of Manaus, Brazil, researchers hypothesized that noise pollution masking vocal signals can lead to an increase in scent marking (Sobroza et al., 2024). Although rates of vocal communication did not change in response to noise pollution, scent marking rates increased (Sobroza et al., 2024). Increased scent marking in this case could be a way to complement the transfer of information in situations where pied tamarins may sense their calls as being unreliable in noisy conditions. However, the authors also considered whether the increase in scent marking could be due to stress associated to noise pollution and other urban stressors (Sobroza et al., 2024). Similarly, banded mongooses (*Mungos mungo*) scent marked more in urbanized areas than in undeveloped ones (Alexander & Nichols, 2020).

Besides studying scent marking rates, one can assess where animals are placing their chemical signals across urban landscapes. This has been done in a few mammals by determining the spatial distribution of their latrines. Latrines are sites that individuals use repeatedly to place their urine and feces (Tsunoda et al., 2019). Latrine use was investigated in racoon dogs (*Nyctereutes procyonoides*) across a large park in Tokyo. Across this urban area, there are varying levels of human disturbance, which can be assessed by considering the distance to parking lots or paths, where human activity is higher (Tsunoda et al., 2019). The authors hypothesized that latrines that were closer to sites with high levels of human activity would mostly be used at night (when human activity is lower) but that overall latrine use would be maintained due to their importance to communicate socio-spatial information (Tsunoda et al., 2019). Although latrine use indeed was a mostly nocturnal activity, racoon dogs decreased their use of latrines in disturbed areas (Tsunoda et al., 2019). In European rabbits (*Oryctolagus cuniculus*), a distinction can be made between territorial boundary latrines (for between-group communication, mainly territorial defense) and territorial core latrines (for within-group communication) (Ziege et al., 2016). The relative use of these two types of latrines should depend on population densities and social group sizes (two variables that tend to differ between urban and nonurban populations) (Ziege et al., 2016). In rural areas, the largest latrines and the highest latrine densities are typically close to the burrow. In contrast, in urban and suburban areas, there is a higher percentage of latrine use away from the burrow (i.e., there is a higher use of territorial boundary latrines), indicating an increased importance of peripheral marking (Ziege et al., 2016). This was the predicted result, as in urban environments there are increased population densities and smaller social group sizes, both factors requiring an increased investment in territorial defense and thus leading to rabbits increasingly using latrines at the periphery of their territories (Ziege et al., 2016). An opposite result has been reported for urban European badgers (*Meles meles*), with lower rates of boundary marking in urban environments (Cresswell & Harris, 1988; Davison et al., 2009). In rural areas with high population densities, badgers emphasize their peripheral marking (Ziege et al., 2016). However, in urban habitats, badger groups are more isolated. Also, urban badgers still have large territories, and to regularly visit their territory boundaries, they would need to traverse areas with high levels of human disturbance. As badgers are averse to such disturbed areas, they are less inclined than rabbits to travel frequently to their territory boundaries to use their territorial boundary latrines (Cresswell & Harris, 1988; Davison et al., 2009). Even in situations in which the home ranges of two urban groups overlapped, boundary latrines were not detected, which clearly indicates a relaxation of territorial scent marking in urban badgers (Davison et al., 2009).

The processing of chemical signals in the brain of urban animals remains to be investigated. To the extent that urban stressors may lead to neuroethological impairments (Blocker & Ophir, 2013; Lee & Thornton, 2021), we can predict that the processing of chemical signals by urban animals can also be negatively impacted. From the few available studies, we can gather that urban animals respond differently to chemical signals from conspecifics. In cabbage moths (*Mamestra brassicae*), males exposed to artificial night-lighting were less likely to be attracted to a pheromone lure compared to males in non-illuminated areas (van Geffen et al., 2015b). This result was associated to a significant decrease in the proportion of mated females in urban areas illuminated by artificial light compared to non-illuminated areas (van Geffen et al., 2015b). In another study in which dark-eyed juncos were tested in a Y-maze (i.e., a two-way choice test), females from both urban and nonurban populations showed a preference for the odors (preen oil) of urban males (Whittaker et al., 2011). Interestingly, the females from the nonurban population showed the stronger preference for the odors of urban males (Whittaker et al., 2011), suggesting that some aspect about the composition of odors from urban males is particularly attractive to females across different populations. This also highlights the importance of teasing apart the characteristics of the chemical signals emitted by the sender and their processing by the receiver.

In summary, there is some indication that the chemical composition of preen oil in birds can be affected by urbanization (Whittaker et al., 2010; Díez-Fernández et al., 2021). However, we need additional information from more species to determine whether there are any consistent patterns across taxa. We also need data from species in which chemical signals differ among individuals not only in the relative abundance of a constant set of compounds (as is normally the case for the preen oil of bird species) but also in the differential mix of such compounds across individuals. The lack of studies on the chemical composition of odors in urban animals is thus obvious. Compared to the composition of chemical signals, we have more evidence for changes in scent marking behavior in urban animals (Cresswell & Harris, 1988; Davison et al., 2009; Ziege et al., 2016; Tsunoda et al., 2019; Alexander & Nichols, 2020; Sobroza et al., 2024). The direction of those changes differs across species, but overall, urban animals tend to scent mark at higher rates compared to nonurban animals, and they can change their scent marking patterns in response to urban disturbance and changes in social structure. We know very little about the preference of individuals for the odors of conspecifics in urban populations (Whittaker et al., 2011; van Geffen et al., 2015b). And we know nothing about how urbanization can affect the processing of chemical signals in the receivers of those signals.

As stated above, there are many opportunities to conduct novel research on chemical communication in urban populations. I propose the following research directions, with the understanding that there are many others available. First, we need more studies across a representative set of species in order to gain a generalized understanding of chemical communication in urban environments. Second, it would be important to know whether the responses within a given species are consistent or not across different cities (delBarco-Trillo & Putman, 2023). Third, we should conduct more studies to determine whether any observed changes associated to chemical communication are driven by phenotypic plasticity or by genetic or epigenetic changes. Research on the mechanisms underlying chemical signal changes in vertebrates remains scarce, especially within an urban framework. Fourth, we should investigate the hypothesis that higher population densities in urban populations, requiring an increased transfer of information among conspecifics, will lead to more complex chemical signals. Fifth, we should assess whether interindividual variation increases in urban populations. In urban areas, each individual is likely to be exposed to a very specific combination of stressors during their lifetime (e.g., in terms of soil or water pollutants or diet items), and this should lead to a higher degree of interindividual variation in the composition, transfer, and processing of chemical signals. Sixth, if chemical signals are disrupted in random ways depending on different milieus of urban stressors, the informative value and reliability of these chemical signals may be drastically reduced, resulting in suboptimal mating decisions and thus in reduced reproductive success (Candolin, 2019). Finally, we are likely to advance our understanding of chemical signals in urban environments considerably when all the different components of chemical communication (i.e., production, transmission, reception, processing, and response to chemical signals) are investigated in any given urban species in a comprehensive and interrelated manner.

References

Alexander, K., & Nichols, C. (2020). Behavior-landscape interactions may create super-spreader environments: Vigilance-olfactory interactions across land type and disease transmission potential in the banded mongoose. *Frontiers in Ecology and Evolution, 8*, 47.

Birnie-Gauvin, K., Peiman, K. S., Gallagher, A. J., de Bruijn, R., & Cooke, S. J. (2016). Sublethal consequences of urban life for wild vertebrates. *Environmental Reviews, 24*, 416–425.

Blocker, T., & Ophir, A. (2013). Cryptic confounding compounds: A brief consideration of the influences of anthropogenic contaminants on courtship and mating behavior. *Acta Ethologica, 16*, 105–125.

Bokony, V., Uveges, B., Verebelyi, V., Ujhegyi, N., & Moricz, A. M. (2019). Toads phenotypically adjust their chemical defences to anthropogenic habitat change. *Scientific Reports, 9*, 3163.

Candolin, U. (2019). Mate choice in a changing world. *Biological Reviews, 94*, 1246–1260.

Cornelis, J., Nordberg, E. J., & Schwarzkopf, L. (2019). Antipredator behaviour of invasive geckos in response to chemical cues from snakes. *Ethology, 125*, 57–63.

Crawford, J. C., Boulet, M., & Drea, C. M. (2011). Smelling wrong: Hormonal contraception in lemurs alters critical female odour cues. *Proceedings of the Royal Society of London, Series B: Biological Sciences, 278*, 122–130.

Cresswell, W. J., & Harris, S. (1988). Foraging behavior and home range utilization in a suburban badger (*Meles meles*) population. *Mammal Review, 18*, 37–49.

Cronin, A. D., Smit, J. A. H., Muñoz, M., Poirier, A., Moran, P. A., Jerem, P., & Halfwerk, W. (2022). A comprehensive overview of the effects of urbanisation on sexual selection and sexual traits. *Biological Reviews, 97*, 1325–1345.

Davison, J., Huck, M., Delahay, R. J., & Roper, T. J. (2009). Restricted ranging behaviour in a high-density population of urban badgers. *Journal of Zoology, 277*, 45–53.

delBarco-Trillo, J., & Putman, B. J. (2023). Not all cities are the same: Variation in animal phenotypes across cities within urban ecology studies. *Urban Ecosystems, 26*, 1725–1737.

Derryberry, E. P., & Luther, D. (2021). What is known-and not known-about acoustic communication in an urban soundscape. *Integrative and Comparative Biology, 61*, 1783–1794.

Díez-Fernández, A., Martínez-de la Puente, J., Martín, J., Gangoso, L., López, P., Soriguer, R., & Figuerola, J. (2021). Sex and age, but not blood parasite infection nor habitat, affect the composition of the uropygial gland secretions in European blackbirds. *Journal of Avian Biology, 52*, e02630.

Donihue, C. M., Herrel, A., Martin, J., Foufopoulos, J., Pafilis, P., & Baeckens, S. (2020). Rapid and repeated divergence of animal chemical signals in an Island introduction experiment. *Journal of Animal Ecology, 89*, 1458–1467.

Gaynor, K. M., Hojnowski, C. E., Carter, N. H., & Brashares, J. S. (2018). The influence of human disturbance on wildlife nocturnality. *Science, 360*, 1232–1235.

Gomes, L., Solé, M., Sousa-Lima, R. S., & Baumgarten, J. E. (2022). Influence of anthropogenic sounds on insect, anuran and bird acoustic signals: A meta-analysis. *Frontiers in Ecology and Evolution, 10*, 827440.

Halfwerk, W., Blaas, M., Kramer, L., Hijner, N., Trillo, P. A., Bernal, X. E., Page, R. A., Goutte, S., Ryan, M. J., & Ellers, J. (2019). Adaptive changes in sexual signalling in response to urbanization. *Nature Ecology & Evolution, 3*, 374–380.

Heinen-Kay, J. L., Kay, A. D., & Zuk, M. (2021). How urbanization affects sexual communication. *Ecology and Evolution, 11*, 17625–17650.

Henneken, J., & Jones, T. M. (2017). Pheromones-based sexual selection in a rapidly changing world. *Current Opinion in Insect Science, 24*, 84–88.

Hutton, P., & McGraw, K. J. (2016). Urban impacts on oxidative balance and animal signals. *Frontiers in Ecology and Evolution, 4*, 54.

Idris, M., & Prakash, I. (1986). Scent marking in the Indian gerbil, *Tatera indica* in response to conspecific odours. *Proceedings of the Indian Academy of Sciences (Animal Sciences), 95*, 89–96.

Idris, M., & Prakash, I. (1987). Scent marking activity in the Indian gerbil, *Tatera indica* in relation to population density. *Animal Behaviour, 35*, 920–922.

Johnson, M. T. J., & Munshi-South, J. (2017). Evolution of life in urban environments. *Science, 358*, eaam8327.

Lee, V. E., & Thornton, A. (2021). Animal cognition in an urbanised world. *Frontiers in Ecology and Evolution, 9*, 633947.

Leveau, L. (2021). United colours of the city: A review about urbanisation impact on animal colours. *Austral Ecology, 46*, 670–679.

Masui, N., Shiojiri, K., Agathokleous, E., Tani, A., & Koike, T. (2023). Elevated O_3 threatens biological communications mediated by plant volatiles: A review focusing on the urban environment. *Critical Reviews in Environmental Science and Technology, 53*, 1982–2001.

Narango, D. L., & Rodewald, A. D. (2016). Urban-associated drivers of song variation along a rural–urban gradient. *Behavioral Ecology, 27*, 608–616.

O'Donnell, K., & delBarco-Trillo, J. (2020). Changes in the home range sizes of terrestrial vertebrates in response to urban disturbance: A meta-analysis. *Journal of Urban Ecology, 6*, juaa014.

Park, D., Hempleman, S. C., & Propper, C. R. (2001). Endosulfan exposure disrupts pheromonal systems in the red-spotted newt: A mechanism for subtle effects of environmental chemicals. *Environmental Health Perspectives, 109*, 669–673.

Partan, S. R., Fulmer, A. G., Gounard, M. A. M., & Redmond, J. E. (2010). Multimodal alarm behavior in urban and rural gray squirrels studied by means of observation and a mechanical robot. *Current Zoology, 56*, 313–326.

Perugini, M., Di Serafino, G., Giacomelli, A., Medrzycki, P., Sabatini, A. G., Oddo, L. P., Marinelli, E., & Amorena, M. (2009). Monitoring of polycyclic aromatic hydrocarbons in bees (*Apis mellifera*) and honey in urban areas and wildlife reserves. *Journal of Agricultural and Food Chemistry, 57*, 7440–7444.

Prakash, I., & Idris, M. (1982). Scent marking by the female of the Indian gerbil, *Tatera indica*, from 2 distinct desert habitats during estrus. *Indian Journal of Experimental Biology, 20*, 915–916.

Proppe, D. S., Sturdy, C. B., & St. Clair, C. C. (2011). Flexibility in animal signals facilitates adaptation to rapidly changing environments. *PLoS One, 6*, e25413.

Rosenthal, G. G., & Stuart-Fox, D. (2012). Environmental disturbance and animal communication. In U. Candolin & B. B. M. Wong (Eds.), *Behavioural responses to a changing world: Mechanisms and consequences* (pp. 16–31). Oxford University Press.

Rubene, D., Urhan, U., Ninkovic, V., & Brodin, A. (2022). Great tits learn odors and colors equally well, and show no predisposition for herbivore-induced plant volatiles. *Frontiers in Ecology and Evolution, 9*, 800057.

Rubene, D., Low, M., & Brodin, A. (2023). Birds differentially prioritize visual and olfactory foraging cues depending on habitat of origin and sex. *Royal Society Open Science, 10*, 221336.

Schmidt, R., Morrison, A., & Kunc, H. P. (2014). Sexy voices – No choices: Male song in noise fails to attract females. *Animal Behaviour, 94*, 55–59.

Secondi, J., Hinot, E., Djalout, Z., Sourice, S., & Jadas-Hécart, A. (2009). Realistic nitrate concentration alters the expression of sexual traits and olfactory male attractiveness in newts. *Functional Ecology, 23*, 800–808.

Sepp, T., McGraw, K. J., & Giraudeau, M. (2020). Urban sexual selection. In M. Szulkin, J. Munshi-South, & A. Charmantier (Eds.), *Urban evolutionary biology* (pp. 234–252). Oxford University Press.

Slabbekoorn, H., & Peet, M. (2003). Birds sing at higher pitch in urban noise. *Nature, 424*, 267.

Sobroza, T. V., Dunn, J. C., Gordo, M., & Barnett, A. A. (2024). Do pied tamarins increase scent-marking in response to urban noise? *Ethology Ecology & Evolution, 36*, 136–149.

Tsunoda, M., Kaneko, Y., Sako, T., Koizumi, R., Iwasaki, K., Mitsuhashi, I., Saito, M. U., Hisano, M., Newman, C., Macdonald, D. W., & Buesching, C. D. (2019). Human disturbance affects latrine-use patterns of raccoon dogs. *Journal of Wildlife Management, 83*, 728–736.

Van Geffen, K. G., Groot, A. T., Van Grunsven, R. H. A., Donners, M., Berendse, F., & Veenendaal, E. M. (2015a). Artificial night lighting disrupts sex pheromone in a noctuid moth. *Ecological Entomology, 40*, 401–408.

Van Geffen, K. G., van Eck, E., de Boer, R. A., van Grunsven, R. H. A., Salis, L., Berendse, F., Veenendaal, E. M., Stewart, A., & Sait, S. (2015b). Artificial light at night inhibits mating in a geometrid moth. *Insect Conservation and Diversity, 8*, 282–287.

Vom Saal, F. S., Nagel, S. C., Palanza, P., Boechler, M., Parmigiani, S., & Welshons, W. V. (1995). Estrogenic pesticides: Binding relative to estradiol in MCF-7 cells and effects of exposure during fetal life on subsequent territorial behaviour in male mice. *Toxicology Letters, 77*, 343–350.

Walsh, S., Goulet, C. T., Wong, B. B. M., & Chapple, D. G. (2018). Inherent behavioural traits enable a widespread lizard to cope with urban life. *Journal of Zoology, 306*, 189–196.

Whittaker, D. J., Soini, H. A., Atwell, J. W., Hollars, C., Novotny, M. V., & Ketterson, E. D. (2010). Songbird chemosignals: Volatile compounds in preen gland secretions vary among individuals, sexes, and populations. *Behavioral Ecology, 21*, 608–614.

Whittaker, D. J., Richmond, K. M., Miller, A. K., Kiley, R., Burns, C. B., Atwell, J. W., & Ketterson, E. D. (2011). Intraspecific preen oil odor preferences in dark-eyed juncos (*Junco hyemalis*). *Behavioral Ecology, 22*, 1256–1263.

Winandy, G. S. M., Japyassú, H. F., Izar, P., & Slabbekoorn, H. (2021). Noise-related song variation affects communication: Bananaquits adjust vocally to playback of elaborate or simple songs. *Frontiers in Ecology and Evolution, 8*, 570431.

Yeh, P. J. (2004). Rapid evolution of a sexually selected trait following population establishment in a novel habitat. *Evolution, 58*, 166–174.

Ziege, M., Bierbach, D., Bischoff, S., Brandt, A. L., Brix, M., Greshake, B., Merker, S., Wenninger, S., Wronski, T., & Plath, M. (2016). Importance of latrine communication in European rabbits shifts along a rural-to-urban gradient. *BMC Ecology, 16*, 29.

3-Mercapto-3-Methyl-1-Butanol, A Component of Leopard Urine, Repels Leopards and Spotted Hyaenas, and Protects Livestock from Leopard Attacks; A Series of Case Studies

Peter Apps[1,2] (✉), Peter Makumbe[3,4], and J. Weldon McNutt[1]

[1] Botswana Predator Conservation, Wild Entrust, Maun, Botswana
[2] Department of Zoology, Nelson Mandela University, Port Elizabeth, South Africa
Peterapps@wildentrust.org
[3] Oppenheimer Generations Research and Conservation, Shangani Ranch, Shangani, Zimbabwe
[4] Sustainability Research Unit, Nelson Mandela University, Port Elizabeth, South Africa

Abstract. Mammals' semiochemicals could provide tools to control pests, deter unwanted behaviors, enhance the survival of imperiled species, and improve the productivity and welfare of domestic animals, but because producing facsimile copies of whole odors is unrealistic in terms of technical challenge and cost, the real-world application of mammals' chemical signals depends on mammals responding to simple subsets of the components of their social odors. Experiments in southern Africa have shown that the odor of 3-mercapto-3-methyl-1-butanol (3M3MB), a component of leopard urine, can protect livestock from leopards, and repel spotted hyaenas. Releasing 3M3MB at a livestock enclosure on a cattle ranch in Botswana, reduced visits by leopards from seven to zero and visits by spotted hyaenas from seven to one over a period of 4 months. At another ranch, 3M3MB was deployed at increasing release rates at a gateway, and the number of leopard records decreased from three per month to zero as the release rate increased. When sheep and goats were fitted with collars emitting 3M3MB, losses to leopards declined from seven per month to zero for the 5 months that the collars were in operation. On a third Botswana ranch, leopards killed 0.6 calves per month when no 3M3MB was present and 0.2 calves per month when 3M3MB was released. The value of the livestock saved was approximately six times the cost of the repellent. At Shangani, in Zimbabwe, camera traps captured images of two leopards just outside shade-cloth kraals where 3M3MB was being released, but no livestock were attacked despite the shade-cloth fence providing no physical barrier. A leopard killed four calves in 5 days at a traditional thorn-branch kraal, and then when 3M3MB was deployed for a month, no more calves were killed. After the 3M3MB was removed, two calves were killed within 10 days. 3M3MB deterred spotted hyaenas from crossing a veterinary cordon fence in Botswana, and its efficacy rose with release rate. These results show that a single scent-mark component affects predators' behavior and protects livestock in real-life trials, and that artificial mammalian chemical signals have practical applications as minimally invasive, ecologically benign, economically sustainable wildlife management tools.

L. M. Schulte (ed.), *Chemical Signals in Vertebrates 16*,
https://doi.org/10.1007/978-3-032-30752-1_17

Keywords: Semiochemical · Camera trapping · Human-wildlife conflict · Pheromone · Carnivore · Predator · Repellent · Deterrent

1 Introduction

Just as with insects, mammals' behavior and physiology are influenced by chemical signaling between individuals. The potential application of these chemical signals to pest control, breeding, or conservation was recognized in the early 1980s (summary in Apps & McNutt, 2023) but has remained unfulfilled, in large part because the entrenched orthodoxy that mammalian chemical signals are coded by irreducibly complex mixtures (Albone, 1984; Pickett et al., 2014; Müller-Schwarze, 2016) has been a barrier both to the identification of signaling compounds (Apps, 2013) and to their practical application, because such complex mixtures would be prohibitively costly to produce and apply. To bring chemical signaling into practical use depends on mammal behavior or physiology being manipulated by single compounds or simple mixtures of compounds.

Insect chemical signals (pheromones) have been effective in the control of insect pests and had a market value of nearly $5 billion in 2024, which is projected to grow to nearly $18 billion by 2032 (Agricultural Pheromones Market, Report, 2025). In stark contrast, the potential of mammal chemical signals for reducing the damage caused by mammalian crop pests and livestock-predators, and protecting endangered wildlife remains unrealized, and commercial mammal pheromones are limited to artificial boar odor to test estrus in sows (*Sus scrofa domesticus*) and "appeasing pheromones" for pets and farm animals. There are currently no commercial applications of mammal chemical signals to reduce the billions of dollars in crop losses inflicted by rodents (which are also vectors and reservoirs of zoonotic diseases) (Hohenberger et al, 2022; Ro Diagne et al., 2023) or to protect the billions of dollars' worth of livestock killed by predators.

Here we report on experimental field tests of the effects of a single component of leopard (*Panthera pardus*) urine, 3-mercapto-3-methylbutanol (3M3MB) (Apps et al., 2014), on leopard predation of livestock and the movements of spotted hyaenas (*Crocuta crocuta*) on commercial livestock ranches in Botswana and Zimbabwe and subsistence cattle posts in northern Botswana. The results provide proof of the concept that a technically simple formulation of a single component of a mammal chemical signal has a practical, commercially viable application.

1.1 Human-Predator Conflict

There is a global decline in the numbers of wild predators and the viability of their populations, as their habitats are lost to expanding human populations with their livestock and crops (Fernandez-Sepulveda & Martín, 2022). Conflict with humans is the single biggest threat to predator populations worldwide, but at the same time, depredation on livestock has significant negative impacts on rural livelihoods, especially on subsistence pastoralists with small flocks of sheep (*Ovis aries*) and goats (*Capra hircus*). In South Africa, predators kill livestock worth ZAR1 billion each year (Kerley et al., 2017), and in the USA, losses of small stock to predators cost $US40 million and of cattle $US98.5 million per annum (USDA, 2010, 2011). In Shorobe, a traditional livestock village in NW Botswana, 67% of livestock owners reported annual losses of up to 14% of goats and 3% of cattle to predators (McNutt et al., 2017), and in the Kalahari in Botswana some cattle posts suffer from more than six depredations per month (Rutina et al., 2017).

As long as it is easier and cheaper for farmers to kill predators than to use nonlethal methods such as herding, predator-proof bomas, or early-warning systems, lethal predator control will continue. Globally, habitat loss and the lethal control that accompanies it have extirpated some species of large carnivores from more than 90% of their historical range (Wolf & Ripple, 2017), and small and mesocarnivores at regional scale (Marneweck et al., 2021). Indiscriminate lethal control methods such as gin-traps, snaring, and poisoning often kill nontarget species (McManus et al., 2015), and in Africa, nontarget poisoning of vultures is a particular concern (Nyirenda et al., 2024).

Official records of lethal control underestimate its true scale because "shoot, shovel, and shut up" is common. Lethal control is the first option for ≥21% of cattle posts in Shorobe (McNutt et al., 2017), and a large majority of the Shorobe farmers report that they would "remove" (in other words kill) a predator in response to loss of livestock. Livestock owners' aversion to predators is due to predators killing livestock, and where attacks on livestock are fewer, tolerance for predators is higher (Lindsey et al., 2013). Nonlethal methods of preventing predation and protecting rural livelihoods without adverse ecological effects (McManus et al., 2015) need to be economically sustainable and applicable to local circumstances, but hardly any existing nonlethal interventions meet these criteria. Early-warning systems based on satellite tracking of collared predators, intensified herding and husbandry (Hawkins et al., 2023), and compensation (Littlewood et al., 2020; Montag, 2003) all fall short due to costs (McManus et al., 2015; Weise et al., 2019, 2018), culturally embedded practices (McNutt et al., 2017; Mmopelwa & Mpolokeng, 2008), and lack of motivation (Mmopelwa & Mpolokeng, 2008). In Botswana, local-breed livestock guarding dogs (LGDs) (*Canis familiaris*) are very effective protection for goats and sheep (Van der Weyde et al., 2020; Horgan et al., 2021) and are the only nonlethal intervention with any spontaneous uptake in the subsistence sector. In South Africa, LGDs are more effective and more cost-effective than lethal control (McManus et al., 2015).

Predator conflict prevention and mitigation are stuck in a rut of ineffective compensation schemes, unsustainable subsidies to livestock husbandry, and high-tech interventions that need continuous external management and funding. The adverse impacts of lethal control and the shortcomings of existing nonlethal measures create the need for novel nonlethal interventions that have minimal adverse impacts, are within the financial and technical reach of subsistence livestock owners, and are easier and less expensive to apply than lethal control (Shivik, 2006). The widespread application of insect pheromones in pest management shows that, once their active ingredients have been identified and formulated, deterrents based on chemical signals are low-tech, low-cost, and quick and easy to apply. The results presented here are the first evidence that mammal chemical signals can be formulated and applied in the same way as those from insects.

2 Methods

2.1 Study Areas

Tests were conducted in northern Botswana and central Zimbabwe (Fig. 1):

- Traditional subsistence livestock (locally referred to as cattle posts) around the settlements of Gogomogo, Tsutsubeka, and Xhoo (GTX) to the east of Maun (−19.913961°23.223117°). This area suffers severe impacts from wildlife because it is separated from the NG/30 and NG/32 Wildlife Management Areas only by a veterinary cordon fence that is in serious disrepair.
- Two commercial cattle ranches in the northern Hainaveld (HR1-20.396460°23.709497°, HR2-20.275596°23.670785°) running cattle and a few sheep and goats, and a commercial cattle ranch in the Motopi area (MR) (−20.163434°24.037963°). All three ranches lie in the Kalahari Acacia-Baikiaea savanna and in addition to livestock have low densities of large and medium-sized wild herbivores, spotted and brown hyaenas (*Parahyaena brunnea*), leopards, African wild dogs (*Lycaon pictus*), and smaller predators.
- Shangani Holistic Ranch (−19.752018°29.399550°) (ZR) runs Nguni cattle with management aimed at restoration from serious bush encroachment caused by prolonged overgrazing. There are resident leopards and spotted hyaenas and smaller predators.

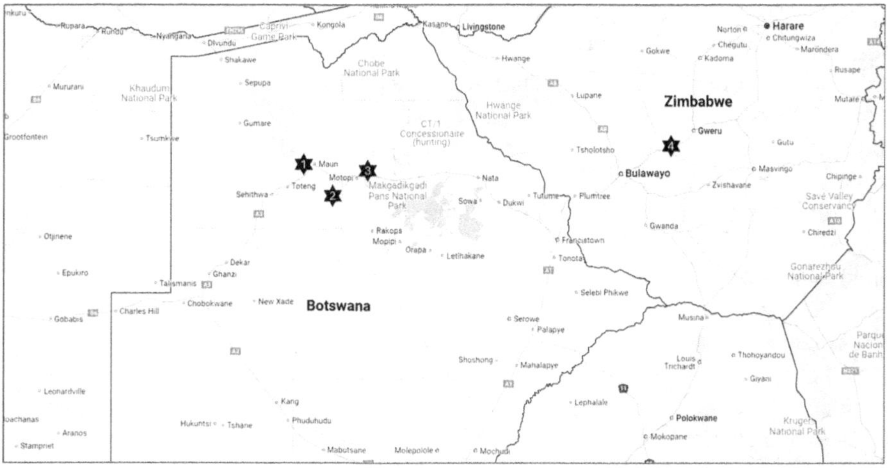

Fig. 1. Locations of study sites in Botswana and Zimbabwe. (1) Gogomoga, Tstsubega, and Xhoo subsistence cattle posts (GTX); (2) Hainaveld livestock ranches (HR1 and HR2); (3) Motopi cattle ranch (MR); (4) Shangani ranch (ZR).

On the cattle posts and ranches, normal operations continued, including lethal control, trapping and translocation of predators, improvements to fences and enclosures, and creation of new enclosures.

2.2 Scent Dispensing

3-Mercapto-3-methyl-1-butanol (3M3MB) was obtained from TCI (https://www.tcichemicals.com/JP/en), Carbosynth (https://www.biosynth.com/), and BLDPharm (https://www.bldpharm.com/). Controlled release was achieved by diffusion through silicone septa (Supelco SU860078, Supelco 27020) with the Teflon layer shaved off with a scalpel, on 4 ml (Supelco 27115-U) or 15 ml (Supelco 27088-U) open-capped vials (at HR1, MR, ZR), and polyethylene kitchen cling wrap (Glad Cling Wrap, Clorox Africa, Bryanston) on 15 ml open-capped vials (at MR, HR2, GTX) (Fig. 2). On 15 ml vials, silicone septa and cling-film both generated emission rates of approximately 20 ng/s at 40 °C, which is close to the maximum daytime temperature. The odor of 3M3MB was clearly detectable to the human nose 5–10 cm from the vial, but not detectable (to humans) 1 m away.

Increments in release rate were created by increasing membrane area (15 ml vs. 4 ml vials) and increasing the number of dispensers at a site.

To release 3M3MB at kraals, fences, and gates, the controlled-release vials were protected by metal housings made from 25 mm galvanized pipe connectors. The housings with vials inside were wired to fences about 30–50 cm off the ground (Fig. 2). For release from animal-borne collars, the vials were inserted into holders made from electrical trunking or CPVC water pipe, that were tied around the animals' necks with cotton rope (Fig. 3).

2.3 Camera Trapping

The presence of predators and their immediate reactions to 3M3MB were monitored with camera traps. In Botswana, we used Reconyx XR6s, Browning Strike Force Pro XDs (model BTC-5PXD), Browning Patriots (model BTC-Patriot-FHD), and Browning Spec Ops Elites (model BTC-BE-HP5) (TrailcamPro, https://www.trailcampro.com/) mounted 30–60 cm off the ground and set for maximum PIR sensitivity, maximum infrared illuminator power, and 30 s videos for Reconyx and 20 s videos for Brownings. At Shangani, still images were captured by Cuddeback X-Change™ Color (model 1279) camera traps attached to the outer fences of the kraal about 1 m above the ground, aimed away from the fence.

2.4 Livestock Losses

Livestock losses were reported by the owners.

2.5 Deployments

After a kraal owner in the GTX subsistence area reported having to chase a leopard away from a calf kraal on four consecutive nights, seven silicone-topped dispensers in housings were wired to the kraal fence, and the immediate surroundings of the kraal were monitored with four camera traps aimed outward from the corners of the kraal.

Fig. 2. (**a**) Controlled-release vials, back row 15 ml, front row 4 ml. Left to right with silicone septum, cling-film, cling-film and gauze fabric. (**b**) Housing and vial before assembly. (**c**) Housing with vial inside wired to a kraal fence.

Dispensers were attached to the GTX veterinary cordon fence where spotted hyaenas used gaps under the wire to cross between the wildlife area and the livestock area. Camera traps were mounted on the fence poles, aimed along the fence toward a gap, or on trees approximately 10 m from the fence, aimed at gaps. At the most frequently used gap, release rates were increased at approximately monthly intervals over a period of 6 months.

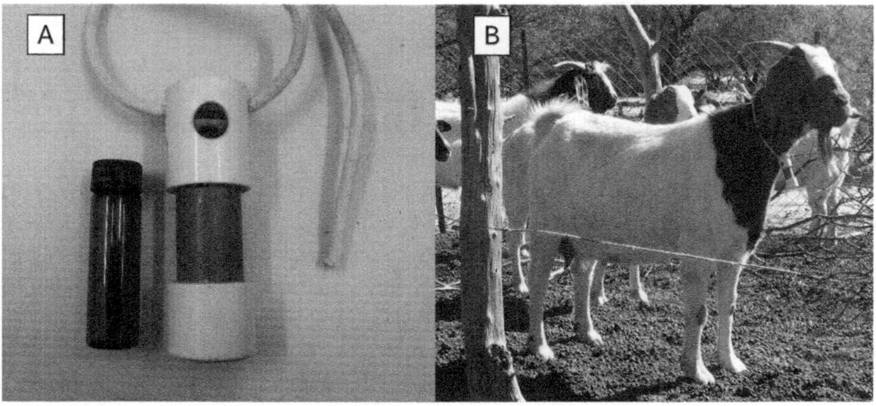

Fig. 3. (**a**) 15 ml dispenser vial and collar for sheep and goats. (**b**) A goat wearing a predator deterrent collar.

At the Hainaveld 1 ranch, predator activity was monitored for 4 months without 3M3MB and for a further 4.5 months with 3M3MB being released from dispensers positioned on both sides of two gateways.

At the Hainaveld 2 ranch, predators passing through a gateway were monitored for 1 month with two empty dispenser housings on each side of the gate, and then 3M3MB was deployed with incrementally increasing release rates over a period of 5 months. In June 2024, leopards killed seven sheep and goats out of a flock of about 30, and six of the survivors were fitted with collars emitting 3M3MB in July 2024 (Fig. 3).

At the Motopi cattle ranch, 3M3MB was released from dispensers on kraal fences, initially in response to a female leopard and her grown cub killing four calves in three nights in August 2019. Two days later, eight 3M3MB dispensers with 15 ml vials were hung approximately 15 m apart at the corners and midpoints of the kraal fence. The dispensers were moved around as the kraals were rebuilt and calves were moved, and left in place for varying periods determined by the availability of 3M3MB, vehicle breakdowns, and COVID movement restrictions.

At Shangani, in June 2022, 3M3MB was dispensed from eight points (corners and middles of sides) on the perimeters of two shade-cloth kraals with a history of leopard kills and recent sightings of leopards or their spoor. In October 2022, 15 3M3MB dispensers were deployed about 15 m apart around a thorn-branch kraal where a leopard was repeatedly breaching the wall and killing calves.

Except for the subsistence kraal, Motopi, and Shangani, where deployments were reactive, empty dispenser housings were in place during the control periods to reduce neophobic responses to their mere presence.

3 Results

At the subsistence cattle post, no videos of leopards were captured following deployment of 3M3MB, but a leopard was shot nearby 10 days later in an incident that was not connected with livestock predation.

3M3MB deterred spotted hyaenas from crossing the veterinary cordon fence between the GTX subsistence area and the neighboring wildlife areas. The strength of the deterrent effect increased with increasing release rate (Figs. 4 and 5).

On Hainaveld Ranch 1, camera trapping for 4 months without 3M3MB yielded seven records of leopards, seven records of spotted hyaenas, and one record of four African wild dogs, and a leopard killed a calf but was not recorded on video. Over a further 4.5 months with 3M3MB being released, there were no records of leopards and only one record each of spotted hyaenas and wild dogs, with no losses of calves.

At Hainaveld Ranch 2, during the 1-month control period, three leopards entered through the gate. As the 3M3MB release rate was increased, the number of leopards recorded per month decreased to zero and then increased again (Figs. 6 and 7). Reported losses of sheep and goats to leopards dropped from seven in the month before the goats and sheep were given deterrent collars to zero for the next 5 months.

At the calf kraals on Motopi cattle ranch, 3M3MB was intermittently released for approximately 18 months over a period of slightly longer than 3 years (Table 1).

From November 2020, the rate of camera-trapped leopard visits declined to one per month or less, with no livestock losses, irrespective of whether 3M3MB was being released, and so the 3M3MB was removed in August 2022 to reestablish the baseline. In September 2022, the rate of leopard detections at the kraal increased to three per month, and in October 2022 to seven per month, but with no attacks on livestock. In November 2022, the number of leopard records fell again to zero and remained there

Fig. 4. A frame from a video of spotted hyaenas deterred from crossing a veterinary cordon fence by the odor of 3M3MB. The full video is at https://www.youtube.com/watch?v=f8tkQkksLsA

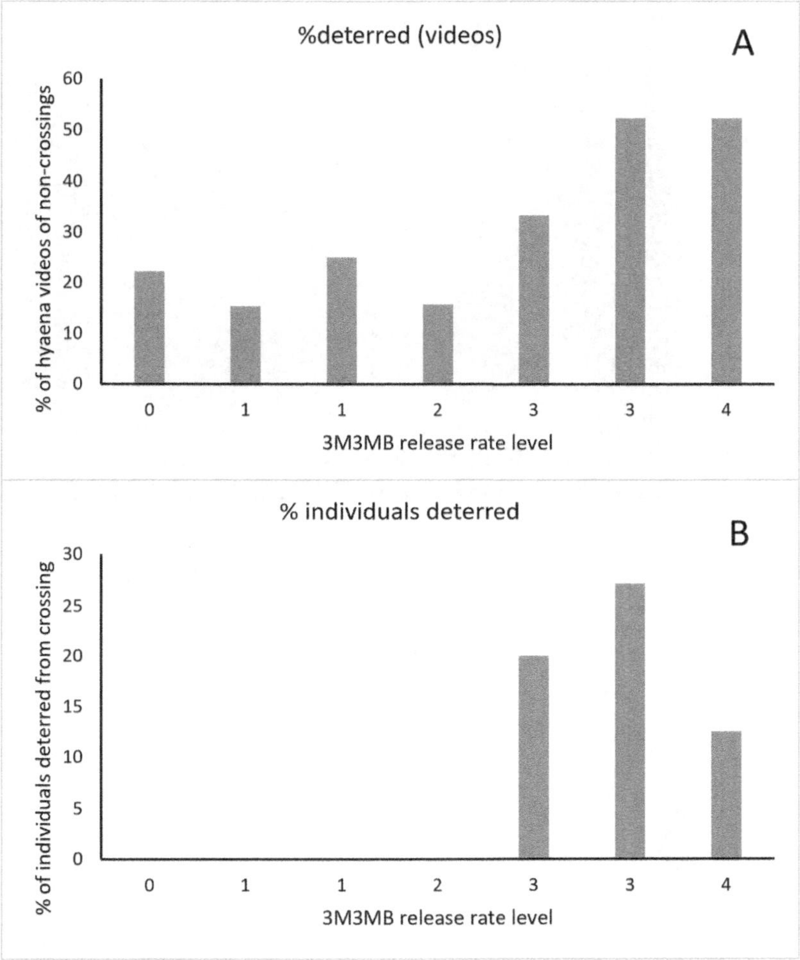

Fig. 5. (**a**) The percentage of videos of spotted hyaenas at a gap under a veterinary cordon fence that showed hyaenas not crossing the fence; (**b**) the percentage of individual hyaenas that did not cross the fence, with increasing release rates of 3M3MB.

until February 2023 with no 3M3MB being released. In March 2023, a leopard killed two calves, and there was a spike to 14 for the month in camera trap videos of leopards at the kraal (background levels were 0 or 1 per month). Two days after the second calf was killed, eight 3M3MB dispensers were deployed on the fences of the calf kraals, and six extra camera traps were added to the four already at the kraal. At the same time, the farmer reinforced the perimeters of the calf kraal with diamond mesh. Over the following 31 days, a leopard was captured on 237 videos of 54 incidents, but despite this high level of activity, no calves were killed. The leopard killed a baby donkey in a kraal with no repellent on April 21, 2023, and was trapped and relocated on April 24, 2023. The repellent and extra cameras were removed on May 16, 2023. On September 8, 2023, a leopard killed two more calves, but killing stopped although no 3M3MB was available.

Fig. 6. A frame from a video of a leopard sniffing the 3M3MB dispensers before turning back from the gateway. The full video is at https://www.youtube.com/watch?v=PpoTyVJVmDU

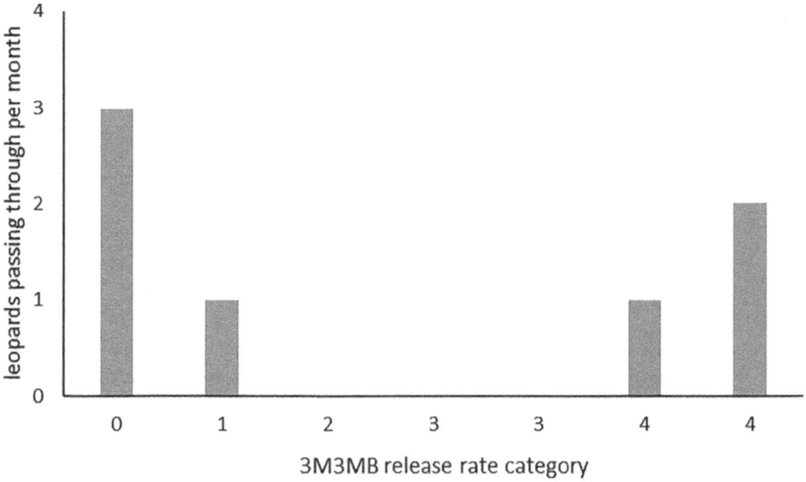

Fig. 7. Numbers of leopards passing through a livestock ranch gateway with incrementally increasing rates of 3M3MB release. Levels 0, 1, and 2 were deployed for 1 month each, followed by 2 months at level 3 and 2 months at level 4. 0 = no 3M3MB, 4 = the highest release rate.

Overall, leopards made 0.2 kills per month when 3M3MB was present and 0.6 kills per month when 3M3MB was absent. The value of the livestock saved was approximately six times the cost of the repellent.

When 3M3MB was present, spoor tracking by Motopi ranch staff showed leopards turning back from the kraal or skirting around it approximately 50 m away. Kraal staff

Table 1. Control and 3M3MB release periods in chronological order, with numbers of calves taken from a kraal by leopards.

	Duration	Calves killed
Pretrial	5 days	4
3M3MB	21 weeks	2
Control	42 weeks	6
3M3MB	12 weeks	1
Control	4 weeks	1
3M3MB	10 weeks	0
Control	7 weeks	0
3M3MB	10 weeks	0
Control	32 weeks	0
3M3MB	21 weeks	0

also reported that there were fewer hyaena tracks and less hyaena activity when 3M3MB was present.

At Shangani, in June 2022, camera traps captured images of two leopards just outside shade-cloth kraals where 3M3MB had been deployed, but no livestock were attacked despite the shade-cloth fence providing no physical barrier. In September–October 2022, a leopard killed four calves in 5 weeks at a thorn-branch kraal, and 11 days after the fourth calf was killed, 3M3MB and camera traps were deployed for the following month, during which no leopard images were captured or calves killed. After the 3M3MB was removed, two calves were killed within 10 days.

4 Discussion

Camera traps are the best available means of recording the presence of predators and their immediate reactions to deterrents, but the strongest repellent effects will result in animals not being recorded at all because they will not enter the cameras' detection zones, which extend a maximum of about 20 m in front of the camera over a field of view about 45° wide (Apps & McNutt, 2018a, b). Notwithstanding this shortcoming, camera trap videos of predators turning back after sniffing 3M3MB are the most convincing evidence of its effects; neither coincidence nor regression to the mean (see below) can explain animals doing U-turns when they encounter an odor (Figs. 4 and 6).

Odor deterrents in collars on goats or sheep provided protection from leopards for sheep and goats (this study) and from caracals for goats (Apps unpubl.), with losses stopping immediately and completely, even though on the Hainaveld 2 ranch at least three leopards continued to be captured by camera traps, and leopards killed uncollared calves within 100 m of the collared sheep and goats. Odor deterrents protected whole flocks, even though only about one-third of the animals had a collar. Eyespots or crosses painted onto cow's rumps protect the painted animals, but not other herd members (Radford et al., 2020). In Europe, pilot trials are being run on sheep collars that emit wolf "pheromones" (De Boer, 2023; IRSEA, n.d.; Samuel, 2023), but at the time of writing (March 2025), no results on their effectiveness have been published.

The repellent effect of 3M3MB on spotted hyaenas was unexpected, because 3M3MB has not been found in spotted hyaena odors (Hofer et al., 2001; Apps unpubl.), but it may well be an interspecific cue that alerts the hyaenas to the presence of a potentially dangerous competitor. Interspecific responses to odors are widespread and common, but seriously neglected (Apps et al., 2019).

Any of the quantitative results presented here could have been coincidence; a leopard might have moved away from livestock just before or just after 3M3MB was released for reasons unconnected with the 3M3MB, including control measures by livestock owners such as the use of diamond-mesh fence by the rancher at Motopi. Nonetheless, coincidence becomes less and less convincing the more instances there are, and at some point the evidence from case studies, with its limitations, becomes sufficient to justify investment in rigorously designed, large-scale trials.

Overall, livestock predation is rare, and when 3M3MB was deployed in reaction to predation, the subsequent declines in livestock losses could have been due to regression toward a mean of close to zero kills per day from exceptionally high rates of one or two per day, although the coincidences in timing between 3M3MB deployment and predation stopping suggest otherwise. To evaluate the relevance of regression to the mean would need comparable series of data on losses of livestock and camera trap detections of predators in the absence of deterrents, and these are not currently available.

At both of the sites where the release rate of 3M3MB was systematically increased, the strength of its effects increased with release rate, at one site with leopards and at the other with spotted hyaenas. Dose-response studies are common in work in insect pheromones, and there is some work on lures and baits for mammals (e.g., Jackson et al., 2018), but they are very scarce for mammal semiochemicals, even for laboratory and farm species (Dorries et al., 1995; Leinders-Zufall et al., 2000; He et al., 2010; Li et al., 2013), and these results are the first for a chemical signal from a wild mammal in the wild. In epidemiology, dose-response relationships are evidence of cause and effect (e.g., Shimonovich et al., 2021), but mammal semiochemistry, especially in the wild, is still at the level of John Snow's tracing of a cholera outbreak to a public water pump in London's Broad Street, and Percivall Pott's recognition that scrotal cancer in chimney sweeps was caused by soot (Bradford Hill, 1965).

The reason why the number of leopards passing through the HR2 gate increased at the highest rate of 3M3MB release is unknown. Habituation or another form of learning is plausible, because 3M3MB had been continuously released there for 4 months. There may also be individual differences in sensitivity to the deterrent, or in motivation, as was apparently the case with spotted hyaenas at the veterinary cordon fence (Fig. 5b), where some individuals were deterred from crossing and others were not. While there are a few mammalian chemical signals whose activity depends on concentration (reviewed in Apps, 2013), only two—a putative rat (*Rattus norvegicus*) estrus signal (Nielsen et al., 2011) and the rabbit (*Oryctolagus cuniculus*) mammary pheromone (Coureaud et al., 2004)—have been shown to have effects that increase and then decrease with increasing concentration. No peak in response has been demonstrated for the other species where different concentrations have been tested: galagos (*Galago crassicaudatus*) (Katsir & Crewe, 1980), tree shrews (*Tupaia belangeri*) (Von Stralendorff, 1982, 1987), and mice (*Mus musculus*) (Ninomiya & Kimura, 1990; Li et al., 2013). As far as we know, there is no relevant background information on individual differences in response to odors or on how long it might take a wild animal to habituate to intermitted exposure to an odor or learn that it is irrelevant.

In domestic cats (*Felis silvestris catus*), 3M3MB's metabolic precursor is produced as a spillover from the cholesterol pathway (Miyazaki, 2026, this volume), which may make 3M3MB an honest signal of a felid's body condition, and if it is, a higher release rate would signal the presence of a more dangerous competitor to both other leopards and to spotted hyaenas.

Laboratory animals respond to single components of social odors (reviewed in Apps, 2013), and wild African predators and red foxes (*Vulpes vulpes*) have been shown to scent-mark in response to simplified mixtures or single components of their social odors (Whitten et al., 1980; Apps et al., 2017; Apps & McNutt, 2023), but this is the first study that shows wild animals' movements and hunting being affected by a single signal component, and that it offers protection against livestock predation.

As the success of insect pheromones shows, the field application of deterrents based on chemical signals is low-tech, low-cost, and quick and easy. Formulation, packaging, marketing, and distribution are within the technical reach of small local businesses, so that wildlife conflict amelioration can be embedded into local economies. If predator deterrents reduce livestock losses to predation by an extremely conservative 10% (reduction in losses at the Motopi calf kraal was around 66%, and with collared sheep and goats 100%), then the savings in South Africa will be ZAR100 million in direct costs and ZAR450 million total (US$5.4 million and US$24.3 million, respectively) per annum. The costs of deterrents would be recovered in months rather than the years required for predator-proof kraals (Weise et al., 2018). Savings in the USA would be US$14 million per annum. Because the value of the livestock saved will be higher than the cost of the deterrents, deterrent-based conflict reduction will not need continual financial subsidies, unlike other nonlethal measures. Odor-based deterrents are the only nonlethal conflict intervention with any prospect of being cheaper and easier to apply than bullets, snares, and poisons and the only one likely to be spontaneously adopted by African farmers and integrated into local economies rather than depending permanently on inputs of foreign money and expertise. This demonstration that predators' behavior is affected by a single component of their odors, and that this compound protects livestock in real-life trials, opens the door to the practical application of artificial mammalian chemical signals as minimally invasive, ecologically benign, economically sustainable wildlife management tools.

Acknowledgments. Funding and donations in kind for the work reported here were provided by the Leopardess Foundation, WWF Netherlands InnoFonds via Stichting Spots, the WildIze Foundation, Oppenheimer-Generations Research and Conservation, the Eppley Foundation for Research, Spectral Works, and Ceva Wildlife Research Fund. Dr. Marius Viljoen and Rra Ditso Ntloyathuto allowed us to work on their land, and the Village Development committees and people of Tsutsubega, Gogomoga, and Xhoo provided information on predator activity and access to their cattle posts. Rra Johane Masene provided technical support in the field in Botswana, and the Shangani Ranch Research Team assisted in setting up and monitoring the project in Zimbabwe. The work was carried out under research permit EWT 8/36/4 XXXVIII (15) granted by the Ministry of Environment, Wildlife and Tourism and the Office of the President of Botswana.

References

Agricultural Pheromones Market, Report. (2025). https://www.fortunebusinessinsights.com/industry-reports/agricultural-pheromones-market-100071. Accessed 6 Mar 2025.

Apps, P. J. (2013). Are mammal olfactory signals hiding right under our noses? *Naturwissenschaften, 100*, 487–506.

Apps, P. J., & McNutt, J. W. (2018a). How camera traps work and how to work them. *African Journal of Ecology, 56*, 702–709.

Apps, P. J., & McNutt, J. W. (2018b). Are camera traps fit for purpose? A rigorous, reproducible and realistic test of camera trap performance. *African Journal of Ecology, 56*, 710–720.

Apps, P. J., & McNutt, J. W. (2023). Reactions of free-ranging African carnivores to single components of carnivore odours: Evidence that mammalian semiochemicals are not all irreducibly complex. In B. Schaal, D. Rekow, M. Keller, & F. Damon (Eds.), *Chemical signals in vertebrates* (Vol. 15, pp. 419–434). Springer.

Apps, P. J., Mmualefe, L., Jordan, N. R., Golabek, K. A., & McNutt, J. W. (2014). The "tomcat compound" 3-mercapto-3-methylbutanol occurs in the urine of free-ranging leopards but not in African lions or cheetahs. *Biochemical Systematics and Ecology, 53*, 17–19.

Apps, P. J., Claase, M., Yexley, B., & McNutt, J. W. (2017). Interspecific responses of wild African carnivores to odour of 3-mercapto-3-methylbutanol, a component of wildcat and leopard urine. *Journal of Ethology, 35*, 153–159.

Apps, P., Rafiq, K., & McNutt, J. W. (2019). Do carnivores have a world wide web of interspecific scent signals? In C. Buesching (Ed.), *Chemical signals in vertebrates 14* (pp. 182–202). Springer.

Bradford Hill, A. (1965). The environment and disease: association or causation? *Proceedings of the Royal Society of Medicine, 108*(1), 295–300.

Coureaud, G., Langlois, D., Sicard, G., & Schaal, B. (2004). Newborn rabbit responsiveness to the mammary pheromone is concentration-dependent. *Chemical Senses, 29*(4), 341–350.

De Boer, J. (2023). Swiss sheep given scented collars to help stop wolf attacks. https://www.iamexpat.ch/lifestyle/lifestyle-news/swiss-sheep-given-scented-collars-help-stop-wolf-attacks (viewed 6 March 2025).

Dorries, K. M., Adkins-Regan, E., & Halpern, B. P. (1995). Olfactory sensitivity to the pheromone, androstenone, is sexually dimorphic in the pig. *Physiology & Behavior, 57*(2), 255–259.

Fernandez-Sepulveda, J., & Martín, C. A. (2022). Conservation status of the world's carnivorous mammals (order Carnivora). *Mammalian Biology, 102*(5), 1911–1925.

Hawkins, H. J., Minnie, L., van Niekerk, H. N., de Waal, H. O., Balfour, D., Kerley, G. I., & Twine, M. (2023). Shepherding is not a shot in the dark: Evidence of low predation losses from the northern cape province of South Africa. *African Journal of Range & Forage Science, 40*(4), 373–384.

He, J., et al. (2010). Distinct signals conveyed by pheromone concentrations to the mouse vomeronasal organ. *Journal of Neuroscience, 30*, 7473–7483.

Hofer, H., East, M., Sämmang, I., & Dehnhard, M. (2001). Analysis of volatile compounds in scent-marks of spotted hyenas (*Crocuta crocuta*) and their possible function in olfactory communication. In A. Marchlewska-Koj, J. J. Lepri, & D. Müller-Schwarze (Eds.), *Chemical signals in vertebrates 9* (pp. 141–148). Springer.

Hohenberger, J., Friesen, A., Wieck, S., & Kümmerer, K. (2022). In search of the holy grail of rodent control: Step-by-step implementation of safe and sustainable-by-design principles on the example of rodenticides. *Sustainable Chemistry and Pharmacy, 25*, 100602.

Horgan, J. E., Van Der Weyde, L. K., Comley, J., Klein, R., & Parker, D. M. (2021). Every dog has its day: Indigenous Tswana dogs are more practical livestock guardians in an arid African savanna compared with their expatriate cousins. *Journal of Vertebrate Biology, 69*(3), 20104.

IRSEA. (n.d.). IRSEA's ambition: To make wolves and sheep live in harmony. https://uk.irsea-institute.com/wolf-project/ (viewed 6 March 2025).

Jackson, M. D., Keyzers, R. A., & Linklater, W. L. (2018). Single compounds elicit complex behavioural responses in wild, free-ranging rats. *Scientific Reports, 8*, 12588.

Katsir, Z. & Crewe, R.M. (1980) Chemical communication in *Galago crassicaudatus*: investigation of the chest gland secretion. *African Zoology, 15,* 249–254.

Kerley, G. I. H., Behrens, K. G., Carruthers, J., Diemont, M., Du Plessis, J., Minnie, L., Richardson, P. R. K., Somers, M. J., Tambling, C. J., Turpie, J., Van Niekerk, H. N., & Balfour, D. (2017). Livestock predation in South Africa: The need for and value of a scientific assessment. *South African Journal of Science, 113*(3/4), 1–3.

Leinders-Zufall, T., Lane, A. P., Puche, A. C., Ma, W., Novotny, M. V., Shipley, M. T., & Zufall, F. (2000). Ultrasensitive pheromone detection by mammalian vomeronasal neurons. *Nature, 405*(6788), 792–796.

Li, Q., Korzan, W. J., Ferrero, D. M., Chang, R. B., Roy, D. S., Buchi, M., Lemon, J. K., Angeldeep, W., Kaur, A. W., Stowers, L., Fendt, M., & Liberles, S. (2013). Synchronous evolution of an odor biosynthesis pathway and behavioral response. *Current Biology, 23*, 11–20.

Lindsey, P. A., Havemann, C. P., Lines, R., Palazy, L., Price, A. E., Retief, T. A., Rhebergen, T., & Van der Waal, C. (2013). Determinants of persistence and tolerance of carnivores on Namibian ranches: Implications for conservation on Southern African private lands. *PLoS One, 8*(1), e52458.

Littlewood, N. A., Rocha, R., Smith, R. K., Martin, P. A., Lockhart, S. L., Schoonover, R. F., Wilman, E., Bladon, A. J., Sainsbury, K. A., Pimm, S., & Sutherland, W. J. (2020). *Terrestrial mammal conservation: Global evidence for the effects of interventions for terrestrial mammals excluding bats and primates. Synopses of Conservation Evidence Series*. University of Cambridge, Open Book Publishers.

Marneweck, C., Butler, A. R., Gigliotti, L. C., Harris, S. N., Jensen, A. J., Muthersbaugh, M., Newman, E. A., Saldo, K., Shute, K. L., Titus, S. W., Yu, D. S., & Jachowski, D. S. (2021). Shining the spotlight on small mammalian carnivores: Global status and threats. *Biological Conservation, 255*, 109005.

McManus, J. S., Dickman, A. J., Gaynor, D., Smuts, B. H., & Macdonald, D. W. (2015). Dead or alive? Comparing costs and benefits of lethal and non-lethal human–wildlife conflict mitigation on livestock farms. *Oryx, 49*(4), 687–695.

McNutt, J. W., Stein, A. B., McNutt, L. B., & Jordan, N. R. (2017). Living on the edge: Characteristics of human–wildlife conflict in a traditional livestock community in Botswana. *Wildlife Research, 44*(7), 546–557.

Miyazaki, M. (2026). Chemical communication using the urinary compounds in the domestic cat. In L. M. Schulte (Ed.), *Chemical Signals in Vertebrates 16* (pp. 145–158). Springer.

Mmopelwa, G., & Mpolokeng, T. (2008). Attitudes and perceptions of livestock farmers on the adequacy of government compensation scheme: Human-carnivore conflict in Ngamiland. *Botswana Notes and Records, 2008*, 147–158.

Montag, J. (2003). Compensation and predator conservation: Limitations of compensation. *Carnivore Damage Prevention News, 6*, 2–6.

Müller-Schwarze, D. (2016). Where we stand and what might be next. In B. A. Schulte, T. E. Goodwin, & M. H. Ferkin (Eds.), *Chemical signals in vertebrates 13* (pp. 11–16). Springer.

Nielsen, B. L., Jerôme, N., Saint-Albin, A., Thonat, C., Briant, C., Boué, F., Rampin, O., & Maurin, Y. (2011). A mixture of odorant molecules potentially indicating oestrus in mammals elicits penile erections in male rats. *Behavioural Brain Research, 225*, 584–589.

Ninomiya, K., & Kimura, T. (1990). Female-attracting property of urinary compounds in male mice. Stereospecificity and interaction with other factors. *Naturwissenschaften, 77*, 586–588.

Nyirenda, V. R., Nkhata, B. A., Phiri, D., Nyirenda, W., Phiri, D. N., Malasa, M., Smith Becker, M., Tembo, E. M., Nyirenda, M. A., Simpamba, T., Mwitwa, J., & Chomba, C. (2024). Vulture poisoning in sub-Saharan Africa and its implications for conservation planning: A systematic review. *Heliyon, 10*(3), e25126.

Pickett, J. A., Barasa, S., & Birkett, M. A. (2014). Vertebrate pheromones and other semiochemicals: The potential for accommodating complexity in signaling by volatile compounds for vertebrate management. *Biochemical Society Transactions, 42*(4), 846–850.

Radford, C., McNutt, J. W., Rogers, T., Maslen, B., & Jordan, N. (2020). Artificial eyespots on cattle reduce predation by large carnivores. *Communications Biology, 3*(1), 430.

Ro Diagne, C., Ballesteros-Mejia, L., Cuthbert, R. N., Bodey, T. W., Fante-Lepczyk, J., Angulo, E., Bang, A., Daubigny, G., & Courchamp, F. (2023). Economic costs of invasive rodents worldwide: The tip of the iceberg. *PeerJ, 11*, e14935.

Rutina, L. P., Mower, K. M., Seonyatseng, E., Mpofu, C., & Ntloyathuto, D. (2017). Herders' ecological knowledge and carnivore predation on livestock investigations in Makgadikgadi and Nxai national parks, Botswana. *Koedoe, 59*(2), a1389.

Samuel, H. (2023). Sheep in wolves' clothing scare off predators. https://www.telegraph.co.uk/world-news/2023/12/01/sheep-collar-pheromones-scares-wolf-predators-switzerland/

Shimonovich, M., Pearce, A., Thomson, H., Keyes, K., & Katikireddi, S. V. (2021). Assessing causality in epidemiology: Revisiting Bradford Hill to incorporate developments in causal thinking. *European Journal of Epidemiology, 36*, 873–887.

Shivik, J. A. (2006). Tools for the edge: what's new for conserving carnivores. *Bioscience, 56*(3), 253–259.

USDA. (2010). Sheep and goats death loss. National Agricultural Statistics Service (NASS), Agricultural Statistics Board, U.S. Department of Agriculture. https://downloads.usda.library.cornell.edu/usda-esmis/files/hh63sv88v/mc87pt05z/jw827f62b/sgdl-05-27-2010.pdf

USDA. (2011). Cattle death loss. National Agricultural Statistics Service (NASS), Agricultural Statistics Board, U.S. Department of Agriculture. https://downloads.usda.library.cornell.edu/usda-esmis/files/vh53wv75j/xp68kk00g/v405sd14m/CattDeath-05-12-2011.pdf

Van Der Weyde, L. K., Kokole, M., Modise, C., Mbinda, B., Seele, P., & Klein, R. (2020). Reducing livestock-carnivore conflict on rural farms using local livestock guarding dogs. *Journal of Vertebrate Biology, 69*(3), 20090–20091.

Von Stralendorff, F. (1982). A behaviorally relevant components of the scent signals of male *Tupaia belangeri*: 2,5-dimethylpyrazine. *Behavioral Ecology and Sociobiology, 11*, 101–107.

Von Stralendorff, F. (1987). Partial chemical characterization of the urinary signaling pheromone in tree shrews (Tupaia belangeri). *Journal of Chemical Ecology, 13*, 655–679.

Weise, F. J., Hayward, M. W., Aguirre, R. C., Tomeletso, M., Gadimang, P., Somers, M. J., & Stein, A. B. (2018). Size, shape and maintenance matter: A critical appraisal of a global carnivore conflict mitigation strategy – Livestock protection kraals in northern Botswana. *Biological Conservation, 225*, 88–97.

Weise, F. J., Hauptmeier, H., Stratford, K. J., Hayward, M. W., Aal, K., Heuer, M., Tomeletso, M., Volker, W., Somers, M. J., & Stein, A. B. (2019). Lions at the gates: Trans-disciplinary design of an early warning system to improve human-lion coexistence. *Frontiers in Ecology and Evolution, 6*, 242.

Whitten, W. K., Wilson, M. C., Wilson, S. R., Jorgenson, J. W., Novotny, M., & Carmack, M. (1980). Induction of marking behavior in wild red foxes (*Vulpes vulpes* L.) by synthetic urinary constituents. *Journal of Chemical Ecology, 6*, 49–55.

Wolf, C., & Ripple, W. J. (2017). Range contractions of the world's large carnivores. *Royal Society Open Science, 4*(7), 170052.

Using Semiochemical Analogs to Help Pigs Cope with Farming Practices

Manon Chasles[1(✉)], Miriam Marcet-Rius[2], Alessandro Cozzi[3], and Patrick Pageat[3]

[1] Ethology, Physiology and Chemical Communication Department, IRSEA, Institute of Research in Semiochemistry and Applied Ethology, Apt, France
m.chasles@irsea-institute.com

[2] Ethics, Legislation and Animal Welfare Department, IRSEA, Institute of Research in Semiochemistry and Applied Ethology, Apt, France

[3] Research and Education Board, IRSEA, Institute of Research in Semiochemistry and Applied Ethology, Apt, France

Abstract. Pork is one of the most widely consumed meat across the world, and most pigs are raised under intensive methods, which can negatively impact animal welfare. Efforts have been made to improve the lives of these animals, but husbandry practices still present challenges that they must adapt to. To minimize the physiological and behavioral consequences, maternal appeasing pheromones or their analogs can be used to promote coping mechanisms, helping animals adapt more quickly and effectively to new situations. This review aimed to centralize information on the use of Pig Appeasing Pheromone (PAP) in both farm and experimental conditions. Overall, the compiled results validate the efficacy of PAP in helping pigs adopt appropriate responses when facing challenges. A decrease in aggressiveness has been observed during mixing or feeding, and the inhibitory action of PAP on cortisol secretion has been noted in various studies and contexts. This opens up the possibility of using appeasing pheromones in pig farming to mitigate the negative consequences of intensive husbandry practices on animal welfare.

Keywords: Pigs · Appeasing pheromone · Animal welfare · Adaptation · Farming practices

1 Farming Practices and Pig Welfare

Pork ranks as the world's second most consumed meat, with the European Union, South Korea, and China being the biggest consumers, with more than 29 kg per person per year (OECD 2025). Half of this pork meat is produced in China, with 57.94 million tons produced in 2023/2024 (USDA 2025). Pigs can be raised using a wide range of farming methods, from free-range organic systems to intensive production. Most of the pigs raised in high-income countries are on intensive farms, making it the dominant system (Maes et al., 2019). Intensive livestock farming is characterized, among other things, by higher stocking density, minimal space per animal, indoor housing, and standardized

L. M. Schulte (ed.), *Chemical Signals in Vertebrates 16*,
https://doi.org/10.1007/978-3-032-30752-1_18

procedures such as precocious weaning or mixing with unfamiliar pigs to maximize profitability. It often involves labor specialization, with some farms dedicated to breeding and piglet production, while others focus on fattening weaners. Intensive pig farms typically operate in batches, using an all-in-all-out system that ensures proper sanitary downtime between groups. While this approach optimizes profitability and labor organization, some husbandry practices can be challenging for pigs, negatively impacting their health and welfare (Maes et al., 2019).

Natural behaviors of pigs include rooting and exploration, behaviors that intensive housing conditions do not allow them to express as they have insufficient space and are usually on slatted floors leading to frustration and stress in pigs. Indeed, pigs provided with straw and rooting material exhibit more exploratory behaviors and are less likely to bite other pigs' tails (Guy et al., 2002; Rosvold et al., 2024). If natural weaning in pigs occurs around 4 months of age, under intensive farming conditions, weaning occurs abruptly at around 1 month of age causing a significant amount of stress in piglets (Campbell et al., 2013; Jensen & Recén, 1989). After weaning, they will be mixed with unfamiliar piglets—allotment, which consists in establishing new groups of pigs according to age and/or weight, is performed in young as well as in adults to keep homogeneous groups of pigs. Hence, mixing (or allotment) is done routinely in pig farms, even if it is known to negatively impact animal growth and performances. This impact on growth is mainly caused by the increased number of fights between pigs at mixing (Tan et al., 1991; Erhard et al., 1997; D'Eath, 2002; Stukenborg et al., 2011). Indeed, when unfamiliar animals encounter each other, social conflict typically occurs, often manifesting as aggression. Pigs are a hierarchical species, and agonistic interactions are essential for establishing a new dominance hierarchy between newly introduced pigs (Meese & Ewbank, 1972, 1973). Hence, if having animals in social groups can be seen as beneficial for their welfare, deleterious housing conditions, such as regular mixing with unfamiliar pigs and a high density of animals, can compromise the benefit of being housed in social groups, especially when conflicts escalate to aggression (Aureli et al., 2002). Moreover, even in a socially stable group of pigs, the introduction of a resource conflict can lead to agonistic interactions that can degenerate into aggression. Feeding method is a good example, with group-fed pigs being involved in more agonistic interactions than pigs fed individually in stalls (Barnett et al., 1992). While not exhaustive, these examples illustrate farming practices that pose challenges to animal welfare and, as a result, impact their zootechnical performance.

Efforts have been made to improve housing conditions to better meet animals' needs. For example, since 2013, members of the European Union have been required to house pregnant sows in groups from the fourth week of gestation until the week of farrowing, as stated in Directive 2001/88/EC. In EU, Directive 2008/120/EC also states that tail docking in pigs cannot be a routine procedure and should only be performed if other solutions have first been considered. Nevertheless, most EU pigs still have their tails docked as housing conditions promote aggressive behaviors and thus tail biting (D'Eath et al., 2016). Intensive farming practices have been designed to improve herd performance—they are not totally effective, but they are sufficiently profitable and efficient to make their eradication difficult. In response to these practices, some individuals may exhibit undesirable behaviors, such as stereotypies or aggression, which can significantly impact their fitness compared to more behaviorally flexible animals that adapt more effectively. This ability to adapt to stressors by adopting a behavioral response enabling a decrease of its physiological consequences is called coping (Schouten & Wiepkema, 1991; Wechsler, 1995). Hence, one strategy to help animals to limit the stress-related consequences of husbandry practices on their health, behavior, and welfare is to promote coping.

2 Appeasing Pheromones: A Tool to Improve Animals' Coping Abilities

Pheromones vary widely in chemical structure, from small, volatile molecules to steroid derivatives or even large protein-ligand complexes. They can be found in various secretions such as saliva, urine, or tears (Liberles, 2014). Factors like age, sex, sexual activity, and endocrine status can influence the type of pheromonal secretion. Pheromones play an important role across different contexts, including reproduction, territoriality, or maternal behavior. Indeed, chemical communication is important between mother and young, playing a role in identification and mutual recognition promoting mother-offspring bonding (Broad et al., 2006; Keller & Lévy, 2012). Among the different semiochemicals involved in the mother-young relationship, a family of pheromones secreted by the mothers to attract young, stimulate suckling, and create a secure environment have been discovered and called "appeasing pheromones" (Pageat 2002; Brennan, 2010). Although those pheromones are naturally secreted to appease young, researchers found that they can have a similar effect in adults (Pageat & Gaultier, 2003; Mills et al., 2006; Vitale, 2018). In mammals, appeasing pheromones are secreted by mammary glands and are composed of different fatty acids, such as palmitic acid, oleic acid, and linoleic acid. The composition and proportions of each fatty acid depend on the species (Pageat, 2002). Even if most of the appeasing pheromones characterized have been found in mammals, it is important to notice that mother hens also produce a semiochemical that is secreted by the uropygial gland and that have the same action as appeasing pheromones (Madec, 2008).

Behavioral and physiological effects of appeasing pheromones have been characterized in pets and farm animals but also wild animals. In pets, the pheromonatherapy is now widely used to facilitate or prevent the display of certain species-specific behaviors (Pageat & Gaultier, 2003; Mills, 2005). The Dog and Cat Appeasing Pheromones, DAP and CAP, respectively, are known to help animals to cope with stressful situations—indeed, they are described as "learning facilitators." It is thought that by reducing anxiety, they alleviate its inhibitory effect on animals' learning abilities (Mills, 2005; Denenberg & Landsberg, 2008; DePorter et al., 2019). This effect on cognitive performances has also been observed in rabbits (Coureaud et al., 2010) as well as in horses. Indeed, horses treated with the Equine Appeasing Pheromone (EAP) had better scores in a learning task than the horses that received the placebo (Mengoli et al., 2014). Learning is a strategy that enables individuals to adapt to environmental changes by modifying their behavior accordingly. Therefore, by promoting learning, appeasing pheromones also facilitate the coping process. In dogs, the use of DAP products reduces behavioral signs of stress in both adults and young exposed to a potentially stressful situation (Sheppard & Mills, 2003; Tod et al., 2005, Mills et al., 2006; Taylor & Mills, 2007; Landsberg et al., 2015;

Puglisi et al., 2022). In cats, providing a CAP treatment decreases tensions and conflicts between individuals and thus facilitates cohabitation (Cozzi et al., 2010; DePorter et al., 2019). It is really interesting to see how conservative the effects of appeasing pheromones are among different species. Indeed, the effects observed in two carnivores, dogs and cats, are similar to what has been observed in rabbits. Indeed, the use of the Rabbit Appeasing Pheromone (RAP) helps rabbits to better cope when subjected to an unknown environment (Codecasa et al., 2022) but also when submitted to a potentially stressful situation, a veterinary examination (Asproni et al., 2024). All these effects observed in domestic animals are desirable in farm species. As mentioned above, farm animals are exposed to various situations and procedures that challenge their adaptive capacities. Using pheromones to promote learning and facilitate coping processes should be highly appealing to farmers, as it helps reduce the negative effects of stress, such as decreased productivity, weakened immune function, and behavioral issues (Dinand Ekkel et al., 1996; Maes et al., 2019). The potential of using pheromonatherapy in farms has been reviewed recently. In this review, Kekan et al. (2017) listed the different sources and types of pheromones, as well as how those semiochemicals could be used in farms to improve management, productivity, and reproductive parameters in pigs, ruminants, and horses. This article emphasizes the versatility of pheromone applications in farming systems.

3 Modalities of Application of Pheromonal Products

When it comes to pheromone treatment, there are two main options: treating the environment or treating the animal itself (Fig. 1). Choosing between these two modalities of application depends on several parameters.

Objectives of the treatment determine the application method. If the goal is to help animals cope with a specific situation (e.g., transportation, physical examination), a short-term application directly on the animal is preferred. Conversely, if the aim is to help animals adapt to new housing or manage recurrent undesirable behaviors, long-term diffusion in the animal's environment is recommended. In this case, a limitation of environmental diffusion is the surface area to be treated. Indeed, the efficacy of both active and passive diffusers is known to be restricted by the size of the treated area. Therefore, under extensive conditions, where animals are housed in larger pens, direct application of the pheromone to the animal can be recommended to ensure continuous exposure within its effective range. Indeed, the environmental conditions in which the animal lives are also to be taken into account to choose the best treatment. To spread the pheromone message in the environment, several strategies can be used, such as spraying the product at specific locations, for example, to attract or deter animals from a particular area (Menuge et al., 2023). Another strategy is to use active or passive diffusers to ensure the effective distribution of the pheromonal message throughout the animal's environment. While the use of active plug-in diffusers is easy to set up and provides good results for pets (Sheppard & Mills, 2003; DePorter et al., 2019; Pereira et al., 2023), their use is more complicated on farms. Therefore, a passive diffusion device has been developed, allowing the pheromonal message to be distributed without the need for access to a power outlet. This solution is particularly well-suited for indoor intensive farming, as it allows the treatment of a large number

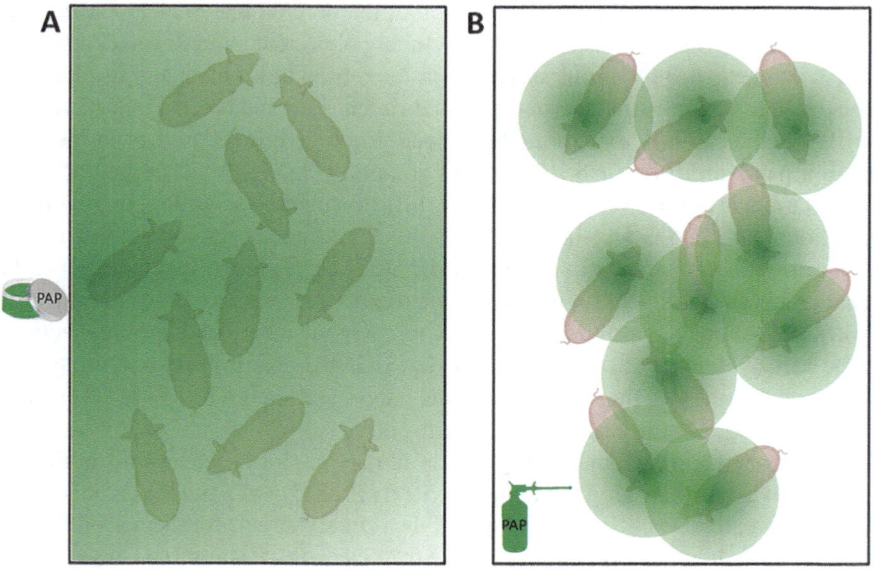

Fig. 1. Schematic representation of two diffusion modalities available for PAP products. (**a**) Passive diffusion through a solid block that allows the pheromonal message to spread in the pigs' environment. (**b**) A cutaneous form in which each treated pig becomes a vector of the pheromonal information for the other pigs in the same group.

of animals simultaneously (Fig. 1). This formulation has shown significant improvements in both welfare and production parameters in farmed rabbits and chickens (Madec et al., 2006, 2008; Bouvier & Jacquinet, 2008; Asproni et al., 2020). Because they continuously release active compounds, passive diffusion devices provide long-term effects without the need for reapplication. However, they require a 24-h delay after installation to ensure optimal diffusion throughout the space.

An alternative solution that has been developed for both pets and farm animals is direct application on the animal. This enables the animal to carry the pheromonal message wherever it goes, ensuring continuous exposure to the treatment for maximal efficacy. Moreover, by carrying the message for itself, the animal also transmits it to its conspecifics, acting simultaneously as both the target and the vector of the pheromonal message (Fig. 1). In both dairy and beef cattle, a cutaneous application of the Bovine Appeasing Pheromone (BAP) leads to improved welfare and production parameters (Osella et al., 2018; Colombo et al., 2020; Garcia-Alvarez et al., 2025).

4 The Pig Appeasing Pheromone (PAP) and Its Applications in Farms

The PAP was first isolated by comparing the secretions from sebaceous glands around the intermammary groove of pregnant and lactating sows from a mixed breed of Chinese and European strains (Pageat, 1998). A gas chromatography-mass spectrometry

(GC/MS) analysis revealed the composition of the porcine maternal appeasing pheromone, which is a mixture of fatty acids, including oleic acid, palmitic acid, and linoleic acid. A synthetic analog of this secretion has then been developed, and its efficacy has been tested in various situations and pigs of different ages and breeds. Indeed, regarding the PAP products, different forms have been developed to be used in farms, among which a passive diffuser and a cutaneous solution (Fig. 1). As described above, this allows farmers to choose the most suitable diffusion method according to their housing conditions and, above all, their expectations regarding the treatment (long-term or short-term effect). For example, a block diffuser of the product SecurePig® (SIGNS Labs, France) is sufficient to diffuse the pheromonal message in 25 square meters for 6 weeks and should be installed 24 h in advance. On the other hand, the cutaneous form of the same reference, SecurePig® FLASH (SIGNS Labs, France), when applied as 5 mL on the withers of pigs, lasts for around 5 days and should be applied 4–6 h before the challenging event (transportation, mixing, etc.).

As described in part 1, the mixing done routinely in pigs is deleterious for their welfare as well as for their growth (Tan et al., 1991; Erhard et al., 1997; Stukenborg et al., 2011). Hence, this practice has been one of the primary targets for the use of PAP products. Most studies on mixing have been conducted on weaners (i.e., piglets around 3–4 weeks of age that have just been weaned from their mother). Indeed, right after weaning, piglets are introduced to non-related conspecifics for the first time. Weaners from commercial breeds that received PAP exhibited less agonistic behaviors up to 48 hours post-mixing than weaners from the control groups (McGlone & Anderson, 2002; Guy et al., 2009; Temple et al., 2016). Moreover, a decrease in the number of lesions on the body of PAP-treated weaners has been observed (McGlone & Anderson, 2002; Guy et al., 2009) sign of more appeased interactions between individuals. This is consistent with the observation made by Temple et al. (2016) highlighting an increase in positive interactions between pigs up to 24 h after the mixing. The author hypothesizes that PAP could promote the formation of stable social interactions, facilitating the rapid establishment of a new hierarchy and thus reducing fights. Similar results were found in adult commercial sows, with PAP-treated females engaging in fewer aggressive interactions until 8 days post-mixing. It is important to note that the PAP diffuser block was removed on the seventh day (Plush et al., 2016). It is also worth noting that these studies indicate that, regardless of the galenic formulation, PAP treatment appears to modify pigs' feeding behavior. Hence, compared to control animals, treated pigs spent more time eating and ate more frequently (McGlone & Anderson, 2002; Plush et al., 2016; Temple et al., 2016). McGlone and Anderson (2002) even observed at 1 month post-weaning that PAP-treated weaners were 1 kg heavier than the control pigs with an average daily gain (ADG) increased by 23%. Some researchers investigated the effect of the Pig Appeasing Pheromone in the context of social encounters in more controlled conditions using minipigs as a model for commercial pigs. In a first study involving eight adult females, authors observed that 2 weeks of exposure to the PAP inhibited the cortisol rise that is observed following a social stress (Yonezawa et al., 2009). They made encounters in an unknown place between pigs, first behind a fence, then without restriction. Control pigs exhibited an increase in salivary cortisol at each step of the protocol—this was not observed for PAP-treated pigs in which the cortisol levels remained low. They also observed that the latency to the first agonistic behavior was longer in PAP-treated pigs, which is consistent with the results presented above. Another study confirmed this effect of PAP on the dynamics of pig interactions. Conducted on minipigs, this study demonstrated that when PAP is used, encounters between unfamiliar pigs are less likely to result in a fight (36.36% vs. 63.64%) compared to when a placebo is used (Marcet-Rius et al., 2022). They also found that agonistic interactions were 1.9 times less frequent during encounters between pigs that had previously received PAP. Additionally, PAP-treated pigs tended to spend more time observing each other. The authors hypothesize that this could represent a form of pro-social behavior, which would be consistent with the findings of Temple et al. (2016), who reported more positive interactions in a mixing context with PAP treatment.

As evoked in the first part of this review, encounters with unfamiliar pigs are not the only situations that can provoke agonistic interactions between individuals—introduction of a resource conflict, such as food, can degenerate into aggression (Barnett et al., 1992). In a recent study, pigs were treated or not with a cutaneous form of PAP before being submitted to collective feeding, and their behaviors were monitored (Chasles et al., 2024). When treated with the pheromone analog, pigs performed less head-knock, an agonistic behavior, but also, they were less likely to flee from the feeding area. Interestingly, during the meals with PAP, pigs were more likely to stop eating when another pig was approaching, even though agonistic behaviors were reduced (Chasles et al., 2024). It can be seen as a conflict avoidance strategy—hence, it was hypothesized that the PAP helped the pig adapt its behavior to this new situation, which aligns with studies highlighting appeasing pheromones as learning facilitators and coping promoters.

Like mixing, transportation is routinely done in pig farms, to send weaners from the breeding facility to the fattening one or to take pigs to the slaughterhouse. In pigs, transportation has been linked to a cortisol increase which is a sign of stress (Bradshaw et al., 1996). Hence, transport has been another target to investigate the efficacy of PAP to promote coping. A study performed on commercial finishing pigs highlighted that PAP inhibited the cortisol rise observed in control pigs when loading into the lorry but also that this effect is maintained up to 30 min after arrival in the slaughterhouse (Wöhr et al., 2003). Those results are perfectly in line with the observations of Yonezawa et al. (2009) in another context, social encounters. Both of those studies revealed that the PAP can decrease the physiological reaction to a stressor which is the definition of coping. Moreover, another study on PAP and transportation showed that PAP-treated pigs had a lower heart rate during the first 2 h of a transport simulation (Driessen et al., 2008).

5 Conclusion

Altogether, those observations on the impact of the application of a Pig Appeasing Pheromone (PAP) on pigs or in their environment highlight the potential of its use in the pig production system. It has proved its efficacy for mixing, transportation, and conflict management. In all those contexts, PAP-treated pigs exhibited a behavioral and physiological response that enabled them to minimize the effects of the challenging situation.

The mechanism of action of PAP could be investigated more in detail to understand better how it affects the behavior and physiology of animals. A recent study yielded preliminary results in this regard, showing an increase in oxytocin in the saliva of minipigs as early as 1 h after a single application of PAP, with this effect persisting for up to 24 h (unpublished data). This first result is especially noteworthy as oxytocin is known to be secreted during social interactions after stimulation of the supraoptic nucleus (SON) and paraventricular nucleus (PVN) (Froemke and Young 2021). Hence, at the brain level, the hypothalamus seems to be at least one of the possible targets of appeasing pheromones which seems logical regarding its central role in the hypothalamic-pituitary-adrenal (HPA) axis.

Appeasing pheromone acts as a coping promoter in this species as it has been shown in other species as well. It has been suggested that appeasing pheromones could act as a social buffer, mimicking the calming presence of a conspecific (Kikusui et al., 2006). This is interesting as for social mammals, the presence of a conspecific helps them recover from aversive experiences (for review: Mulej Bratec et al., 2020). This opens up a wide range of uses for appeasing pheromones, both in livestock and companion animals. They should be considered not only as a tool to respond to stressful events but also as a preventive measure to help animals better cope with the challenges they may face: chemical communication is always part of the life of animals and not only to manage stressful contexts.

References

Asproni, P., Bienboire-Frosini, C., Barthélémy, H., Mechin, V., Teruel, E., Leclercq, J., Cozzi, A., & Pageat, P. (2020). Single fluff-spray application of mother hen uropygial secretion analogue positively influences bursa of Fabricius development and the heterophil-to-lymphocyte ratio in ROSS 308 chicks. *Poultry Science, 99*(12), 6300–6306. https://doi.org/10.1016/j.psj.2020.08.076

Asproni, P., Codecasa, E., Marcet-Rius, M., Demellier, J., Descout, E., Verbaere, M., Vinck, O., Pageat, P., & Cozzi, A. (2024). Helping rabbits cope with veterinary acts and vaccine-related stress: The effects of the rabbit appeasing pheromone (RAP). *Animals, 14*(23), 3549. https://doi.org/10.3390/ani14233549

Aureli, F., Cords, M., & Van Schaik, C. P. (2002). Conflict resolution following aggression in gregarious animals: A predictive framework. *Animal Behaviour, 64*(3), 325–343. https://doi.org/10.1006/anbe.2002.3071

Barnett, J. L., Hemsworth, P. H., Cronin, G. M., Newman, E. A., McCallum, T. H., & Chilton, D. (1992). Effects of pen size, partial stalls and method of feeding on welfare-related behavioural and physiological responses of grouphoused pigs. *Applied Animal Behaviour Science, 34*(3), 207–220. https://doi.org/10.1016/S0168-1591(05)80116-9

Bouvier, A. C., & Jacquinet, C. (2008). Pheromone in rabbits: Preliminary technical results on farm use in France. In *Proceedings of the 9th world rabbit congress* (pp. 10–13). World Rabbit Science Association.

Bradshaw, R. H., Parrott, R. F., Forsling, M. L., Goode, J. A., Lloyd, D. M., Rodway, R. G., & Broom, D. M. (1996). Stress and travel sickness in pigs: Effects of road transport on plasma concentrations of cortisol, beta-endorphin and lysine vasopressin. *Animal Science, 63*(3), 507–516. https://doi.org/10.1017/S135772980001540X

Brennan, P. (2010). Pheromones and mammalian behavior. In A. Menini (Ed.), *The neurobiology of olfaction*. CRC Press/Taylor & Francis.

Broad, K. D., Curley, J. P., & Keverne, E. B. (2006). Mother–infant bonding and the evolution of mammalian social relationships. *Philosophical Transactions of the Royal Society B: Biological Sciences, 361*(1476), 2199–2214. https://doi.org/10.1098/rstb.2006.1940

Campbell, J. M., Crenshaw, J. D., & Polo, J. (2013). The biological stress of early weaned piglets. *Journal of Animal Science and Biotechnology, 4*(1), 19. https://doi.org/10.1186/2049-1891-4-19

Chasles, M., Marcet-Rius, M., Chou, J. Y., Teruel, E., Pageat, P., & Cozzi, A. (2024). Cutaneous application of SecurePig® FLASH, a pig appeasing pheromone analogue, facilitates adaptation and manages social behavior during feeding in semi-extensive conditions. *Porcine Health Management, 10*(1), 13. https://doi.org/10.1186/s40813-024-00363-z

Codecasa, E., Pageat, P., Marcet-Rius, M., Asproni, P., Chabaud, C., Arroub, S., Francois, O., Monneret, P., Kuentz, V., & Cozzi, A. (2022). The use of a synthetic analogue of the rabbit appeasing pheromone on rabbits experiencing the arrival in a novel environment: A pilot study. In *Proceedings of the 4th annual meeting of the European veterinary congress of Behavioural medicine and animal welfare*. Mallorca.

Colombo, E. A., Cooke, R. F., Brandão, A. P., Wiegand, J. B., Schubach, K. M., Duff, G. C., Gouvêa, V. N., & Cappellozza, B. I. (2020). Administering an appeasing substance to optimize performance and health responses in feedlot receiving cattle. *Journal of Animal Science, 98*(11), skaa339. https://doi.org/10.1093/jas/skaa339

Coureaud, G., Charra, R., Datiche, F., Sinding, C., Thomas-Danguin, T., Languille, S., Hars, B., & Schaal, B. (2010). A pheromone to behave, a pheromone to learn: The rabbit mammary pheromone. *Journal of Comparative Physiology A: Neuroethology, Sensory, Neural, and Behavioral Physiology, 196*(10), 779–790. https://doi.org/10.1007/s00359-010-0548-y

Cozzi, A., Monneret, P., Lafont-lecuelle, C., Bougrat, L., Gaultier, E., & Pageat, P. (2010). The maternal cat appeasing pheromone: Exploratory study of the effects on aggressive and affiliative interactions in cats. *Journal of Veterinary Behavior, 5*(1), 37.

D'Eath, R. B. (2002). Individual aggressiveness measured in a resident-intruder test predicts the persistence of aggressive behaviour and weight gain of young pigs after mixing. *Applied Animal Behaviour Science, 77*(4), 267–283. https://doi.org/10.1016/S0168-1591(02)00077-1

D'Eath, R. B., Niemi, J. K., Vosough Ahmadi, B., Rutherford, K. M. D., Ison, S. H., Turner, S. P., Anker, H. T., Jensen, T., Busch, M. E., Jensen, K. K., Lawrence, A. B., & Sandøe, P. (2016). Why are most EU pigs tail docked? Economic and ethical analysis of four pig housing and management scenarios in the light of EU legislation and animal welfare outcomes. *Animal, 10*(4), 687–699. https://doi.org/10.1017/S1751731115002098

Denenberg, S., & Landsberg, G. M. (2008). Effects of dog-appeasing pheromones on anxiety and fear in puppies during training and on long-term socialization. *Journal of the American Veterinary Medical Association, 233*(12), 1874–1882. https://doi.org/10.2460/javma.233.12.1874

DePorter, T. L., Bledsoe, D. L., Beck, A., & Ollivier, E. (2019). Evaluation of the efficacy of an appeasing pheromone diffuser product vs placebo for management of feline aggression in multi-cat households: A pilot study. *Journal of Feline Medicine and Surgery, 21*(4), 293–305. https://doi.org/10.1177/1098612X18774437

Dinand Ekkel, E., Savenije, B., Schouten, W. G., & Tielen, M. J. (1996). Health, welfare, and productivity of pigs housed under specific-stress-free conditions in comparison with two-site systems. *Journal of Animal Science, 74*(9), 2081. https://doi.org/10.2527/1996.7492081x

Driessen, B., Peeters, E., & Geers, R. (2008). Influence of olfactory substances on the heart rate and lying behaviour of pigs during transport simulation. *Animal Welfare, 17*, 155–160.

Erhard, H. W., Mendl, M., & Ashley, D. D. (1997). Individual aggressiveness of pigs can be measured and used to reduce aggression after mixing. *Applied Animal Behaviour Science, 54*, 337–151.

Froemke, R. C., & Young, L. J. (2021). Oxytocin, neural plasticity, and social behavior. *Annual Review of Neuroscience, 44*(1), 359–381. https://doi.org/10.1146/annurev-neuro-102320-102847

Garcia-Alvarez, J., Teruel, E., Cozzi, A., Harris, E., Rutter, S. M., & Beaver, A. (2025). Effects of a synthetic analog of the bovine appeasing pheromone on the overall welfare of dairy calves from birth through weaning. *Journal of Dairy Science, 108*(2), 1964–1977. https://doi.org/10.3168/jds.2024-25452

Guy, J. H., Rowlinson, P., Chadwick, J. P., & Ellis, M. (2002). Behaviour of two genotypes of growing–finishing pig in three different housing systems. *Applied Animal Behaviour Science, 75*(3), 193–206. https://doi.org/10.1016/S0168-1591(01)00197-6

Guy, J. H., Burns, S. E., Barker, J. M., & Edwards, S. A. (2009). Reducing post-mixing aggression and skin lesions in weaned pigs by application of a synthetic maternal pheromone. *Animal Welfare, 18*, 249–255.

Jensen, P., & Recén, B. (1989). When to wean – Observations from free-ranging domestic pigs. *Applied Animal Behaviour Science, 23*(1–2), 49–60. https://doi.org/10.1016/0168-1591(89)90006-3

Kekan, P. M., Ingole, S. D., Sirsat, S. D., Bharucha, S. V., Kharde, S. D., & Nagvekar, A. S. (2017). The role of pheromones in farm animals – A review. *Agricultural Reviews, 38*(02), 83–93. https://doi.org/10.18805/ag.v38i02.7939

Keller, M., & Lévy, F. (2012). The main but not the accessory olfactory system is involved in the processing of socially relevant chemosignals in ungulates. *Frontiers in Neuroanatomy, 6*, 1–9. https://doi.org/10.3389/fnana.2012.00039

Kikusui, T., Winslow, J. T., & Mori, Y. (2006). Social buffering: Relief from stress and anxiety. *Philosophical Transactions of the Royal Society B: Biological Sciences, 361*(1476), 2215–2228. https://doi.org/10.1098/rstb.2006.1941

Landsberg, G. M., Beck, A., Lopez, A., Deniaud, M., Araujo, J. A., & Milgram, N. W. (2015). Dog-appeasing pheromone collars reduce sound-induced fear and anxiety in beagle dogs: A placebo-controlled study. *Veterinary Record, 177*(10), 260. https://doi.org/10.1136/vr.103172

Liberles, S. D. (2014). Mammalian pheromones. *Annual Review of Physiology, 76*(1), 151–175. https://doi.org/10.1146/annurev-physiol-021113-170334

Madec, I. (2008). *Effets du sémiochimique MHUSA (Mother Hens' Uropygial Secretion Analogue) sur le stress des poulets de chair*. Doctoral dissertation, Institut National Polytechnique (Toulouse).

Madec, I., Pageat, P., Bougrat, L., Saffray, D., Falewee, C., Gervasoni, M. A., Bollart, A., & Gabarrou, J. F. (2006). Influence of a semiochemical analogue on growing performances and meat quality of broilers. *Poultry Science, 85*(12), 2112–2116. https://doi.org/10.1093/ps/85.12.2112

Madec, I., Pageat, P., Bougrat, L., Lecuelle-Lafont, C., Saffray, D., Falewee, C., Bollart, A., Chabrol, P., & Gabarrou, J. F. (2008). Influence of a preen gland secretion on growth and meat quality of heavy broilers. *Animal, 2*(4), 631–635. https://doi.org/10.1017/S1751731107001504

Maes, D. G. D., Dewulf, J., Piñeiro, C., Edwards, S., & Kyriazakis, I. (2019). A critical reflection on intensive pork production with an emphasis on animal health and welfare. *Journal of Animal Science, 98*, S15–S26. https://doi.org/10.1093/jas/skz362

Marcet-Rius, M., Mendonça, T., Pageat, P., Arroub, S., Bienboire-Frosini, C., Chabaud, C., Teruel, E., & Cozzi, A. (2022). Effect of wither application of an analogue of pig appeasing pheromone on encounters between unfamiliar mini-pigs. *Porcine Health Management, 8*(1), 50. https://doi.org/10.1186/s40813-022-00294-7

McGlone, J. J., & Anderson, D. L. (2002). Synthetic maternal pheromone stimulates feeding behavior and weight gain in weaned pigs. *Journal of Animal Science, 80*(12), 3179–3183. https://doi.org/10.2527/2002.80123179x

Meese, G. B., & Ewbank, R. (1972). A note on instability of the dominance hierarchy and variations in level of aggression within groups of fattening pigs. *Animal Production, 14*(3), 359–362. https://doi.org/10.1017/S0003356100011090

Meese, G. B., & Ewbank, R. (1973). The establishment and nature of the dominance hierarchy in the domesticated pig. *Animal Behaviour, 21*(2), 326–334. https://doi.org/10.1016/S0003-3472(73)80074-0

Mengoli, M., Pageat, P., Lafont-Lecuelle, C., Monneret, P., Giacalone, A., Sighieri, C., & Cozzi, A. (2014). Influence of emotional balance during a learning and recall test in horses (*Equus caballus*). *Behavioural Processes, 106*, 141–150. https://doi.org/10.1016/j.beproc.2014.05.004

Menuge, F., Pageat, P., Arroub, S., Teruel, E., Marcet-Rius, M., Monneret, P., & Cozzi, A. (2023). Can a semiochemical deter cats from trash cans? *Journal of Feline Medicine and Surgery, 25*(1), 1098612X221140855. https://doi.org/10.1177/1098612X221140855

Mills, D. S. (2005). Pheromonatherapy: Theory and applications. *In Practice, 27*(7), 368–373. https://doi.org/10.1136/inpract.27.7.368

Mills, D. S., Ramos, D., Estelles, M. G., & Hargrave, C. (2006). A triple blind placebo-controlled investigation into the assessment of the effect of Dog Appeasing Pheromone (DAP) on anxiety related behaviour of problem dogs in the veterinary clinic. *Applied Animal Behaviour Science, 98*(1–2), 114–126. https://doi.org/10.1016/j.applanim.2005.08.012

Mulej Bratec, S., Bertram, T., Starke, G., Brandl, F., Xie, X., & Sorg, C. (2020). Your presence soothes me: A neural process model of aversive emotion regulation via social buffering. *Social Cognitive and Affective Neuroscience, 15*(5), 561–570. https://doi.org/10.1093/scan/nsaa068

OECD. (2025). Meat consumption data. https://www.oecd.org/en/data/indicators/meat-consumption.html. Accessed 30 Jan 2025.

Osella, M. C., Cozzi, A., Spegis, C., Turille, G., Barmaz, A., Lecuelle, C. L., Teruel, E., Bienboire-Frosini, C., Chabaud, C., Bougrat, L., & Pageat, P. (2018). The effects of a synthetic analogue of the bovine appeasing pheromone on milk yield and composition in Valdostana dairy cows during the move from winter housing to confined lowland pastures. *Journal of Dairy Research, 85*(2), 174–177. https://doi.org/10.1017/S0022029918000262

Pageat, P. (1998, January 21). Pig appeasing pheromones to decrease stress, anxiety and aggressiveness. U.S. Patent No US6169113B1. https://patents.google.com/patent/US6169113B1/en. Accessed 4 Feb 2025.

Pageat, P. (2002, May 7). Animal Appeasing Pheromones. U.S. Patent No 6384252B1. https://patents.google.com/patent/US6384252B1/en. Accessed 12 Mar 2025.

Pageat, P., & Gaultier, E. (2003). Current research in canine and feline pheromones. *Veterinary Clinics of North America: Small Animal Practice, 33*(2), 187–211. https://doi.org/10.1016/S0195-5616(02)00128-6

Pereira, J. S., Demirbas, Y. S., Meppiel, L., Endersby, S., da Graça Pereira, G., & De Jaeger, X. (2023). Efficacy of the Feliway® classic diffuser in reducing undesirable scratching in cats: A randomised, triple-blind, placebo-controlled study. *PLoS One, 18*, e0292188. https://doi.org/10.1371/journal.pone.0292188

Plush, K., Hughes, P., Herde, P., & van Wettere, W. (2016). A synthetic olfactory agonist reduces aggression when sows are mixed into small groups. *Applied Animal Behaviour Science, 185*, 45–51. https://doi.org/10.1016/j.applanim.2016.09.011

Puglisi, I., Masucci, M., Cozzi, A., Teruel, E., Navarra, M., Cirmi, S., Pennisi, M. G., & Siracusa, C. (2022). Effects of a novel gel formulation of dog appeasing pheromone (DAP) on behavioral and physiological stress responses in dogs undergoing clinical examination. *Animals, 12*(18), 2472. https://doi.org/10.3390/ani12182472

Rosvold, E. M., Ocepek, M., & Andersen, I. L. (2024). Effects of different rooting materials on behaviour and welfare of finishing pigs. *Applied Animal Behaviour Science, 277*, 106344. https://doi.org/10.1016/j.applanim.2024.106344

Schouten, W. G. P., & Wiepkema, P. R. (1991). Coping styles of tethered sows. *Behavioural Processes, 25*(2–3), 125–132. https://doi.org/10.1016/0376-6357(91)90015-R

Sheppard, G., & Mills, D. S. (2003). Evaluation of dog-appeasing pheromone as a potential treatment for dogs fearful of fireworks. *Veterinary Record, 152*(14), 432–436. https://doi.org/10.1136/vr.152.14.432

Stukenborg, A., Traulsen, I., Puppe, B., Presuhn, U., & Krieter, J. (2011). Agonistic behaviour after mixing in pigs under commercial farm conditions. *Applied Animal Behaviour Science, 129*(1), 28–35. https://doi.org/10.1016/j.applanim.2010.10.004

Tan, S. S. L., Shackleton, D. M., & Beames, R. M. (1991). The effect of mixing unfamiliar individuals on the growth and production of finishing pigs. *Animal Production, 52*(1), 201–206. https://doi.org/10.1017/S0003356100005845

Taylor, K., & Mills, D. S. (2007). A placebo-controlled study to investigate the effect of dog appeasing pheromone and other environmental and management factors on the reports of disturbance and house soiling during the night in recently adopted puppies (*Canis familiaris*). *Applied Animal Behaviour Science, 105*(4), 358–368. https://doi.org/10.1016/j.applanim.2006.11.003

Temple, D., Barthélémy, H., Mainau, E., Cozzi, A., Amat, M., Canozzi, M. E., et al. (2016). Preliminary findings on the effect of the pig appeasing pheromone in a slow releasing block on the welfare of pigs at weaning. *Porcine Health Management, 2*, 1–5. https://doi.org/10.1186/s40813-016-0030-5

Tod, E., Brander, D., & Waran, N. (2005). Efficacy of dog appeasing pheromone in reducing stress and fear related behaviour in shelter dogs. *Applied Animal Behaviour Science, 93*(3–4), 295–308. https://doi.org/10.1016/j.applanim.2005.01.007

USDA. (2025). Pork production data. https://www.fas.usda.gov/data/production/commodity/0113000. Accessed 30 Jan 2025.

Vitale, K. R. (2018). Tools for managing feline problem behaviors: Pheromone therapy. *Journal of Feline Medicine and Surgery, 20*(11), 1024–1032. https://doi.org/10.1177/1098612X18806759

Wechsler, B. (1995). Coping and coping strategies: A behavioural view. *Applied Animal Behaviour Science, 43*, 123–134.

Wöhr, A.-C., Meier, C., Hollwich, P., Mertens, P., Unshelm, J., & Erhard, M. (2003). Porcine pheromones: A novel method to improve the well-being of fattening pigs during transportation to the slaughterhouse? International Congress in Animal Hygiene, Mexico.

Yonezawa, T., Koori, M., Kikusui, T., & Mori, Y. (2009). Appeasing pheromone inhibits cortisol augmentation and agonistic behaviors during social stress in adult miniature pigs. *Zoological Science, 26*(11), 739–744. https://doi.org/10.2108/zsj.26.739

The Vomeronasal Organ in Domestic Carnivores: From the Physiology to the Pathology, from the Diagnostic to the Clinical Perspectives

Pietro Asproni[1(✉)], Alessandro Cozzi[2], Violaine Mechin[1], Tolga Karaali[1], and Patrick Pageat[2]

[1] Tissue Biology and Chemical Communication Department, IRSEA, Institute of Research in Semiochemistry and Applied Ethology, Apt, France
p.asproni@irsea-institute.com

[2] Research and Education Board, IRSEA, Institute of Research in Semiochemistry and Applied Ethology, Apt, France

Abstract. The vomeronasal organ (VNO) plays a key role in animal behavior, thanks to its capability in detecting semiochemicals, among them pheromones. The VNO has been extensively investigated in the murine model, while less widely in companion animals. This chapter aims to review the literature that investigated the VNO, its physiology, and its pathology in domestic carnivores, in order to better characterize its role and the possible perspectives and applications in clinics. Both cats and dogs possess two VNOs, located in the nasal cavity, which open into the oral cavity. Semiochemicals are detected by the vomeronasal sensory epithelium (VNSE), which is composed of bipolar neurons that transmit the signal to the accessory olfactory bulb. Thanks to pheromone detection, the VNO is crucial for the induction and the display of social, sexual, and maternal intraspecific behavior. We previously described the presence of spontaneous inflammation of the VNSE in cats presenting significantly more intraspecific aggressive behaviors than healthy ones. Other individual observations revealed that vomeronasalitis can also affect dogs. Moreover, in this species, recently a case of VNO agenesis associated with intraspecific aggressive behavior has been reported. These findings are in line with what has been described in farm pigs and strongly emphasize the VNO role also in companion animal behavior and life. These studies also pave the way for further studies that should investigate the possibility of diagnosing VNO changes in living animals to treat them in order to prevent and reduce the behavioral troubles that they can induce.

Keywords: Cat · Chemical communication · Dog · Domestic carnivores · Pathology · Vomeronasal organ

L. M. Schulte (ed.), *Chemical Signals in Vertebrates 16*,
https://doi.org/10.1007/978-3-032-30752-1_19

1 The Vomeronasal Organ of Domestic Carnivores: Anatomy and Role

Different chemosensory organs are able to detect pheromones; nevertheless, the vomeronasal organ (VNO) is the main peripheral structure involved in sociosexual communication (Tirindelli, 2021). In mammals, the VNO is a bilateral, blind-ended, fluid-filled tubular chemosensory structure located in the nasal cavity which opens into the nasal or oral cavity (Swaney & Keverne, 2009; Salazar & Sánchez-Quinteiro, 2011; Dzięcioł et al., 2020). The VNO is connected to the accessory olfactory bulb (AOB) by the vomeronasal nerves, and, contrary to the mouse, in domestic carnivores it is surrounded by the vomeronasal cartilage (Salazar & Quinteiro, 2009).

As in other species, the chemosensory properties of the VNO are provided by the vomeronasal sensory epithelium (VNSE), which in dogs (*Canis lupus familiaris*) and cats (*Felis catus*) is located in the medial wall of the VNO (Salazar & Quinteiro, 2009; Salazar & Sánchez-Quinteiro, 2011). The lateral side of the vomeronasal cavity is lined by a typical ciliated and pseudostratified respiratory epithelium, which is a non-sensory epithelium (NSE) (Salazar et al., 1996). Finally, the rostral part of the VNO is characterized by a transition from respiratory to stratified squamous epithelium. The VNSE is composed of different layers of different types: a layer of few basal cells, a stratum formed by several layers of bipolar neurons, a layer of epithelial supporting cells, and, on the top, a luminal layer composed of extensions of the neuronal cells, with their microvilli that flow into the vomeronasal cavity (Salazar et al., 1996; Salazar & Quinteiro, 2009). In domestic carnivores, only the vomeronasal type 1 receptors (V1Rs) have been identified as functional, which are coupled with the $G\alpha i2$ protein, as in other species (Salazar & Sánchez-Quinteiro, 2011; Salazar et al., 2013; Tirindelli, 2021). More precisely, in dogs and cats, until now, 9 and 30 V1R functional genes were identified, respectively (Tirindelli, 2021). In both species, no functional vomeronasal type 2 receptors (V2Rs) nor the presence of the $G\alpha o$ protein has been described so far (Tirindelli, 2021). Interestingly, a recent study reported the immunohistochemical presence of the V2Rs in wild canids, specifically wolves and foxes. However, using the same antibody on domesticated canine VNO samples, no immunohistochemical signal was observed (Ortiz-Leal et al., 2024). This interesting finding was explained by the authors as the consequence of the regression of the vomeronasal system (VNS) in dogs, compared to their wild counterparts (Ortiz-Leal et al., 2024).

Despite this possible effect of domestication on the VNS of domestic carnivores, the VNO still plays an important role in the behavior of dogs and cats (Pageat & Gaultier, 2003). Both species are macrosmatic mammals, and, as all the macrosmatic species, they possess the three different structures that define a true VNS: the VNO, the accessory olfactory bulb (AOB), and the vomeronasal amygdala (VNAg) (Halpern & Martínez-Marcos, 2003; Salazar & Quinteiro, 2009). The receptors located in the VNSE project through the vomeronasal nerves to the AOB; from the AOB, the accessory olfactory tract transfers the information to the VNAg, which treats the semiochemical information, modulating various social behaviors, such as the male and female mating behavior, aggression, territorial marking, and maternal behavior (Halpern & Martínez-Marcos, 2003). The VNS is also connected to different hypothalamic structures, which are an important component of the social behavior network (Kelliher, 2007; Trives et al., 2025).

Semiochemicals can enter into the vomeronasal cavity using different mechanisms. In both dogs and cats, the two vomeronasal ducts open into the oral cavity in the incisive papilla (Salazar et al., 1996, 2013). In cats, the surrounding air is aspirated into the VNO, thanks to the flehmen behavior, which consists of the elevation of the upper lip and the opening of the mouth for several seconds (Hart & Leedy, 1987). During the emission of this behavior, semiochemicals are transported into the vomeronasal cavity by the inhaled air or also by fluids, thanks to a naso-oral contact with the source of the stimuli (Hart & Leedy, 1987; Pageat & Gaultier, 2003). Moreover, the aspiration during the flehmen is also produced by a pumping/sucking mechanism, caused by the vasoconstriction of the VNO walls that increases the diameter of the lumen of the organ, generating the fall in pressure needed for the suction (Pageat & Gaultier, 2003; Torres et al., 2023).

In dogs, the existence of the flehmen behavior is still controversial; nevertheless, in this species, an analogous behavior that facilitates the entrance of the semiochemicals into the VNO is well described: the tonguing. During the emission of this behavior, the dog pants, elevates the upper lip, folds the nostrils, and rapidly flicks the tongue against the incisive papilla during exploration of feces, urine, or proestral blood (Pageat & Gaultier, 2003). The emission of these behaviors facilitates the entrance of semiochemicals into the vomeronasal cavity, as well as their binding to the vomeronasal receptors and the activation of the VNS.

Even if the VNS of companion carnivores is less studied than that of laboratory species, several evidences of its role in behavior are reported also in dogs and cats. In the domestic cat, studies from the 1970s described that male cats need the VNO to detect female pheromones from urine and cheek gland secretion, which contain information about the females' hormonal phase (Verberne and de Boer, 1976). Different semiochemical secreting structures have been identified in dogs and cats, and the behavior associated to the detection of these secretions has been also described (Pageat & Gaultier, 2003). Six semiochemical sources have been described in domestic carnivora: the facial area, the pedal complex, the perianal complex, the genital complex, the mammary complex, and urine and feces (Pageat & Gaultier, 2003).

The facial complex is composed of glands located under the perioral and cheek skin, and the secretions of these glands are deposited on surfaces by rubbing the cheeks, the chin, the vibrissae, and the lips (Pageat & Gaultier, 2003). In cats, facial secretions are divided into five fractions (F1, F2, F3, F4, and F5), but only for the synthetic analogs of the F2, F3, and F4 the behavioral effects are known (MacDonald, 1985). The F2 is associated with sexual facial marking in male cats; the F3 with the facial marking on items, as antagonist of urine marking and scratching; and finally, the F4 with allomarking, as antagonist of territorial or irritative aggression (MacDonald, 1985). In dogs, the facial complex produces secretions which are mainly involved in the management of social relationships (Bekoff, 1979a; MacDonald, 1985; Mills, 2002).

The pedal complex is composed of various structures located both in the plantar pads and in the skin of the interdigital region. Nevertheless, in dogs, the presence of glands seems to be limited to the interdigital area, while in the plantar pad is not as evident as in the cat (Pageat & Gaultier, 2003). In both species, the secretions produced and released by this complex are involved in territorial marking and in producing alarm pheromones; during territorial marking, the pedal secretions are associated with scratching behavior, which produces a persistent visual signal for conspecifics (Bekoff, 1979a, b; Stoddart, 1980; MacDonald, 1985; Pageat, 1995).

The perianal complex is composed of three distinct types of glands: the supracaudal gland, located on the top of the tail; the circumanal glands, surrounding the anus; and the anal sac, located internally and possessing ducts that open into the anal canal (Pageat & Gaultier, 2003; Hepper & Wells, 2015). The supracaudal glands, which are well developed in cats and less developed in dogs, are sebaceous glands that produce secretions involved in the recognition of males by females in estrus (MacDonald, 1985; Pageat & Gaultier, 2003). The circumanal glands are sebaceous, hepatoid, and modified sweat glands that are interspersed around the anus (Shabadash & Zelikina, 2002; Hepper & Wells, 2015). These structures are more developed in dogs than in cats, in particular in males, and they are involved in social interactions (Pageat & Gaultier, 2003). The anal sacs are two symmetric sacs that open between the anus and the skin in dogs and into the rectum in cats (Pageat & Gaultier, 2003; Hepper & Wells, 2015). They are composed of sweat and sebaceous glands, and their secretions trigger different types of behaviors in dogs, such as fear-induced marking and sexual behavior (Pageat & Gaultier, 2003). In cats, the sebaceous component is predominant, which makes the secretion lipid-rich. Moreover, in this species, the sacs open into the rectum; therefore, the secretion is released with the feces. Finally, feline anal sacs also secrete large amounts of Fel d 1 protein, which is able to bind semiochemicals (Dornelas de Andrade et al., 1996; Bienboire-Frosini et al., 2020).

The genital area possesses sebaceous or mucous type glands, situated in the prepuce, in the vulva, and in the urethra. The secretions produced and released by this complex contribute in both social and sexual behavior. In dogs, this area is deeply explored during social contacts (Gorman & Trownbridge, 1989). In bitches, during the estral phase, this complex secretes a secretion containing methyldihydroxybenzoate, which is highly attractive to male dogs and triggers sexual behavior (Bekoff, 1979a; Goodwin et al., 1979; MacDonald, 1985).

The mammary complex is composed of sebaceous glands located in the intermammary sulcus, and it permits the release of appeasing pheromones by the nursing mother (Kim et al., 2010; Bertoni et al., 2022). These are not produced immediately after the parturition but 3–4 days after birth and continue 2–5 days after the weaning of the puppies (4 months) or kittens (6–12 weeks) (Pageat & Gaultier, 2003). As for the other mammals, the appeasing pheromones produced by the mammary complex are able to appease and reassure the litter during the first period of life (Mills et al., 2013).

Finally, urine and feces are both a complex source of chemical cues in domestic carnivores. Their release is often associated with marking behaviors, both in dogs and cats (Pageat & Gaultier, 2003). The chemical components of urine are produced by the glands located in the different parts of the urinary tract and are secreted into the urinary tract lumen, while the components of feces are released into the anal duct and are modified by the saprophytic bacterial flora (Albone et al., 1978; Bekoff, 1979a; MacDonald, 1985). Urine and feces cues are associated with territorial and sexual behavior in both species.

The role of the VNO in these species is also demonstrated by the use of pheromone analogs both in dogs and cats to handle different behavioral problems. In dogs, the dog appeasing pheromone (DAP)—the synthetic analog of the maternal appeasing pheromone produced and released by the intermammary glands in the nursing bitch—is used to help dogs to cope with stressful situations like transport, separation, sound sensitivity, and novel environment adaptation and in the improvement of dog conditions in the shelter (Pierotti, 2016; Hargrave, 2021). In cats, the cat appeasing pheromone (CAP) is used to manage problems associated with the agonistic behavior in domestic cats and European wildcats (Cozzi et al., 2010; Amat et al., 2009; DePorter et al., 2019; Bertoni et al., 2022). Also, feline facial pheromones have been studied, and their analogs are commonly used. For example, the synthetic analogs of the fraction F3 are able to handle various behavioral problems and to improve cat adaptation to a novel environment (Pageat & Gaultier, 2003). Moreover, the use of pheromones to help pets cope with veterinary procedures has been previously demonstrated, both in dogs (Mills et al., 2006, 2013; Siracusa et al., 2010; Hargrave, 2021; Puglisi et al., 2022) and cats (Kronen et al., 2006; Pereira et al., 2016).

This substantial literature confirms that the pheromonal communication and the VNS play a key role in dog and cat behavior, as also reported for the other mammals.

2 The Spontaneous Alterations of the VNO in Cats and Dogs

The role of the VNO in animals has been elucidated also, thanks to several works focused on the induction of experimental alterations of this organ, in order to reduce or suppress its function and to observe the consequences on animal behavior. From the second half of the twentieth century, different experimental methods have been used to decrease or eliminate VNO functionality, such as the resection of the vomeronasal nerves (Powers & Winans, 1975), the occlusion of the vomeronasal canals (Johns et al., 1978), and the VNO surgical removal (Wysocki et al., 1983). All these techniques have been applied to different species (mouse, hamster, Guinea pig, rat, goat and sheep) and showed that animals with a non-functional VNO presented clear changes in sexual, maternal and social behavior (Powers & Winans, 1975; Winans & Powers, 1977; Johns et al., 1978; Kaneko et al., 1980; Wysocki et al., 1983; Beauchamp et al., 1985; Booth & Katz, 2000; Kiyokawa et al., 2007; Pankevich et al., 2006; Booth & Webb, 2010). Despite these interesting findings, the presence of spontaneous alterations of the VNO has only recently been investigated.

2.1 VNO Inflammation in Cats and Its Effect on Behavior

The first description of VNO spontaneous alteration was the presence of chronic inflammation in the VNO of domestic cats from private owners (Asproni et al., 2016). In this study, using histology, we detected that 14 of the 20 studied cats (70%) presented an inflammation of the VNO. In 13 cats (65%), the inflammation was located in the NSE part, while in 11 (55%) it was located in the VNSE. From a histopathological point of

Table 1. Histopathological findings in the vomeronasal organ of 20 cats.

	VNO		VNSE		NSE	
Histopathological finding	Number of cases	%	Number of cases	%	Number of cases	%
No alterations	6	30	9	45	7	35
Inflammation	14	70	11	55	13	65
Chronic	5	25	3	15	5	25
Chronic-active	6	30	5	25	5	25
Pyogranulomatous	2	10	2	10	2	10
Vasculitis	1	5	1	5	1	5

VNO vomeronasal organ, *VNSE* vomeronasal sensory epithelium, *NSE* non-sensory epithelium

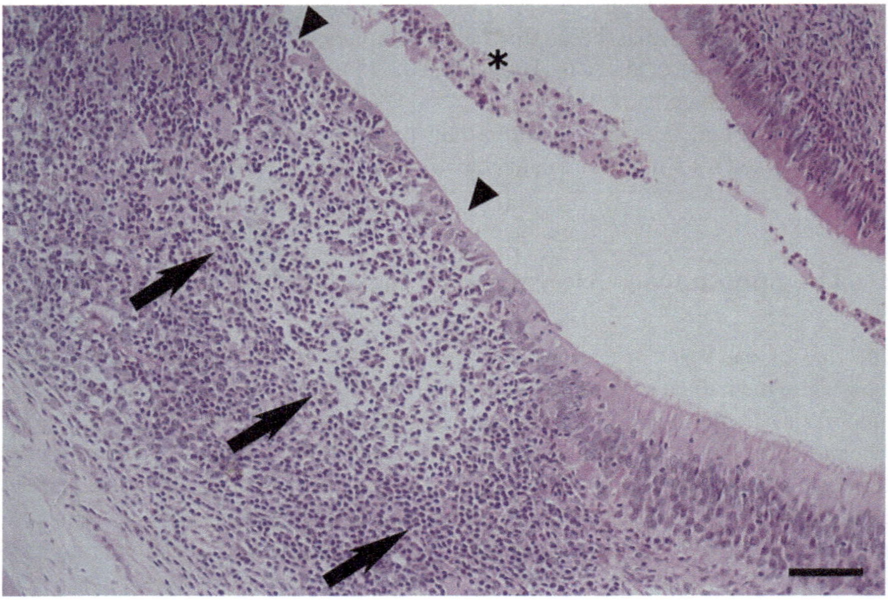

Fig. 1. Vomeronasal organ of a domestic cat showing a feline chronic inflammation located in the VNSE. The inflammatory infiltrate (arrows) is mainly composed of lymphocytes and, to a lesser degree, of macrophages and plasma cells. The VNSE is partially destroyed (arrowhead) by the presence of inflammatory cells, and two aggregates of necrotic VNSE cells are observable in the VNO cavity (asterisk). The tissue is stained with hematoxylin and eosin; the scale bar is 50 μm.

view, inflammations were all of chronic type, and some of them were also subclassified as chronic-active processes, pyogranulomatous inflammations, or vasculitis, according to the distribution reported in Table 1.

The chronic vomeronasalitis was characterized by an infiltrate mainly composed of lymphocytes and, to a lesser degree, macrophages and plasma cells (Figs. 1 and 2). The chronic-active inflammation was characterized by the same cellular infiltrate of the chronic inflammation, with the concurrent involvement of mature neutrophils, which is

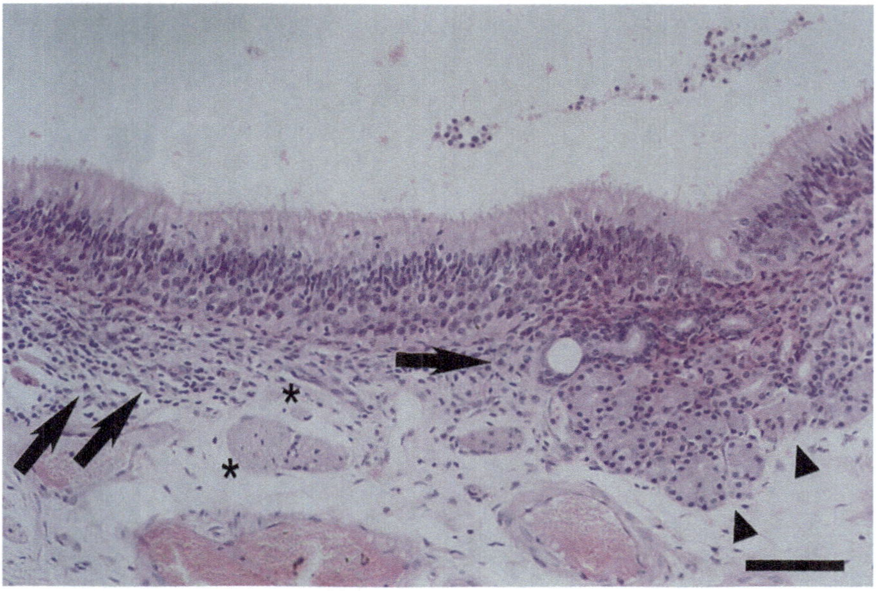

Fig. 2. Vomeronasal organ of a domestic cat showing a feline chronic inflammation located in the VNSE. Lymphocytes and plasma cells were interspersed (arrows) around the vomeronasal nerves (asterisks) and glands (arrowheads). The tissue is stained with hematoxylin and eosin; the scale bar is 50 μm.

a sign of the reactivation of the inflammatory process. When the inflammation was very severe, the infiltrate partially invaded and destroyed the VNSE, with clusters of necrotic cells that were detached into the vomeronasal lumen (Fig. 1). Even if in weak inflammations the VNSE was not invaded or destroyed, inflammatory cells were interspersed in the soft tissue under the VNSE, around vomeronasal nerves and glands (Fig. 2). In the VNOs affected by pyogranulomatous inflammation, the cellular infiltrate was dense and composed of macrophages and neutrophils, and the VNSE was strongly decreased in thickness (Fig. 3). The vasculitis was observed in only one case, in a cat that died from an effusive form of feline infectious peritonitis (FIP), which microscopic lesions were characterized by the infiltration of inflammatory cells around the blood vessels and vascular congestion.

Interestingly, the statistical analysis revealed that the presence of the VNSE inflammation was associated with the presence of intraspecific aggressive behavior ($p = 0.038$; Fisher's exact test), which was assessed, thanks to a questionnaire filled by the owners. On the contrary, no statistical associations were observed between NSE inflammation and intraspecific aggressive behavior ($p = 0.42$) or between NSE or VNSE inflammation and aggression against humans ($p = 0.55$ and $p = 0.67$, respectively).

This article was the first description of a spontaneous alteration of the VNO and its possible effects on animal behavior in any species. Up to now, it remains also the only case describing this condition in companion carnivores. This first report showed that also spontaneous changes of the VNO may have an impact on animal behavior, as previously shown by the studies that induced experimental changes (see above). This

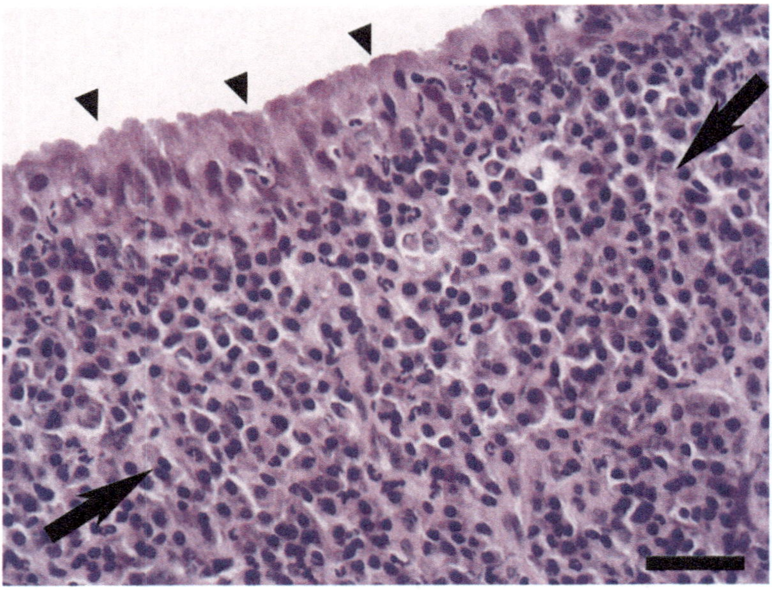

Fig. 3. Vomeronasal organ of a domestic cat showing a feline pyogranulomatous inflammation located in the VNSE. Macrophages and neutrophils massively infiltrate the soft tissue under the VNSE (arrows), which is reduced in thickness (arrowheads). The tissue is stained with hematoxylin and eosin; the scale bar is 25 μm.

report also confirms that the VNO plays a significant role in the behavior and life of cats, in particular in the management of inter-cat relationships, rather than human-cat relationships. The intraspecific aggression is the most common behavioral problem in cats, and it can be caused by several conditions (Amat et al., 2009; Moesta & Crowell-Davis, 2011). This report suggested that among these various causes of inter-cat aggression, also the alteration of the VNO could be taken into account during the diagnostic process.

In cats, no studies have investigated the molecular and cellular changes that affect an inflamed VNSE, but some interesting data have been recently published in farm pigs (Mechin et al., 2022). In fact, also this species can be affected by vomeronasalitis, which is associated with aggressive behaviors in the pen (Asproni et al., 2022). In pigs, we observed that an inflamed VNO is less thick and possesses less Gαi2-positive cells than a healthy VNO, which means that the inflammatory process induces a neuronal loss and a reduction of V1Rs. This loss seems to limit the functioning of the VNO; thus, an animal affected by vomeronasalitis cannot properly use chemical communication to interact with their conspecifics and to manage the onset of possible conflicts, as observed in cats and pigs (Asproni et al., 2016, 2022).

In our report being the first to describe this condition in cats, several implications were revealed and still need to be explored, such as the causes of this alteration in pets, the possibility to diagnose them in living patients, and the treatments that could be used to reduce its effects on behavior.

2.2 VNO Spontaneous Alteration in Dogs

In dogs, no reports of vomeronasalitis have been published up to now. Nevertheless, some personal observations were collected in our Institute on five dogs (three from owners and two from laboratory dogs from our facility). In these five cases, a chronic inflammation of the VNSE has been observed, always composed of a lymphocytic infiltrate, as commonly reported in cats (Fig. 4).

An interesting case report has been recently published in this species, which described the congenital agenesis of the VNO in a 1.5-year-old, entire, male mongrel dog (Muñiz-de Miguel et al., 2023). This dog was referred for a behavioral consultation since it presented aggressive behaviors toward unknown dogs, both males and females, even though it had an adequate socialization and did not have previous traumatic experiences. The ethological and pharmacological treatment did not show effects; therefore, the dog was submitted to medical imaging analyses, i.e., a computed tomography (CT) scan and magnetic resonance imaging (MRI). These analyses revealed a defect of the complete rostroventral osseous nasal septum, which affected also the vomer bone and the palatine-maxillary bone, with the absence of the VNO. These analyses also showed a complete absence of the septum pellucidum (SP), which is a component of the limbic system. Obviously, the authors of this report stated that it was not possible to distinguish the individual role of the VNO agenesis and the SP agenesis in the presence of the behavioral troubles of this dog. Nevertheless, this first report of VNO congenital agenesis gives a hint that an alteration of the VNS might have an impact on the behavior of the affected animals, in particular in the onset of intraspecific aggressive behaviors, in dogs and, as previously reported, in cats and pigs.

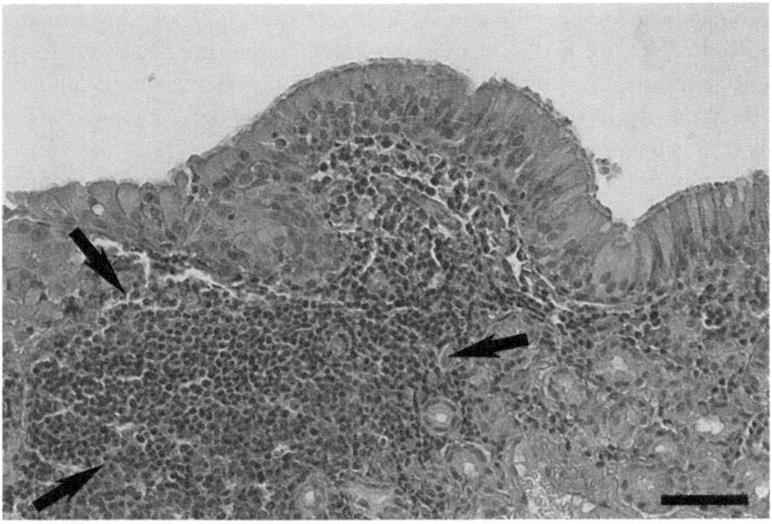

Fig. 4. Vomeronasal organ of a dog with a chronic inflammation located in the VNSE, showing a dense inflammatory infiltrate which is mainly composed of mature lymphocytes infiltrating the soft tissue under the VNSE (arrows). The tissue is stained with hematoxylin and eosin; the scale bar is 50 μm.

3 Diagnostic and Clinical Perspectives

Dogs and cats are the two most common companion animals worldwide, and the interest in their health and well-being is constantly increasing (AVMA, 2024; Contalbrigo et al., 2024). Behavioral troubles are commonly reported in these species, and aggression is the most common problem in both species (Fatjò et al., 2007; Amat et al., 2009; Moesta & Crowell-Davis, 2011). The first descriptions of spontaneous alterations of the VNO in domestic carnivores suggest that these conditions can contribute to the onset of intraspecific aggression, which is a complex condition that can be induced by various factors. In cats, it can be status- or conflict-related, play-related, fear-related, or redirected or caused by the arrival of a new cat into a preexisting group of cats (Moesta & Crowell-Davis, 2011). In dogs, the aggression toward other dogs can be fear-related, protective, resource guarding, redirected, territorial, and pathophysiological or caused by miscommunication problems (Landsberg et al., 2013).

In both species, it is therefore important to identify the cause of the intraspecific aggression, in order to be able to treat it and to improve the well-being of the animals and owners. It is also important to distinguish if the origin of the cause is behavioral or organic, which implies a completely different clinical approach. Among the organic causes, various alterations should be taken into account during the diagnostic process, such as anatomical abnormalities of the central nervous system (CNS), CNS tumors, metabolic disease states (endocrine disorders, hepatic encephalopathy), CSN trauma, infections (rabies), and toxicological conditions (Landsberg et al., 2013). Beyond these well-known causes, the studies on VNO alterations suggest that also these conditions should be taken into account during the clinical behavioral examination of intraspecific aggressive pets.

Unfortunately, in our previous studies, in order to histologically diagnose the inflammation of the VNO, we had to sample the VNOs during the necropsy. To perform a histological analysis, the sample of the whole organ is necessary to obtain a diagnosis. Even if for some organs or lesions it is possible to sample only a biopsy, the position of the VNO makes the access very complicated. Moreover, the size of the VNO suggests that the sampling of a portion could induce important damages and a reduction of its functionality. For these reasons, it is crucial to find a noninvasive method that can allow the diagnosis of vomeronasalitis *in vitam*.

Interestingly, Dzięcioł et al. (2020) were able to introduce a 24G intravenous cannula into the VNOs of 13 dogs, 5 cadavers, and 8 living anesthetized dogs, also injecting into the nasal cavity a contrast medium to perform a MRI analysis. These significant findings open an important path in order to find a method to sample cells or fluids from the vomeronasal cavity to be then submitted to further laboratory analysis. In fact, the cytology of cavity fluids is commonly used in veterinary medicine, since it allows the sampling of exfoliated cells or inflammatory cells, which can help in the diagnosis of an inflammatory condition and in its classification (Rebar & Thompson, 2010). Furthermore, the study of Dzięcioł and colleagues also showed that it is possible to inject a liquid into the VNO cavity, suggesting that the flushing of this cavity to collect cells for cytology is feasible, which is another sampling technique commonly used in clinics (Meyer et al., 2010). VNO fluids and washes could be also submitted to other biochemical or molecular analyses, in order to measure the levels of some inflammatory markers. Moreover, all the 13 dogs were beagles, which is a small-sized hound breed (Fédération Cynologique Internationale, 2011), thus suggesting that this technique is also applicable to small individuals.

Future studies should investigate the possibility to collect materials from the VNO to be analyzed by cytology or other biochemical and molecular methods, in order to offer to the clinician some rapid, non-expensive, and noninvasive tools to be used with pets that present intraspecific aggressive behaviors. This analysis could be useful to confirm or exclude the vomeronasalitis as potential cause of this behavioral trouble, helping the clinician to correctly choose the treatment, avoiding to prescript a behavioral therapy when the cause is of a different nature.

4 Conclusions

The literature presented here demonstrates the importance of the VNO and pheromonal communication in domestic carnivores. The VNO allows the detection of semiochemicals and the transmission of messages to the AOB. The VNS is involved in different aspects of dog and cat behavior, such as the sexual, social, and maternal behavior. Recent studies showed that the VNO can be affected by spontaneous alterations, of inflammatory or developmental nature, which can contribute to the onset of aggressive behaviors, both in cats and dogs. These evidences demonstrate that the chemical communication and its dysfunction have an important impact in animal life and thus in veterinary clinical practice. For this reason, further studies should explore other kinds of VNO alterations and their effects on animals and investigate the possibility of diagnosing VNO pathologies in living animals, in order to be able to detect the presence of these changes *in vitam* and to treat them properly.

References

Albone, E. S., Gosden, P. E., Ware, G. C., Macdonald, D. W., & Hough, N. G. (1978). Bacterial action and chemical signalling in the red fox, *Vulpes vulpes*, and other mammals. *Proceedings of the American Chemical Society, 67*, 78–91.

Amat, M., de la Torre, J. L. R., Fatjo, J., Mariotti, V. M., van Wijk, S., & Manteca, X. (2009). Potential risk factors associated with feline behaviour problems. *Applied Animal Behaviour Science, 121*, 134–139.

American Veterinary Medical Association. (2024). *2024 AVMA pet ownership and demographics sourcebook*. Publisher AVMA, Veterinary Economics Division.

Asproni, P., Cozzi, A., Verin, R., Lafont-Lecuelle, C., Bienboire-Frosini, C., Poli, A., & Pageat, P. (2016). Pathology and behaviour in feline medicine: Investigating the link between vomeronasalitis and aggression. *Journal of Feline Medicine and Surgery, 18*, 997–1002.

Asproni, P., Mainau, E., Cozzi, A., Carreras, R., Bienboire-Frosini, C., Teruel, E., & Pageat, P. (2022). Is there a link between vomeronasalitis and aggression in stable social groups of female pigs? *Animals, 12*, 303.

Beauchamp, G. K., Wysocki, C. J., & Wellington, J. L. (1985). Extinction of response to urine odor as a consequence of vomeronasal organ removal in male Guinea pigs. *Behavioral Neuroscience, 99*, 950–955.

Bekoff, M. (1979a). Scent marking by free ranging domestic dogs: Olfactory and visual components. *Behavioral Biology, 4*, 123–139.

Bekoff, M. (1979b). Ground scratching by male domestic dogs: A composite signal. *Journal of Mammalogy, 60*, 847–848.

Bertoni, V., Spiezio, C., Regaiolli, B., Cozzi, A., Valsecchi, P., & Normando, S. (2022). Group Reunion in zoo European wildcats using cat appeasing pheromone (CAP) and gradual release of the animals in the exhibit – A case study. *Animals, 12*, 1302.

Bienboire-Frosini, C., Durairaj, R., Pelosi, P., & Pageat, P. (2020). The major cat allergen Fel d 1 binds steroid and fatty acid semiochemicals: A combined *in silico* and *in vitro* study. *International Journal of Molecular Sciences, 21*, 1365.

Booth, K. K., & Katz, L. S. (2000). Role of the vomeronasal organ in neonatal offspring recognition in sheep. *Biology of Reproduction, 63*, 953–958.

Booth, K. K., & Webb, E. C. (2010). Effect of blockage of the ducts of the vomeronasal organ on LH plasma levels during the "Whitten effect" in does. *Veterinary Medicine International, 2010*, 305468.

Contalbrigo, L., Normando, S., Bassan, E., & Mutinelli, F. (2024). The welfare of dogs and cats in the European Union: A gap analysis of the current legal framework. *Animals, 14*, 2571.

Cozzi, A., Monneret, P., Lafont-Lecuelle, C., Bougrat, L., Gaultier, E., & Pageat, P. (2010). The maternal cat appeasing pheromone: Exploratory study of the effects on aggressive and affiliative interactions in cats. *Journal of Veterinary Behavior, 5*, 37–38.

DePorter, T. L., Bledsoe, D. L., Beck, A., & Ollivier, E. (2019). Evaluation of the efficacy of an appeasing pheromone diffuser product vs placebo for management of feline aggression in multi-cat households: A pilot study. *Journal of Feline Medicine and Surgery, 21*, 293–305.

Dornelas de Andrade, A., Birnbaum, J., Magalon, C., Magnol, P., Lanteaume, A., Charpin, D., & Vervloet, D. (1996). Fel d 1 levels in cat anal glands. *Clinical & Experimental Allergy, 26*, 178–180.

Dzięcioł, M., Podgórski, P., Stańczyk, E., Szumny, A., Woszczyło, M., Pieczewska, B., Niżański, W., Nicpoń, J., & Wrzosek, M. A. (2020). MRI features of the vomeronasal organ in dogs (*Canis familiaris*). *Frontiers in Veterinary Science, 7*, 159.

Fatjò, J., Amat, M., Mariotti, V. M., Ruiz-De-La-Torre, J., & Manteca, X. (2007). Analysis of 1040 cases of canine aggression in a referral practice in Spain. *Journal of Veterinary Behavior: Clinical Applications and Research, 2*, 158–165.

Fédération Cynologique Internationale. (2011). FCI breeds nomenclature FCI-Standard N° 161- Beagle.

Goodwin, M., Gooding, K. M., & Reginer, F. (1979). Sex pheromone in the dog. *Science, 203*, 559–561.

Gorman, L., & Trownbridge, B. (1989). The role of odor in the social lives of carnivores. In J. L. Gittleman (Ed.), *Carnivore behavior, ecology and evolution* (pp. 57–88). Cornell University Press.

Halpern, M., & Martínez-Marcos, A. (2003). Structure and function of the vomeronasal system: An update. *Progress in Neurobiology, 70*, 245–318.

Hargrave, C. (2021). Pheromones and 25 years of pheromonotherapy: What are they and how do they work? *The Veterinary Nurse, 12*, 116–122.

Hart, B. L., & Leedy, M. G. (1987). Stimulus and hormonal determinants of flehmen behavior in cats. *Hormones and Behavior, 21*, 44–52.

Hepper, P., & Wells, D. (2015). Olfaction in the order Carnivora: Family Canidae. In R. L. Doty (Ed.), *Handbook of olfaction and gustation* (pp. 591–604). Wiley.

Johns, M., Feder, H., Komisaruk, B. et al. Urine-induced reflex ovulation in anovulatory rats may be avomeronasal effect. Nature 272, 446–448 (1978). https://doi.org/10.1038/272446a0

Kaneko, N., Debski, E. A., Wilson, M. C., & Whitten, W. K. (1980). Puberty acceleration in mice. II. Evidence that the vomeronasal organ Isa receptor for the primer pheromone in male mouse urine. *Biology of Reproduction, 22*, 873–878.

Kelliher, K. R. (2007). The combined role of the main olfactory and vomeronasal systems in social communication in mammals. *Hormones and Behavior, 52*, 561–570.

Kim, Y. M., Lee, J. K., et al. (2010). Efficacy of dog-appeasing pheromone (DAP) for ameliorating separation-related behavioral signs in hospitalized dogs. *The Canadian Veterinary Journal, 51*, 380–384.

Kiyokawa, Y., Kikusui, T., Takeuchi, Y., & Mori, Y. (2007). Removal of the vomeronasal organ blocks the stress-induced hyperthermia response to alarm pheromone in male rats. *Chemical Senses, 32*, 57–64.

Kronen, P. W., Ludders, J. W., Erb, H. N., Moon, P. F., Gleed, R. D., & Koski, S. (2006). A synthetic fraction of feline facial pheromones calms but does not reduce struggling in cats before venous catheterization. *Veterinary Anaesthesia and Analgesia, 33*, 258–265.

Landsberg, G., Hunthausen, W., & Ackerman, L. (2013). *Behavior problems of the dog and cat* (3rd ed.). Saunders Ltd.

MacDonald, D. W. (1985). The carnivores: Order Carnivora. In R. E. Brown & D. W. MacDonald (Eds.), *Social odours in mammals* (pp. 619–722). Clarendon Press.

Mechin, V., Asproni, P., Bienboire-Frosini, C., Cozzi, A., Chabaud, C., Arroub, S., Mainau, E., Nagnan-Le Meillour, P., & Pageat, P. (2022). Inflammation interferes with chemoreception in pigs by altering the neuronal layout of the vomeronasal sensory epithelium. *Frontiers in Veterinary Science, 9*, 936838.

Meyer, D. J., Connolly, S. L., & Heng, H. G. (2010). The acquisition and management of cytology specimens. In R. E. Raskin & D. J. Meyer (Eds.), *Canine and feline cytology, a color atlas and interpretation guide* (pp. 1–14). Saunders Ltd.

Mills, D. S. (2002). Pheromonotherapy – An integral part of modern companion animal practice. *UK Vet, 7*, 1–3.

Mills, D. S., Ramos, D., Estelles, M. G., & Hargrave, C. (2006). A triple blind placebo-controlled investigation into the assessment of the effect of dog appeasing pheromone (DAP) on anxiety related behaviour of problem dogs in the veterinary clinic. *Applied Animal Behaviour Science, 98*, 114–126.

Mills, D., Dube, M. B., & Zulch, H. (2013). Stress and pheromonatherapy in the veterinary clinic. In D. Mills, M. B. Dube, & H. Zulch (Eds.), *Stress and pheromonatherapy in small animal clinical behaviour* (pp. 242–256). Wiley-Blackwell.

Moesta, A., & Crowell-Davis, S. (2011). Intercat aggression – General considerations, prevention and treatment. *Tierärztliche Praxis. Ausgabe K, Kleintiere/Heimtiere, 39*, 97–104.

Muñiz-de Miguel, S., Barreiro-Vázquez, J. D., Sánchez-Quinteiro, P., Ortiz-Leal, I., & González-Martínez, Á. (2023). Behavioural disorder in a dog with congenital agenesis of the vomeronasal organ and the septum pellucidum. *Veterinary Record Case Reports, 11*, e571.

Ortiz-Leal, I., Torres, M. V., López-Beceiro, A., Fidalgo, L., Shin, T., & Sanchez-Quinteiro, P. (2024). First immunohistochemical demonstration of the expression of a type-2 vomeronasal receptor, V2R2, in wild canids. *International Journal of Molecular Sciences, 25*, 7291.

Pageat, P. (1995). Properties of cat's facial pheromones. European Patent EP 0 724 832 B1.

Pageat, P., & Gaultier, E. (2003). Current research in canine and feline pheromones. *Veterinary Clinics of North America: Small Animal Practice, 33*, 187–211.

Pankevich, D. E., Cherry, J. A., & Baum, M. J. (2006). Effect of vomeronasal organ removal from male mice on their preference for and neural Fos responses to female urinary odors. *Behavioral Neuroscience, 120*, 925–936.

Pereira, J. S., Fragoso, S., Beck, A., Lavigne, S., Varejão, A. S., & da Graça Pereira, G. (2016). Improving the feline veterinary consultation: The usefulness of Feliway spray in reducing cats' stress. *Journal of Feline Medicine and Surgery, 18*, 959–964.

Pierotti, P. (2016). Clinical applications of pheromones in dogs. *Dog Behavior, 2*, 20–26.

Powers, J. B., & Winans, S. S. (1975). Vomeronasal organ: Critical role in mediating sexual behavior of the male hamster. *Science, 187*, 961–963.

Puglisi, I., Masucci, M., Cozzi, A., Teruel, E., Navarra, M., Cirmi, S., Pennisi, M. G., & Siracusa, C. (2022). Effects of a novel gel formulation of dog appeasing pheromone (DAP) on behavioral and physiological stress responses in dogs undergoing clinical examination. *Animals, 12*, 2472.

Rebar, A. H., & Thompson, C. A. (2010). Body cavity fluids. In R. E. Raskin & D. J. Meyer (Eds.), *Canine and feline cytology, a color atlas and interpretation guide* (pp. 171–191). Saunders Ltd.

Salazar, I., & Quinteiro, P. S. (2009). The risk of extrapolation in neuroanatomy: The case of the mammalian vomeronasal system. *Frontiers in Neuroanatomy, 3*, 22.

Salazar, I., & Sánchez-Quinteiro, P. (2011). A detailed morphological study of the vomeronasal organ and the accessory olfactory bulb of cats. *Microscopy Research and Technique, 74*, 1109–1120.

Salazar, I., Sanchez Quinteiro, P., Cifuentes, J. M., & Garcia Caballero, T. (1996). The vomeronasal organ of the cat. *Journal of Anatomy, 188*, 445–454.

Salazar, I., Cifuentes, J. M., & Sánchez-Quinteiro, P. (2013). Morphological and immunohistochemical features of the vomeronasal system in dogs. *The Anatomical Record, 296*, 146–155.

Shabadash, S. A., & Zelikina, T. I. (2002). Detection of hepatoid glands and distinctive features of the hepatoid acinus. *Biology Bulletin, 29*, 559–567.

Siracusa, C., Manteca, X., Cuenca, R., del Mar Alcalá, M., Alba, A., Lavín, S., & Pastor, J. (2010). Effect of a synthetic appeasing pheromone on behavioral, neuroendocrine, immune, and acute-phase perioperative stress responses in dogs. *Journal of the American Veterinary Medical Association, 237*, 673–681.

Stoddart, D. M. (1980). *The ecology of vertebrate olfaction* (1st ed.). Springer.

Swaney, W. T., & Keverne, E. B. (2009). The evolution of pheromonal communication. *Behavioural Brain Research, 200*, 239–247.

Tirindelli, R. (2021). Coding of pheromones by vomeronasal receptors. *Cell and Tissue Research, 383*, 367–386.

Torres, M. V., Ortiz-Leal, I., Ferreiro, A., Rois, J. L., & Sanchez-Quinteiro, P. (2023). Immunohistological study of the unexplored vomeronasal organ of an endangered mammal, the dama gazelle (*Nanger dama*). *Microscopy Research and Technique, 86*, 1206–1233.

Trives, E., Porte, C., Nakahara, T. S., Keller, M., Vacher, H., & Chamero, P. (2025). Social experience is associated with a differential role of aromatase neurons in sexual behavior and territorial aggression in male mice. *Hormones and Behavior, 170*, 105723.

Verberne, G. and de Boer, J. (1976), Chemocommunication among Domestic Cats, Mediated by the Olfactory andVomeronasal Senses. Zeitschrift für Tierpsychologie, 42: 86-109. https://doi.org/10.1111/j.1439-0310.1976.tb00958.x

Winans, S. S., & Powers, J. B. (1977). Olfactory and vomeronasal deafferentation of male hamsters: Histological and behavioral analyses. *Brain Research, 126*, 325–344.

Wysocki, C. J., Katz, Y., & Bernhard, R. (1983). Male vomeronasal organ mediates female-induced testosterone surges in mice. *Biology of Reproduction, 28*, 917–922.

Index

L. M. Schulte (ed.), *Chemical Signals in Vertebrates 16*,
https://doi.org/10.1007/978-3-032-30752-1

Q

R

GPSR Compliance

The European Union's (EU) General Product Safety Regulation (GPSR) is a set of rules that requires consumer products to be safe and our obligations to ensure this.

If you have any concerns about our products, you can contact us on ProductSafety@springernature.com

In case Publisher is established outside the EU, the EU authorized representative is:

Springer Nature Customer Service Center GmbH
Europaplatz 3
69115 Heidelberg, Germany

Batch number: 10918185

Printed by Printforce, the Netherlands